Materials and Chemistry of Flame-Retardant Polyurethanes Volume 2: Green Flame Retardants

ACS SYMPOSIUM SERIES **1400**

Materials and Chemistry of Flame-Retardant Polyurethanes Volume 2: Green Flame Retardants

Ram K. Gupta, Editor

Department of Chemistry, Kansas Polymer Research Center

Pittsburg State University

Pittsburg, Kansas, United States

American Chemical Society, Washington, DC

Library of Congress Cataloging-in-Publication Data

Names: Gupta, Ram K., editor.
Title: Materials and chemistry of flame-retardant polyurethanes / Ram K. Gupta, editor, Department of Chemistry, Kansas Polymer Research Center, Pittsburg State University, Pittsburg, Kansas, United States.
Description: Washington, DC : American Chemical Society, 2021- | Series: ACS symposium series ; 1399, 1400 | Includes bibliographical references and index. | Contents: volume 1. A fundamental approach -- volume 2. Green flame retardants.
Identifiers: LCCN 2021049631 (print) | LCCN 2021049632 (ebook) | ISBN 9780841298002 (hardcover) | ISBN 9780841297999 (ebook other)
Subjects: LCSH: Fire resistant polymers. | Fireproofing agents. | Polyurethanes.
Classification: LCC TH1074.5 . M38 2022 (print) | LCC TH1074.5 (ebook) | DDC 628.9/223--dc23/eng/20211201
LC record available at https://lccn.loc.gov/2021049619
LC ebook record available at https://lccn.loc.gov/2021049620

The paper used in this publication meets the minimum requirements of American National Standard for Information Sciences—Permanence of Paper for Printed Library Materials, ANSI Z39.48n1984.

Foreword

The ACS Symposium Series is an established program that publishes high-quality volumes of thematic manuscripts. For over 40 years, the ACS Symposium Series has been delivering essential research from world leading scientists, including 36 Chemistry Nobel Laureates, to audiences spanning disciplines and applications.

Books are developed from successful symposia sponsored by the ACS or other organizations. Topics span the entirety of chemistry, including applications, basic research, and interdisciplinary reviews.

Before agreeing to publish a book, prospective editors submit a proposal, including a table of contents. The proposal is reviewed for originality, coverage, and interest to the audience. Some manuscripts may be excluded to better focus the book; others may be added to aid comprehensiveness. All chapters are peer reviewed prior to final acceptance or rejection.

As a rule, only original research papers and original review papers are included in the volumes. Verbatim reproductions of previous published papers are not accepted.

ACS Books

Contents

Indexes

Preface

The polyurethane (PU) industry is one of the fastest growing industries, with commercial applications in many areas, such as automotive materials, construction materials, medical items, packaging materials, furniture, thermal materials, electrical materials, and vibration insulation. The high flammability of PU foams is a concern, and it restricts many of their valuable applications. The research and progress in PUs are growing rapidly, as witnessed by a rapid increase in the number of publications in this area and wide industrial applications. However, a lack of any books covering the materials, chemistry, and mechanism of various flame retardants (FRs) for PUs under one title makes this book very unique. This book provides current state-of-the-art knowledge on the materials and chemistry of novel FRs for PUs. The synthesis, characterization, and applications of halogen, nonhalogen, and eco-friendly materials as FRs for PUs and their FR mechanisms are explored to provide in-depth knowledge in this field. The FR mechanisms based on the types of FRs in PUs, along with their eco-friendly synthesis, are explored. Environmentally friendly approaches to recycling PUs and their future aspects are discussed.

This book contains a total of 25 chapters, which are divided into two volumes. In the first volume (entitled *Materials and Chemistry of Flame-Retardant Polyurethanes Volume 1: A Fundamental Approach*), 12 chapters are included. The first chapter introduces basic concepts, materials, chemistry, and various approaches for the synthesis of PUs. A wide range of applications of diverse types of PUs—including rigid, flexible, thermoplastic, ionomer, and waterborne PUs—is discussed. This is followed by a chapter highlighting recent advances in bioderived renewable materials for the synthesis of PUs. The subsequent chapters discuss various types of FRs, synthesis, and their applications in reducing the flammability of PUs. Advantages and disadvantages of halogen-based FRs, along with those of many FRs as substitutes for halogen-based FRs, are covered. The last few chapters in the first volume cover nanomaterials, nanocomposites, industrial FR-PUs, and approaches used to recycle FR-PUs.

This second volume (entitled *Materials and Chemistry of Flame-Retardant Polyurethanes Volume 2: Green Flame Retardants*) deals with eco-friendly FRs for PUs. The first few chapters cover natural resources and polysaccharides as green FRs, followed by some chapters covering carbon-based FRs such as carbon nanotubes and graphite. Metal oxides, metal hydroxides, and nitrogen- and phosphorus-based FRs for PUs are explored in detail. The mechanism of flame retardancy of these materials is discussed to provide better insight for the development of new and efficient FRs. The last couple of chapters cover synergism effects in FRs, as well as future perspectives and challenges of FR-PUs.

I would like to thank the authors for their valuable contributions and the reviewers for their productive comments and recommendations to improve the quality of this book. I also gratefully acknowledge the support from the ACS Books editorial team—particularly Anne Brenner, Amanda Koenig, James Antley, and Tracey Glazener—for their timely responses and expert handling of the submitted chapters.

I would like to dedicate this book to my parents, family, teachers, and students, without whom none of my success would be possible.

Ram K. Gupta, Associate Professor of Polymer Chemistry
Department of Chemistry
Kansas Polymer Research Center
Pittsburg State University
Pittsburg, Kansas 66762, United States

Chapter 1

Natural Resources as Flame Retardants for Polyurethanes

Charles Oluwaseun Adetunji,[*,1] Abel Inobeme,[2] Kshitij R. B. Singh,[3] John Tsado Mathew,[4] Olugbemi T. Olaniyan,[5] Jay Singh,[6] Vanya Nayak,[7] and Ravindra Pratap Singh[7]

[1]Applied Microbiology, Biotechnology and Nanotechnology Laboratory, Department of Microbiology, Edo University Iyamho, PMB 04, Auchi, Edo State 312101, Nigeria

[2]Department of Chemistry, Edo University Iyamho, PMB 04, Auchi, Edo State 312101, Nigeria

[3]Department of Chemistry, Govt. V.Y.T. P.G. Autonomous College, Durg, Chhattisgarh 491001, India

[4]Department of Chemistry, Ibrahim Badamasi University Lapai, Niger State 911101, Nigeria

[5]Laboratory for Reproductive Biology and Developmental Programming, Department of Physiology, Edo University Iyamho, PMB 04, Auchi, Edo State 312101, Nigeria

[6]Department of Chemistry, Institute of Science, Banaras Hindu University, Varanasi, Uttar Pradesh 221005, India

[7]Department of Biotechnology, Indira Gandhi National Tribal University, Amarkantak, Madhya Pradesh 484887, India

[*]Email: adetunji.charles@edouniversity.edu.ng

Several studies have reported the wide utilization of polyurethanes (PUs) in numerous areas such as refrigeration, construction, astronautics, aeronautics, and oil pipelines. This depends on their distinctive physical properties, like high compressibility, light weight, good cohesiveness, low water absorption, and low thermal conductivity. The flame-retardant (FR) properties of constituents can be increased through the integration of various natural resources. One of the major FRs applied in PUs is the phospho-halogenated composite. Improvement of innovative halogen-free FRs, which are currently used for polymer applications, is becoming significant because of the bans on certain halogenated FRs, incompetence of current FR additives, and advanced fire performance necessities for materials. This book chapter provides a general focus on natural resources as FRs for PUs, the chemistry of the composites, and the mode of action in FRs for PUs.

Introduction

Polyurethane (PU) is a remarkable multipurpose polymer encompassing urethane bonds (−NHCOO−) in the chain, which has been effectively applied in various areas (*1*). It is one of the thermosetting materials applied widely in day-to-day life and is also a main associate of the polymer family. PUs have been widely used in several fields such as refrigeration, construction, astronautics, aeronautics, and oil pipelines, depending on their distinctive physical properties. These include high compressible strength, light weight, good cohesiveness, low water absorption, and low thermal conductivity (*2*).

In general, the flame-retardant (FR) properties of constituents might increase through the integration of FRs from natural resources. Some of the usual FRs applied in polyurethanes include phospho-halogenated composites. On the other hand, such retardants mostly cause toxic and dense smoke during burning, which prevents their additional expansion based on the different guidelines of fire protection of materials. Intumescent FRs are part of the fresh group of FRs because of their advantages, which include low production of toxic gases generated through anti-dripping and burning properties (*3*). Improvement of innovative halogen-free FRs currently used for applications in polymers is becoming significant because of the bans on certain current halogenated FRs, incompetence of current FR additives, and advanced fire performance necessities for materials. Also, PU is an indispensable polymer that has found relevance in different areas such as wood coatings, textile coatings, fibers, foams, adhesives, and cables (*4*). This chapter provides (Figure 1) the natural resources available as FRs for PUs, the chemistry of the composites, and the mode of action in FRs for PUs.

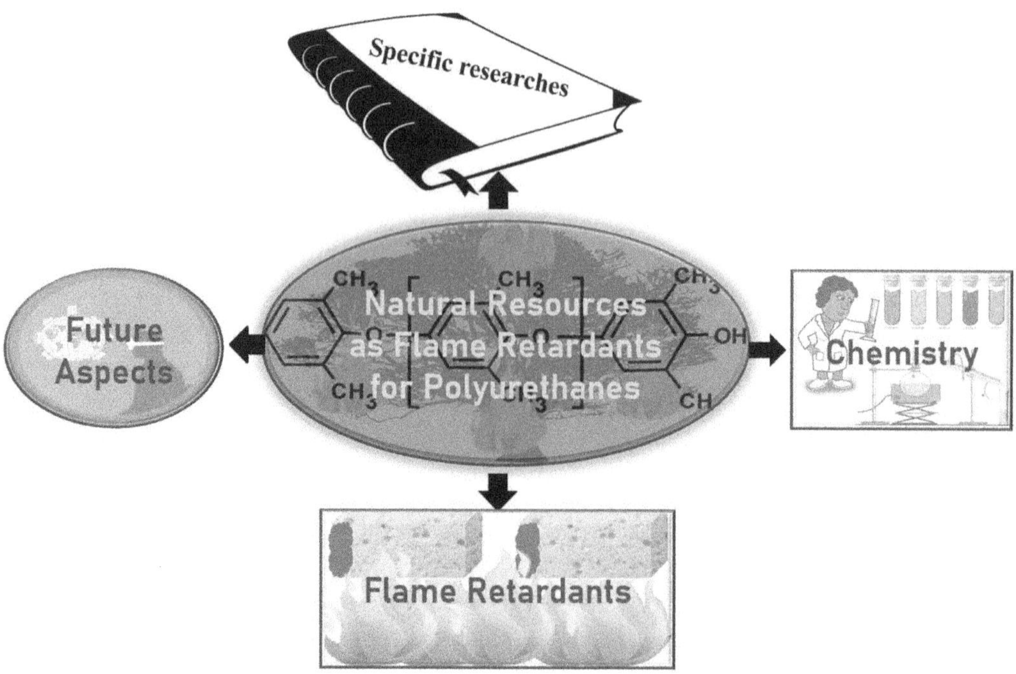

Figure 1. The natural resources available as FRs for PUs, the chemistry of the composites, and the mode of action in FRs for PUs.

Specific Research on the Applications of Natural Resources as FRs for PUs

In their review work, Adetunji et al. (*5*) documented that PUs are formed from the reaction between isocyanate and polyol molecules in the presence of enzymes and light. They also reported that polyols can be obtained from renewable sources such as biomass. Zhang et al. (*6*) reported that lignin-based phosphorus-containing FR was developed and utilized for the production of PU. Their study revealed that a chemical reaction between 9,10-dihydro-9-oxa-10-phosphaphenanthrene-10-oxide, hexamethylene diisocyanate, and lignin produced a novel lignin-based FR (LHD). Different spectral analyses were used for the characterization of the structural components, such as ^{1}H NMR, Fourier transforms infrared (FTIR), and ^{31}P NMR spectroscopy. LHD-PU was generated from lignin, hexamethylene diisocyanate, and polyethylene glycol through a chemical reaction, and it exhibited excellent FR-PU characteristics. It was also noted that the excess of polyphenol biopolymeric material and substantial aromatic structures helped lignin produce a carbon layer, and it showed a certain degree of flame retardancy. To increase the flame retardancy of FR lignin-based PU, the LHDs were added on its surface to entrap the FR lignin-based PU combustible products made in the interior swelling carbon layer. This prevented the transfer of internal and external heat and combustible gases.

Sanket et al. (*7*) reported that many renewable biobased natural resources are available for use as FRs. In their study, they highlighted the roles of orange peels, castor, limonene, vegetable oils, sunflower oil, corn oil, polyethylene, butanediol, neem oil, polypropylene glycol, linseed oil, and soybeans, among other materials, combined with a newly improved phosphorous-based polyol in different amounts in the development of FR-PU foams (PUFs). These composites were reported to have superior improved strength, minimal self-extinguishing time, reduced weight loss, a low peak heat release rate, a reduced total amount of smoke release, a decreased total heat release, and a reduced overall smoke generation rate. They attributed these properties to the formation of a protective char coating over the foam surface by phosphorus compound decomposition. This demonstrated a novel green technological approach for developing high-grade FR-PUFs.

Stapleton et al. (*8*) investigated products for children with PUF to assess the frequency of using PU in the products. Documentation was made on details from the manufacturers and information showing whether the products met international standards. Identification of the nature of the flames and their concentrations in the foam was also done; these samples were selected based on baby products that were commonly used. The findings revealed the presence of identifiable FR additives in 80 samples, which were either brominated or chlorinated. They also revealed that the most commonly employed FRs were phosphate, trisisopropy, and components mostly found in Firemaster 550 in its commercial form.

In their work, Chen et al. (*9*) produced a new FR, which was a phosphorus-based (3-triethoxysilicon) propylamine (DPTP). They also assessed the impact of DPTP on the thermal stability, mechanical strength, and extent of flame retardancy in waterborne rigid PUF (RPUF). They reported that the incorporation of DPTP in the hard PUF could significantly reduce the smoke density and thermal properties that were undesired, bringing about an increase in the limiting oxygen index (LOI). When compared with the pure form, it was also observed that the introduction of DPTP reduced the compressive strength.

Jimenez et al. (*10*) prepared an open-cell flexible PUF, which was coated with a cold plasma-induced graft-polymerization of phosphonate-containing precursors (diethylvinylphosphonate, or DEVP), with or without a cross-linking agent (1,4 butanedioldiacrylate). This could induce the flame retardancy in the PUF. The plasma-treated foam was tested to UL-94 horizontal without sonication.

The surface of PUF was covered with two DEVPs, nongrafted and grafted, with a total weight gain of about 14%. Under the flame test, the nongrafted DEVP quickly vaporized at a low temperature of 100–150 °C, and it was also observed that PO radicals were formed, which scavenged H^- and OH^- radicals, helping them to inhibit the flame for about 10–60 s. The process after was sonication, in which only grafted DEVP was found at the PUF surface with a total gain of 6%. The burning test showed that the flame occurred with a quick charring process caused by the presence of a phosphorous reaction that protected the covered PUF from burning, and the flame was removed as it extinguished itself.

In their study, Jin et al. (*3*) attempted to formulate a green system of flame retardancy that was environmentally compatible. They used expandable graphite (EG) and polyisocyanate PU in combination with aluminum hydroxide (ATH) to efficiently enhance flame retardancy. There was an increase in the level of LOI to 84.2, introducing 24 phr (part per hundred of the matrix) EG and 50 phr ATH into the matrix. Results from the thermogravimetric analysis (TGA) and scanning electron microscopy revealed that ATH could efficiently induce "villi"-shaped substances on the overall surface of EG, thereby making the intumescent char denser.

Weil and Levchik (*11*) did a review covering combustion modifier types of FRs used commercially to formulate water-based PUs and isocyanates that emphasized foams, along with a shorter coverage of coatings and elastomers. In their work, they explored various reaction factors, including catalysts, blowing agents, polyol choices, and other related factors. A detailed discussion on performance factors, such as fogging and scorching, was also done regarding the selected FRs for foam.

Durganala (*12*) investigated novel FRs using a pyrolysis combustion calorimeter, which aids in properly comprehending the potential impact of the reduction in heat released because of phosphorus boron structures. Their work showed that pyrolysis-Combustion Flow calorimetry could be employed to understand the chemistry of flame retardancy and the impact of a new FR before mixing it with the polymer. They also suggested that there is a need for more work to prove such a hypothesis. Vahabi et al. (*13*) did a comprehensive review intending to investigate advances in developing biobased FR-PUs. The chemical structures of the FR-PUs were done based on flame retardancy. Also, various strategies for improving flame retardancy of such PUs were highlighted.

Chang et al. (*14*) investigated the flame retardancy in a poly(butylene terephthalate) engineering plastic, with a biobased FR that was produced from phosphorylated downstream corn oil and phosphorylated linseed oil. The study involved incorporating various phosphorus moieties into the backbone vegetable oil through a reaction that involved reopening. A confirmation of the chemical composition and structure of the phosphorylated oil was done using NMR and FTIR spectroscopy. It was observed that the introduction of about 7.5 wt % of the phosphorylated oil was enough to bring about a change in the UL-94 fire class of poly(butylene terephthalate)s from the initial nonrating to V-0. An evaluation was also done on the mechanism of flame retardancy using TGA. It was concluded that the combination of the effects of the dripping tendency in the blends and the gas-phase mechanism aided in retarding the propagation of the flame.

In a related study, Zhu and Xu (*15*) synthesized a melamine–formaldehyde resin modified with ethylene glycol using melamine, ethylene glycol, and paraformaldehyde; the RPUFs were produced using a combination of polyisocyanate, ethylene glycol–modified melamine, and polyols. The impact of dimethyl methylphosphonate (DMMP) and ammonium polyphosphate on the flame retardancy, stability to heat, mechanical properties, and morphology of the prepared products were also investigated. It was deduced from the study that the FR efficiency of glycol-modified melamine resin was improved through the addition of ammonium polyphosphate and DMMP.

Gong et al. (*16*) carried out a synthesis of phosphaphenanthrene groups that contained soybean oil–derived polyol (DSBP) using dihydro-oxa-phosphaphenanthre-10-oxide and epoxidized soybean oil. The soybean oil-based polyol (HSBP) was produced using water and epoxidized soybean oil. Characterization of the structural composition of HSBP and DSBP was done using NMR and FTIR spectroscopy. Preparation of the RPUF was carried out by mixing HSBP and DSBP. The findings showed a decrease in compression strength and apparent density of the PUF, with an increase in the content of DSBP. An investigation into the thermal degradation and flame retardancy was done using a vertical burning test, LOI, and TGA. This study revealed that the FR impact was basically in the condensed and gas phases.

Kirpluks et al. (*17*) formulated biobased PUFs that were very rigid and investigated the possibility of reducing their flammability by using a sustainable FR. It was observed that various EG-based FRs brought about a remarkable reduction in flammability, while still ensuring a low thermal conductivity in the insulating material.

In their work, Tu et al. (*18*) used phosphate-adsorbed composites obtained as byproducts during the synthesis of magnesium oxide from phosphorus tailings in a hydrothermal approach as efficient FRs for PUs of thermoplastic origin. The results from the study showed a remarkable reduction in the heat released, rate of production of smoke, and rate of release of carbon dioxide. They concluded that this novel approach was efficient as a green tool in environmental protection, recycling resources, and enhancing the fire safety level of polymer materials.

Liu et al. (*19*) investigated a novel composite coating on a PU sponge from sodium lignosulfonate and bioderived phytic acid through a facile soaking approach. It was observed that unlike a pristine sponge, the coated PU could reach a 32% decrease in the rate of heat release during the cone calorimeter experiment. In addition, it brought about prevention of melt dripping after ignition by a butane torch. The composite coating also showed potencies against microorganisms. The coating materials were additionally revealed to be recyclable.

Kapatel and Patel (*20*) improved the flame retardancy of PU materials by incorporating phosphorus moieties into the polymeric materials. Various waterborne FR-PUs were produced through the variation of an NCO:OH mole ratio in the prepolymer mixing process. The water-based PUs produced were coated on glass, wood, and steel panels. NMR and FTIR spectroscopy were used for the characterization step.

In a related work, Zhang et al. (*1*) prepared PUF that was FR in nature from compatible mixtures of phosphorus polyol and HSBP. The synthesis was done through polymerization involving propylene oxide and polyether polyol (THPO-PO). A ring-opening reaction with lactic acid was employed in the production of soybean oil from epoxidized soybean. There was a drop in the density of the foams with an increase in the content of THPO-PO. The overall yield of the PUF was found to decrease first and then rise with the introduction of THPO-PO.

Nontasak et al. (*21*) investigated the formulation of an intumescent coating fire retarding for wood based on modified natural rubber. Their study was the first in the production of a novel modified natural rubber using the solution polymerization method. The intumescent coating fire retarding was produced through the compounding of the synthesized products with the intumescent additive and curing agent, resulting in the coating compound's formation. Cross-linking was achieved through reactions with polyisocyanate under ambient conditions.

Chemistry of FRs for PUs

The overview of a covalently connected FR organophosphonate ester into a PUF was accomplished through the CuAAC "click" reaction of an azidoalkylmonophosphonate and an alkyne–polyol. Many of these functionalized materials were arranged in four different steps. In the first step, a quantity of azidoalkylmonophosphonate complexes was made through nucleophilic substitution of bromoalkyl phosphonates by sodium azide. On the second step, polyols with terminal alkyne groups were made through anionic ring-opening copolymerization among glycidyl propargyl ether and propylene oxide. This was followed by "clicking" the aminoalkyl phosphonate to the polyol and, finally, there was a mixture of the PUF by means of 2.4 wt % of ''click polyol'' (*22*).

In a struggle to produce an environmentally friendly FR scheme for hard polyisocyanate PUFs, ATH combined with EG was primarily applied to increase flame retardancy. Inadequate oxygen index improved to 84.2 by combining 24 parts per hundred of expandable matrix graphite and 50 parts per hundred of matrix ATH into the matrix. However, with TGA and scanning electronic microscopy, it was concluded that ATH would effectually create "villi" like a constituent part on the external part of EG that made the intumescent carbonize thicker. Several of the compressed carbonized coatings possibly would efficiently obstruct the heat transport and bubbles. ATH and EG enhanced the primary degradation, and fleecy carbonize was rapidly produced externally. This resulted in the deceleration of degradation products of the compound and the delayed dispersal of explosive, combustible remains to the flame zone (*3*).

The usual FR was tris(1,3-dichloroisopropyl) phosphate; the frequency of detection (36%), following through constituents characteristically, originated in the Firemaster 550 viable combination. The two chlorinated organophosphate FRs that were not previously noted in the environs were also recognized, were industrially traded as V-6, and comprised tris(2-chloroethyl) phosphate as contamination. In addition, a convenient X-ray fluorescence analyzer was applied to assess the chlorine and bromine constituent of the foam, and also to examine through X-ray fluorescence whether it was a valuable technique for predicting the existence of halo-generated FRs as additives (*8*).

The FRs that have had active development or are in commercial use lead to prospective profitable applications in isocyanurates and PUs. This emphasizes foams and a brief coverage of coatings and elastomers. It also entails the factors of a catalyst, polyol choice, blowing agent, and surfactant that influence the efficiency of FR isocyanurates and PUFs (*11*).

Modes of Action Involved in the Application of Natural Resources as FRs for PUs

Sanket et al. (*7*) reported that the mode of action of some natural resources, like mercaptenized oils, involved the provision of thiol-ene-based reactions together with several structural changes, resulting in a wide range of physical characteristics useful for diverse applications. Zhang et al. (*6*) reported that the mode of interaction in the application of natural resources as FRs in PUs involved valuable π–πinteraction of the aromatic groups between 9,10-dihydro-9-oxa-10-phosphaphenanthrene-10-oxide and benzyl of lignin. The esterification process of the hydroxyl group generated in the decomposition process with the phosphoric acid resulted in the conversion of 15% lignin in LHD to a P–O–C cross-linked structure. LHD has been revealed to facilitate cross-linkage of free lignin during the production of PU. FR application has an important role in saving property and life by decreasing the overall fire risk connected by applying highly flammable raw materials in products, including clothing, textiles, plastics, and composites. There has been an increased demand for fresh FR possibilities because of their promising environmental profiles and cost-effective and durable products (23). Some renewable materials are being increasingly applied in several practical uses instead of short-term application products. Between some other applications, the flame retardancy of those kinds of polymers must be increased for practical uses because of their prospective fire risks and participation in daily life. Several FR complexes based on nonconventional and conventional methods include nitrogen-based FRs, inorganic FRs, nanofillers, and halogenated FRs, which eliminate these potential risks when synthesized. Therefore, they persist ecologically for a long period of time (24).

The types of PUFs that are flexible have widely been used in several industries, including upholstered mattresses, furniture, and motors. The high air permeability, chemical nature of the PU, and high interior surface part of the foam edifice make most of these materials extremely combustible. As a result, the use of FRs in the direction of flexible PUFs is a significant concern. The halogenated FR application is not among the best. Therefore, melamine as a nonhalogenated FR is used almost all of the time. Particularly, the improvement of melamine's high heat capacity and viscosity led to a reduction in rising height and foaming growth through increasing melamine content (25). Materials to be used in the future for thermoplastics should involve a lesser influence on the environment. Authentic thermoplastic materials similarly need "greener" additives for fire, as well as thermal resistance profiles. A study by Konig et al. (25) addressed halogen-free FRs from the perspective of feasible routes with industrial and economical significance. Also, several of the major FRs—such as inorganic, hybrid, or organic ones—can be applied as normal platforms for the ensuing kinds of thermoplastic syntheses, such as synthetic, biobased, and natural ones (26).

Ding et al. (27) prepared phosphorus- and nitrogen-containing FR-PU prepolymers (FRPUPs), which were further cured with castor oil to prepare FR-PU sealants. These enhanced the char residues of PU sealants. Figure 2 shows the possible thermal degradation process, in which initially, the FRPUP is heated from room temperature to 308 °C to break P–O–C and P–C bonds and form phosphoric acid and some H_2O and CO_2. Then, the urethane groups formed between CO and FRPUP are degraded into hydroxyl and isocyanate compounds at 308–378 °C. However, as the temperature increases, more H_2O and CO_2 evolve with the conversion of hydroxyl and isocyanate groups into hydrocarbons and carbodiimide compounds. At the same time, phosphoric acid is formed, which increases the char formation, and finally, a stable layer of char is formed at 378–459 °C. This consists of polyaromatic structures and phosphorous species.

Organic molecules can be achieved from commercial waste streams or natural sources, which can function as active feedstocks for an extensive assortment of polymer products. From epoxies, foams, and mixtures to bulk plastics, thermal stability, mechanical adaptations, and biomolecules, they show processability to accomplish existing material desires. The known biosourced, thermooxidative, environmentally safe, and FR additives are from plant fibers, animal tissues, food waste, and other natural resources. Their integration methods into the environment, environmental safety, mediums by which they are bioderived, flame retardancy, and thermooxidative nature should be appropriate will likewise be made available (28).

A novel two-step process was initiated to produce biopolyol from crude glycerol-based liquefaction of wheat straws, followed by esterification of fatty acid (29). This biopolyol was further utilized to develop FR-PUFs in the presence of 25% FRs composed of DMMP and EG, as shown in Figure 3 (29). These conjugations of DMMP and EG in the biobased PUFs reduced the smoke release rate and considerably enhanced smoke suppression performance. It was observed that the FR-PUFs showed excellent thermal insulation and, therefore, this wheat straw– and crude glycerol–derived PUF could be recognized as a promising material for FR insulation applications.

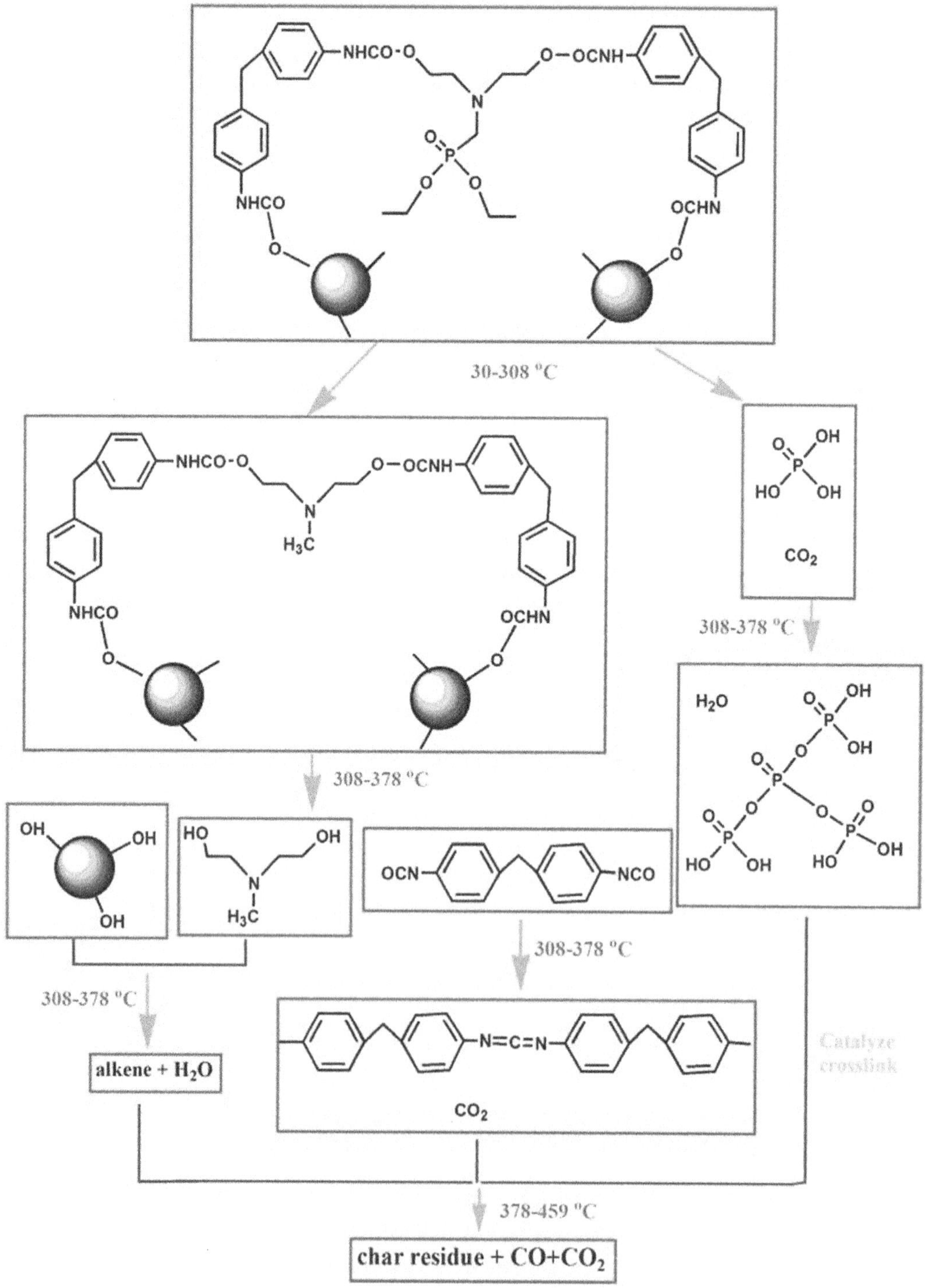

Figure 2. Mechanism of thermal degradation for FR biobased PU. Reproduced with permission from reference (27). Copyright 2016 Springer.

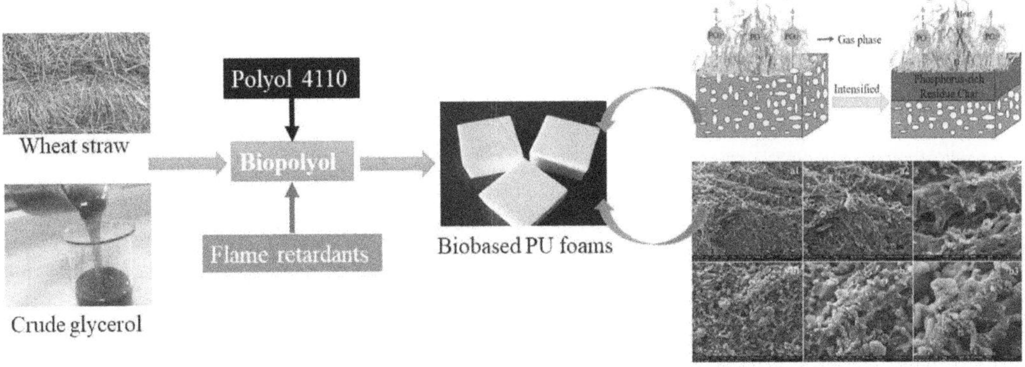

Figure 3. Wheat straw- and crude glycerol-based FR-PUs. Reproduced with permission from reference (29). Copyright 2021 Elsevier.

Most PU materials are quite expansively used in the construction industry as important sensible insulation for dual decker board applications, including roof panels, cladding, and structure insulation. The quest for PU, which has gained popularity, has created significant advantages for reduced energy reduction. In RPUF, foams are flammable materials characterized by quick flame spread, high heat release rates, and the capability to generate large amounts of poisonous gases in an innovative form. FR additives are thus applied to increase the thermal properties of PUs. Some viable additives currently applied are dangerous to the environment and humans, and their wide uses are inadequate because of their adverse effects on polymer mechanical properties. In light of this, obtaining materials that are ecologically harmless and friendly to humans is a pressing demand. Some other unconventional additives that are now in research are fly ash and clay, which are recyclable and natural resources that could improve the FR properties of several polymers (*30*).

Conclusion and Prospects

This chapter has highlighted the important properties of natural polymers as FRs in various PU applications. The chapter has discussed the role of certain natural materials—such as plant fibers, animal tissues, and other natural resources—with tremendous physicochemical properties; for example, they are ecologically harmless and friendly to humans, have mechanical strength, have thermal potential, and are biocompatible. This review also addressed halogen-free FRs from the perspective of feasible routes with industrial and economic significance. In addition, several of the major FRs—such as inorganic, hybrid, or organic ones—can be applied as normal platforms for the ensuing kinds of thermoplastic syntheses. More studies should be carried out to identify organic molecules derived from commercial waste streams or from natural sources that can equally function as active feedstocks for an extensive assortment of polymer products.

References

1. Zhang, K.; Hong, Y.; Wang, N.; Wang, Y. Flame retardant polyurethane foam prepared from compatible blends of soybean oil-based polyol and phosphorus containing polyol. *Journal of Applied Polymer Science* **2017**, 45779 (1), 1–10, DOI: 10.1002/app.45779.
2. Zhaojun, L.; Qianqiong, Z.; Ruilan, F.; Xiaoxue, Y.; Fuli, T. Flame retardancy and thermal properties of rigid polyurethane foam conjugated with a phosphorus–nitrogen halogen-free

intumescent flame retardant. *Journal of Fire Sciences* **2020**, *1* (1), 1–18, DOI: 10.1177/0734904119890685.

3. Jin, J.; Dong, Q.; Shu, Z.; Wang, W.; He, K. Flame Retardant Properties of Polyurethane/Expandable Praphite Composites. *Procedia Engineering* **2014**, *71*, 304–309.
4. Sabyasachi, G.; Matthias, N.; Olivia, C.; Shuyu, L.; Henri, M. Flame retardant polyurethanes based on novel phosphonamidate additives. *International Association for Fire Safety Science* **2014**, *23* (21), 821–831, DOI: 10.3801/IAFSS.FSS.11-821.
5. Adetunji, C. O.; Olaniyan, O. T.; Anani, O. A.; Inobeme, A.; Mathew, J. T. Environmental Impact of Polyurethane Chemistry. American Chemical Society. Polyurethane Chemistry:. *Renewable Polyols and Isocyanates* **2021**, 393–411.
6. Zhang, Y. M.; Zhao, Q.; Li, R.; Zhang, J.; Zhang, Y. M. Synthesis of a lignin-based phosphorus-containing flame retardant and its application in polyurethane. *RSC Adv.* **2018**, *8*, 32252, DOI: 10.1039/c8ra05598j.
7. Sanket, B.; Ionescu, M.; Kahol, P. K.; Gupta, R. K. Sustainable flame-retardant polyurethanes using renewable resources. *Industrial Crops & Products* **2018**, *123*, 480–488.
8. Stapleton, H. M.; Klosterhaus, S.; Keller, A.; Ferguson, L.; Bergen, S. Identification of Flame Retardants in Polyurethane Foam Collected from Baby Products. *Environ. Sci. Technol.* **2011**, *45* (12), 5323–5331. https://doi.org/10.1021/es2007462.
9. Chen, Y.; Jia, Z.; Luo, Y.; Demin Jia; Li, B. Environmentally Friendly Flame-Retardant and Its Application in Rigid Polyurethane Foam. *International Journal of Polymer Science* **2014**, 2014, Article ID 263716. https://doi.org/10.1155/2014/263716.
10. Jimenez, M.; Lesaffre, N.; Bellayer, S.; Dupretz, R.; Vandenbossche, M.; Duquesne, S.; Bourbigot, S. Novel flame retardant flexible polyurethane foam: plasma induced graft-polymerization of phosphonates. *RSC advances* **2015**, *5* (78), 63853–63865.
11. Weil, E.; Levchik, S. V. Commercial Flame Retardancy of Polyurethanes. *Journal of Fire Sciences* **2014**, 22. https://doi.org/10.1177/0734904104040259.
12. Durganala, S. *Synthesis of Non-Halogenated Flame Retardants For Polyurethane Foams*. Master of Science Thesis, School of Engineering, University of Dayton, Aug 2011.
13. Vahabi, H.; Rastin, H.; Movahedifar, E.; Antoun, K.; Brosse, N.; Saeb, M. polymers Review Flame Retardancy of Bio-Based Polyurethanes: Opportunities and Challenges. *Polymers* **2020**, *12*, 1234, DOI: 10.3390/polym12061234.
14. Chang, B. P.; Thakur, S.; Mohanty, A. K. Novel sustainable biobased flame retardant from functionalized vegetable oil for enhanced flame retardancy of engineering plastic. *Sci. Rep.* **2020**, *9*, 15971. https://doi.org/10.1038/s41598-019-52039-2.
15. Zhu, H.; Xu, S. Preparation of Flame-Retardant Rigid Polyurethane Foams by Combining Modified Melamine–Formaldehyde Resin and Phosphorus Flame Retardants. *ACS Omega* **2020**, 5 (17), 9658–9667, DOI: 10.1021/acsomega.9b03659.
16. Gong, Q.; Wang, N.; Zhang, K.; Huang, S.; Wang, Y. Preparation and Characterization of Soybean Oil-based Flame Retardant Rigid Polyurethane Foams Containing Phosphaphenanthrene Groups:. *Letters in Organic Chemistry* **2020**, *17*, 10, DOI: 10.2174/1570178617666200224104336.

17. Kirpluks, M.; Cabulis, U.; Avots, A. Flammability of Bio-Based Rigid Polyurethane Foam as Sustainable Thermal Insulation Material. In *Insulation Materials in Context of Sustainability*; InTech, 2016. https://doi.org/10.5772/62539.
18. Tu, B.; Zhou, K.; Zhou, Q.; Gonga, K.; Hu, D. (2021) Advance Waste to resource: preparation of an efficient adsorbent and its sustainable utilization in flame retardant polyurethane composites. *RSC Advances* **2021**, *17*.
19. Liu, C. Z.; Luo, Y.; Wang, Y.; Li, J.; Wang, H. Multifunctional polyurethane sponge coatings with excellent flame retardant, antibacterial, compressible, and recyclable properties. *Composites Part B: Engineering* **2021**, *215*, 108785.
20. Kapatel, P. M.; Patel, R. H. Green Approach for the Development of Novel Flame Retardant Waterborne Polyurethanes: Synthesis and its Characterizations. *Materials Today Proceedings 2020* **2020**, *23* (2), 389–399.
21. Nontasak, W.; Thongnuanchan, B.; Ninjan, R.; Lopattananon, N.; Wannavilai, P.; Nakason, C. Fire-retardant wood coating based on natural rubber bearing methacrylic functionality. *Polym Eng* **2021**, *41* (1), 44–53. https://doi.org/10.1515/polyeng-2020-0092.
22. Borreguero, A. M.; Sharma, R.; Spiteri, C.; Rodríguez, J. F. A novel click-chemistry approach to flame retardant polyurethanes. *Reactive and Functional Polymers* **2013**, *73* (9), DOI: 10.1016/j.reactfunctpolym.2013.06.003.
23. Kundu, C. K.; Li, Z.; Song, L.; Hu, Y. *Flame Retardant Treatments of Nylon Textiles: A Shift towards Eco-Friendly Approaches* **2020**, DOI: 10.5772/intechopen.94880.
24. Maqsood, M.; Seide, G. Biodegradable Flame Retardants for Biodegradable Polymer. *Biomolecules* **2020**, *10* (7), 1038.
25. Konig, A.; Fehrenbacher, U.; Hirth, T.; Kroke, E. Flexible Polyurethane Foam with the Flame-retardant Melamine. *Journal Of Cellular Plastics* **2008**, *44*, 469–480, DOI: 10.1177/0021955X08095502.
26. Nicoleta, L.; Zina, V.; Mircea, T.; Mihai, C. C. Halogen-free flame retardants for application in thermoplastics based on condensation polymers. *SN Applied Sciences* **2019**, *1*, 422, DOI: org/10.1007/s42452-019-0431-6.
27. Ding, H.; Xia, C.; Wang, J.; Wang, C.; Chu, F. Inherently flame-retardant flexible bio-based polyurethane sealant with phosphorus and nitrogen-containing polyurethane prepolymer. *Journal of Materials Science* **2016**, *51* (10), 5008–5018.
28. Watson, D.; Schiraldi, D. A. Biomolecules as Flame Retardant Additives for Polymers: A Review. *Polymers* **2020**, *12*, 849, DOI: 10.3390/polym12040849.
29. Chang, C.; Liu, L.; Li, P.; Xu, G.; Xu, C. Preparation of Flame Retardant Polyurethane Foam from Crude Glycerol Based Liquefaction of Wheat Straw. *Ind. Crops Prod.* **2021**, *160*, 113098. https://doi.org/10.1016/j.indcrop.2020.113098.
30. Rothon R. Sustainable and Recycled Particulate Fillers. In *Polymers and Polymeric Composites: A Reference Series*; Palsule, S., Ed.; Springer, Berlin, Heidelberg. 2021. https://doi.org/10.1007/978-3-642-37179-0_21-1.

Chapter 2

Fire Retardancy of Polysaccharide-Based Polyurethane Foams

Trinath Biswal[1] and Prafulla K Sahoo[*,2,3]

[1]Department of Chemistry, VSS University of Technology, Burla 768018, India
[2]Department of Chemistry, Utkal University, Vani Vihar, Bhubaneswar 751004, India
[3]School of Applied Sciences, Centurion University of Technology and Management, Jatni 752050, India

[*]Email: psahoochemuu@gmail.com

Polyurethane foams (PUFs) are considered a significant class of polymer-based foam because of their excellent thermal conductivity, low density, amazing mechanical properties, sound-insulation qualities, and very good fire retardancy. Although PUFs are usually extracted from synthetic petroleum-based materials, polysaccharide-based PUFs are now the subject of focus because they are more environmentally friendly. During combustion, PUFs generate a huge quantity of vision-impairing smoke, which is highly toxic because it contains extremely high amounts of CO and HCN. PUFs are made flame retardant (FR) by the addition of suitable additives such as phosphorus-based materials, nitrogen-based additives, halogen-based materials, silicone-enriched materials, and some miscellaneous additives of inorganic and organic compounds. Other fire-resistant groups—such as isocyanates, carbodiimides, and nitrogen-based heterocyclic compounds—also exhibit fire retardancy properties for PUFs. The incorporation of suitable FR additives inhibits the smoke level, flammability, and toxicity of PUFs. New kinds of FR coating materials have been developed and designed for rigid PUFs by utilizing mixtures of different polysaccharides and amorphous sodium polyborate. The fire retardancy of rigid PUF can be enhanced by addition of graphite as a sealant material along with 2-carboxyethyl(phenyl)phosphinic acid and melamine phosphate as additives.

Introduction

Polysaccharide is a macromolecule consisting of many small molecules known as monosaccharides. The monosaccharide molecules are simple sugar molecules like fructose or glucose. There are special kinds of enzymes that bind these small monosaccharide molecules (monomers) together, producing macromolecules of sugar commonly termed polysaccharides. Polysaccharides are also called glycans. Depending upon the kind of monosaccharide molecules, polysaccharides may be classified into two categories: homopolysaccharides, where the monosaccharides are of the same kind, and heteropolysaccharides, where the monosaccharide molecules are of different kinds. The polysaccharides are classified into two different categories depending on the binding of the carbon atom of the monosaccharide units: linear or straight chain polysaccharides, or branched chain polysaccharides (*1*, *2*).

In linear polysaccharides the monosaccharide molecules are simply joined by covalent bonding, whereas in branched chain polysaccharides these are present in the form of turns and arms. Polyurethanes (PUs) are formed by the chemical reaction between the hydroxyl groups (–OH) of a polyol with the isocyanate groups (–NCO) of an isocyanate; the linkage is commonly termed urethane linkage. PU is considered a significant class of polymeric material, which is used in many fields associated with human life (*3*). PU foams (PUFs) contribute 50% of the whole polymer foams in commercial global markets. PUFs are mainly classified into two categories: rigid foams and flexible foams. Again, the PUFs are classified as flexible molded foams, reaction injection molded foams, flexible PU slabs, carpet backing foam, and so forth. The polymer PUs, normally a diversified group of synthetic polymeric materials, were first developed by Otto Bayer in 1937. PUF is an interesting material that is commonly used in various applications such as insulating material in refrigerators and buildings, car cushions, upholstered furniture, seats in trains, and various kinds of mattresses (*4*). PUFs are combustible in nature; therefore, before practical use, it is necessary to make PUF as flame retardant (FR) as possible to avoid fire hazard. The materials extensively used as additives in PUF to enhance fire retardancy include tris(1-chloro-2-propyl)phosphate, tris(2-chloroethyl)phosphate, and tris(1,3-dichloro-2-propyl phosphate (*5*). Halogenated FR materials are usually toxic in nature; therefore, they should be avoided because of health and environmental safety concerns. Although red phosphorus is one of the significant FR materials, it is a hazardous and a highly combustible substance. There are some other important additives, such as graphite, graphene, melamine, and layered silicate materials, which can confer FR properties to PUF. A coating of the mixture of sodium polyborate and starch adds excellent fire retardancy to rigid PU foam (RPUF). The starch acts as a binding agent of the coating, is converted into the carbonized layer upon heating, and exhibits a synergy impact with sodium polyborate (*6*, *7*).

Chemistry of PU

From a chemistry point of view, PUs are recognized as organic polymers containing urethane bonds, and these bonds are formed by the reaction between the polyols and isocyanates. The raw materials used for PU synthesis are isocyanates and polyols, which are obtained from petroleum-based materials. Biopolymers obtained from green, biological, sustainable, and renewable natural resources are now the focus of study because of their sustainability, eco-friendliness, and abundance. These biobased PUs synthesized from natural biomaterials of polyols and isocyanates are now more attractive to researchers and industry. PUs are considered the most useful materials used in the 21st century, serving a variety of different valuable purposes, such as in furnished furniture, medical devices, footwear, wall insulation, sealants, adhesives, roofs, appliances, and coatings. The RPUF is more interesting because of its improved physicochemical properties, including poor thermal conductivity, very good shock absorption, light weight, and high abrasion resistance (*8*, *9*).

Structure of PU

The PUF may be hydrophilic or hydrophobic, if used in wound-healing applications, for example, depending on the composition. However, most PUFs are hydrophobic in nature. The hydrophilic property of PUFs is because of the backbone of polyol, mainly polyalkylene oxide; normally polyester-based polyol or polyethylene oxide and suitable surfactants are added in order to develop the water-wicking properties. The hydrophilicity level and the porosity of PUFs are the measures of its wicking rate. PUF is formed by the reaction between the diisocyanate and ethane-1,2-diol in the presence of amine as catalyst and blowing agent. Scheme 1 represents the general reaction route to the formation of PUF.

The diisocyanate used may be toluene diisocyanate or methylene biphenyl diisocyanate. The blowing agent here is CO_2, which is obtained by the reaction between isocyanate and H_2O. During the process of foaming, two chemical reactions take place simultaneously. The first is the reaction between (ethane-1,2-diol) and diisocyanate, which creates a polymer network. The 1,2-ethanediol is a low-molecular-weight chain extender (in which catalysts are dissolved). The second one is the reaction between H_2O (blowing) and isocyanate forming the foam. The catalysts used are particularly amine compounds, which increase the rate of second reaction (*10, 11*).

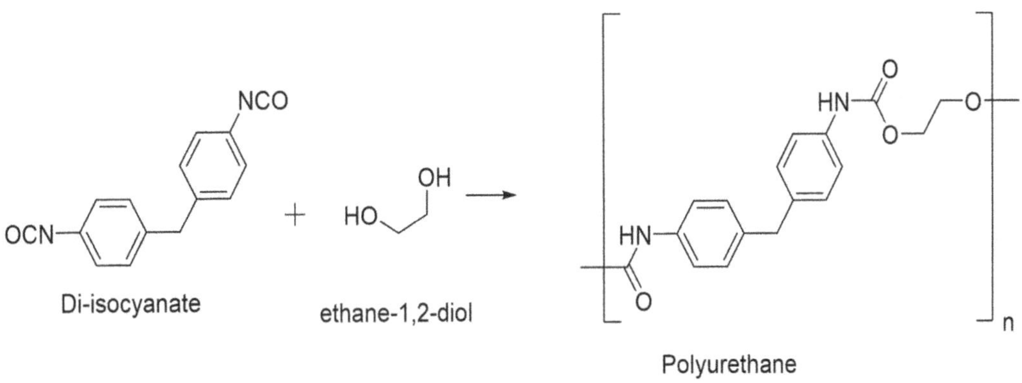

Scheme 1. General reaction route to the formation of PUF.

Properties of PUFs

PUFs are straight chain polymers consisting of organic units successively joined by carbamate linkage. The chemical formula of PUF is $C_{27}H_{36}N_2O_{10}$, which is synthesized in various ways depending on the intended use.

The various properties of the PUF are highly dependent upon its process of synthesis. For example, if the polyol chain (raw material used for synthesis of PU) is flexible and long, then the PUF obtained is elastic, flexible, and soft. Although PUFs are of different categories, the flexible PUF (FPUF) is widely used. Otherwise, if the structure of the PU is that of a three-dimensional network, then it will be of high molecular weight, will be inflexible, and will not easily soften on heating.

PU exhibits a unique foam-producing capability, which is considered the most interesting property of PU. The foam-forming capacity of PU is facilitated by the production of CO_2 gas during the process of polymerization. Another important benefit of PU is the formation of microcellular foams of high-density polymer without blowing agents (*12, 13*).

PUFs are well known, because of their excellent versatility and resilience, which make them a suitable material for use in the automotive and construction industries.

PUFs are usually resistant to a number of solvents. PUFs are normally more advantageous than polyether foams, specifically because of resistance to solvents used for dry cleaning. This is why the polyester foam is used for production of foams for textile laminates, where the solvents used must have excellent resistance to dry cleaning, performance toward flame bonding, and properties of elongation.

PUFs are degradable by the aqueous solution of alkalis, acids, and steam. The amide, urethane, and ester groups are the attacking sites for hydrolytic cleavage. The polyether foams normally have higher resistance to hydrolysis than polyester foams, because the ether groups that are present are not easily attacked.

There is a noticeable distinction in the strength of the foam at the same density, which depends upon the technique of production. The flammability of the foam depends upon the type of the blowing agent used to synthesize the cells in the foam.

The RPUF includes two kinds of polymer. The first one is the formulation of polyisocyanurate and the second one is the PU formulation; however, there is a huge difference observed in the performance, synthesis, and characteristics of both of these.

PUFs are more effective in the growth process of fungi or any other microorganisms than any other polymeric materials. Polyester PU groups are more quickly degraded through the microbial action than polyether PUs, because the ester group is more resistant to hydrolysis. In order to increase the antimicrobial properties, a wide range of appropriate microorganisms are added during the process of manufacturing PUF.

Both polyether and polyester foams gradually discolor and yellow on exposure to UV light, and the rate of yellowing depends upon the intensity of irradiated radiation. The polyether foams yellow more quickly than polyester foams, but the yellowing has no noticeable effect on the physical properties of either kind of foam. Both kinds of foam are fabricated from aromatic diisocyanates. A nonyellowing foam can be synthesized by using aliphatic isocyanates, but this is costly.

FPUFs easily undergo combustion from any source of ignition because they have more surface area and lower density. The combustion resistance of PUFs can be improved by addition of FR materials during synthesis, or after manufacturing, through the addition of FR additives at polymer lattices. The posttreatment process is normally followed if the PUF must meet the strict flammability restrictions necessary in building construction.

PUFs have good mechanical resistance, good behavior with the hydrocarbons, higher sensitivity to hydrolysis and water, good damping and soundproofing properties, and less resistance toward aging (*14, 15*).

Applications of PUFs

The most common application of PU is in the synthesis of foam, and the foams formed are utilized either directly or indirectly in many products, including household furniture, refrigerator sheets, and upholstery fabrics. PUFs are also used in the design of some specific garments.

PUs can be fabricated into door frames and columns and sometimes balusters and window headers. Low-density PUFs are quite flexible and popularly utilized in the manufacture of mattresses; PUFs are used for manufacturing blankets, fabrics, and automobile seats.

FPUFs are used for fabrication of elastic bands and belts. Because of their low density, PUs are widely used in the manufacture of footwear. PUs are also used in the fabrication of kitchen and bathroom sponges, couches, and seat cushions.

In addition, PUs are used in the manufacture of containers with sandwich panels, panels for frigorific rooms, walls of assembled dwellings, sailboards, self-supporting panels, polyester boats, soundproofing screens with foam cores, insulated doors for buildings, and double doors for offices and residences (*16, 17*).

Formulations of PUFs

PUFs are synthesized by the free rise method, and the formulation of the synthesis is represented in Table 1, which was normally in terms of 100 parts of soya-based polyol. In this process the soya-based polyol was amended with suitable additives such as catalysts, a blowing agent, and a surfactant at ambient temperature for 5 min; then isocyanate was added and the mixture was stirred for about 20 s. Then the mixture obtained was immediately transferred into a deep foil pan to facilitate foaming, and PUF was produced. The synthesized soy-based PUF foam was stored for some days for curing. Waterborne PUF was synthesized by using toluene diisocyanate and polyether polyols.

Table 1. Formulation of PUF[a]

Material	*Natural Sources*	*Parts by Weight*
Polyol	Soya-based polyol	100
Surfactant	Copolymer of silicone and furandione	2.5
Catalyst	Tibutyltin dilaurate Tertiary amine	1.5 2.5
Blowing agent	Water	5.0
Polyurea prepolymer	NCO index	130

[a] Source: Data from reference (*18*).

Foaming of PU

In the process of polymerization, solid PU is produced, and gas bubbles are formed in the polymerizing mixture, which is popularly called blowing; then, the foam is synthesized. The foam manufacturing company continuously produces slabstock or laminates and discontinuously some blocks and molded items. The FPUFs can be synthesized comfortably in different shapes and sizes by molding or cutting. The PUFs can be used in mattresses and upholstered furniture. The molding technique of flexible foam is used to make the seats, cushions, and chairs more flexible and comfortable. Because of the cleanness and economy of FPUFs, they are widely used in bedding and upholstery applications. Low-density, strong rigid foams can be produced by using adequate eco-friendly blowing agents, which form closed cell structures with very low thermal conductivity. Because of its excellent thermal insulation, it can be used in building materials, refrigerator cabinets, freezers, and refrigerated transport systems. A simple and quick process of molding can be used to synthesize articles or products of flexible and rigid foam. There are three different kinds of PUFs, which are particularly important on the basis of quality:

- Low-density flexible foams;
- Low-density rigid foams;
- High-density flexible foams (*19*, *20*).

All of these foams are particularly mentioned as integral skin foams of microcellular elastomers. The density of the low-density flexible foams is in the range of 10–80 kg/m^3, which is formed from a less cross-linked polymer with an open cell macrostructure. The low-density rigid foams are formed from extremely cross-linked polymers with basically a closed cell structure and a density ranging from 28 to 50 kg/m^3. The different cells in the foam are separated from each other through a wall of thin polymer, which can stop the rate of flow of the gas within the foam (*21*). These foaming materials provide excellent structural strength associted with their weight, combined with outstanding properties of thermal insulation. The cell normally consists of a mixture of gases and exhibits different thermal insulating properties based on nature and relative percentage of foams. To maintain long-term function, gases of low thermal conductivity are needed within the cellular structure. PUFs with densities of more than 100 kg/m^3 are called high-density flexible foams. The microcellular elastomers and self-skinning foams belong to this category. The systems of integral skin foam or self-skinning foam are normally used for manufacturing the molded parts with a cellular core, which is comparatively dense and used for decorative skin (*22*). These foams are of two kinds. The first one is the open cell core with an average density of approximately 450 kg/m^3; the second one is the microcellular core or completely closed cell with an average density of approximately greater than 500 kg/m^3.

The foam of microcellular elastomers has a complete uniform density of 400–800 kg/m^3, consisting of almost closed cells, which are not suitable for the same applications as low-density PUF. Microcellular elastomers and integral skin are used in vehicle trim, shoe soling, and molded parts for upholstery (*23*). The different PUs and their properties and applications are described in Table 2.

Table 2. Classification, Properties, and Uses of PU[a]

Name of the PU	*Properties*	*Applications*
Thermoplastic PUs	Wide range of service temperature, wide range of hardness, excellent tear resistance, outstanding compression strength, adequate resistance to nonpolar solvents, high tensile strength, good abrasion resistance, very good electrical properties	Laptop keyboard protector, outer cases of mobile and other electronic devices, caster wheels, automobile instrument panels, medical devices, sporting goods, power tools, driving belts, footwear, different extruded films, profile and sheet applications
RPUFs	Highly FR, improved thermal and physical properties, limiting oxygen index (LOI) value is high, highly improved compressive strength and thermal stability, light weight, improved compressive strength, excellent thermal conductivity, good tensile strength, good bending strength	Thermal and sound insulator
FPUFs	Excellent durability, firmness, tearing resistivity, combustibility, and surface elasticity; good flexibility; high degree of cross-linking; adequate crystallinity; reduced flammability	Cushion materials, carpet underlays, interior parts, furniture, bedding, automotive packaging, biomedicine, nanocomposites, automobile interior parts
PU ionomers	Good dispersion in polar solvents because of increased hydrophobicity; improved thermal, mechanical, and shape memory properties; excellent biocompatibility, improved hemocompatibility	Artificial hearts, connector tubes for hearts, pacemakers, hemodialysis tubes

Table 2. (Continued). Classification, Properties, and Uses of PU[a]

Name of the PU	*Properties*	*Applications*
Waterborne PUs	Excellent and versatile mechanical, chemical, and physical properties; outstanding adhesive and bonding properties; high chemical resistivity; excellent drying; low temperature flexibility; adequate scratch resistivity; anticorrosive properties; susceptibility to UV degradation	Coatings, adhesives, sealants, binders

[a] Source: Data from reference (*24*).

Fire Retardancy of Polysaccharide-Based PUFs

The property of fire retardancy means the disturbance of the combustion process at once or in multiple stages, with the whole process ending within a permitted time frame. Normally three different processes are adopted for reduction of flammability property of PUFs:

- The FRs take part in the process of foaming reactions and form chemically into the molecules of polymeric materials in combination with other initiating foam components. This process hinders the bleeding out of the polymeric material and the property of fire retardancy is retained. It does not have a plasticizing effect, nor does it influence the thermal stability of the structural component of the foam. Usually the FR additives are halogen based, nitrogen based, or phosphorus based.
- The phosphorous-based additives are phosphate, phosphonate, phosphite, organophosphorous polyols, phosphinate, and phosphonitride. The halogens and halogen-based compounds are highly effective in chlorinated or brominated states or in combination with the derivatives of both compounds. The inert FR materials do not take part in the foaming reaction and offer the property of fire retardancy based on weight. If these additive materials are compatible with the polymers, they function as plasticizers. If they are not compatible, they are used as fillers or reinforcing material.
- These materials tend to bleed and are always volatile in nature. Therefore, the FR property may progressively decrease during the process of aging. Various kinds of nonreactive additives based on nitrogen, phosphorus, boron, halogens, antimony, sulfur, silicones, aluminum, and carbon are used to enhance the FR property of PUFs. However, the combination of both additive and reactive FR materials forms a synergistic effect. Synergists are usually more significant, because they are cheaper compared with the actual FRs, and their addition as reactive FR materials highly decreases the existence of a synergist without affecting the property of fire retardancy (*25*, *26*). The different additives that are effective against PUFs are given as follows.

Polysaccharide-Based FR Materials for PUFs

Biopolymers can be classified into the categories found in Figure 1.

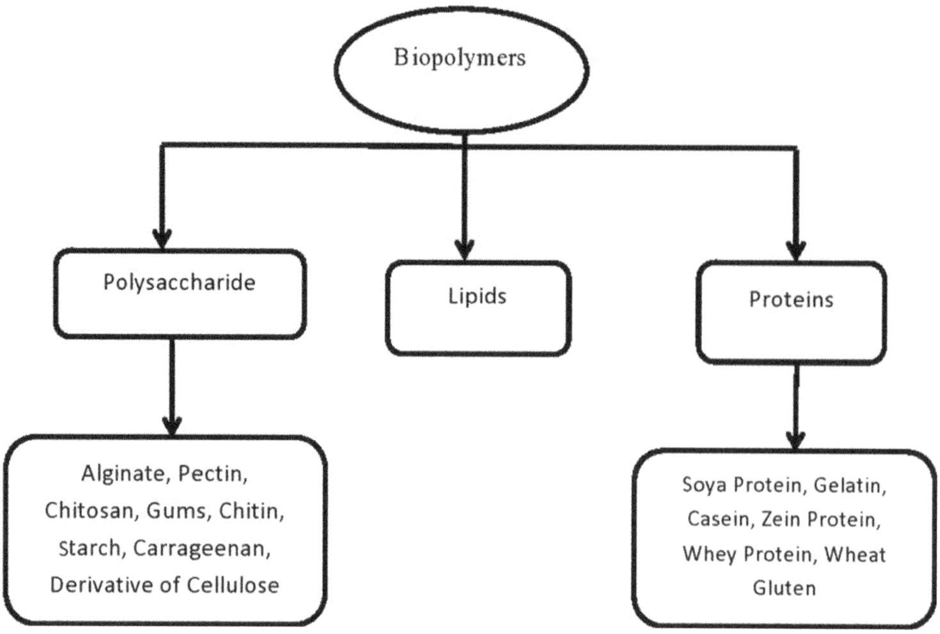

Figure 1. Classification of biopolymer.

Alginate-Based Polymeric Material for Fire Retardancy of PUFs

Alginate is a well-known biopolymer, popularly known as alginic acid; it is an anionic polysaccharide obtained from seaweed or algae and is hydrophilic in nature. Alginates have different chemical structures and vary in physical and chemical properties. Alginates are mainly derived in the form of Na alginate. They possess the capability of forming a gel-like cross-linked structure or hydrogels and can absorb water at two to three times their original weight. The alginate hydrogel becomes FR if it is cross-linked by the exchange of the Na^+ ion with other metal ions such as Ca, Zn, and Ba. The polymeric materials and their composites that are commonly utilized in automobiles, construction materials, and packaging applications possess comparatively less resistance toward degradation through reductive pyrolysis or oxidative decomposition during heating. Here, the addition of alginate hydrogels in polymeric materials improves FR property and is considered an emerging technological application. The bacterial alginate derived from the *Azotobacter* spp. has a higher content of G-blocks, and the gel produced from it is comparatively stiffer. Figure 2 shows the structural formula of sodium alginate and alginic acid (*27, 28*).

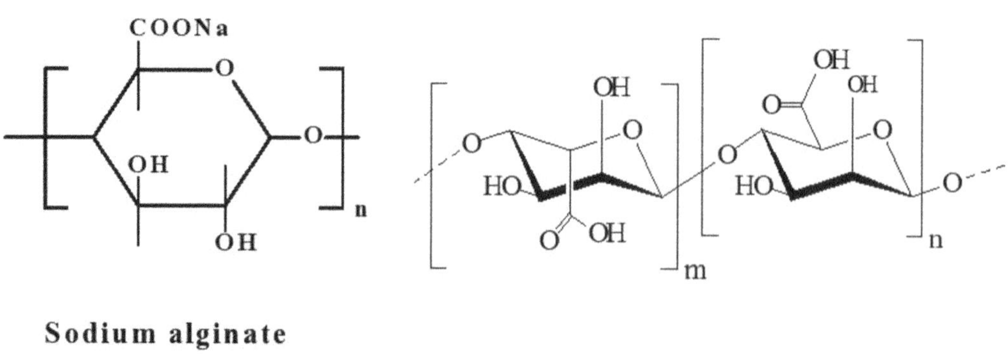

Figure 2. Structure of sodium alginate and alginic acid.

All of the forms of alginates contain three types of blocks in their chemical structure such as G-block (poly α-L-guluronic acid), M-block (ß-D-mannuronic acid), and MG-block (with both the polyuronic acids). The blocks are bonded with each other by between one and four glycosidic linkages (Figure 3) (29).

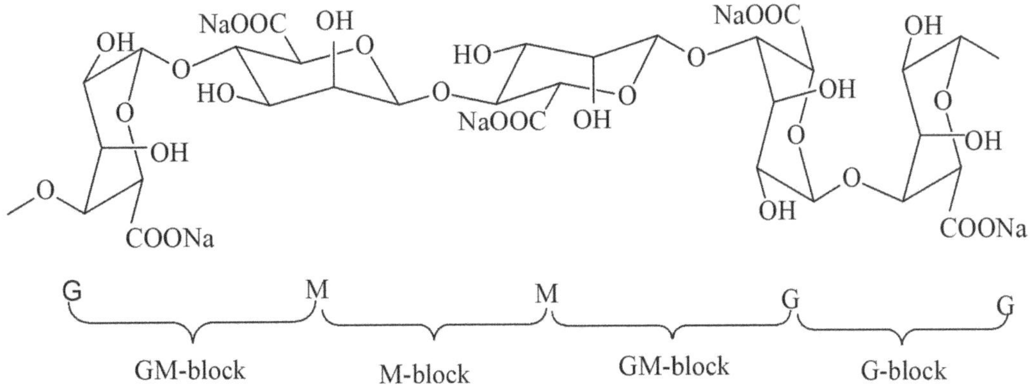

Figure 3. Different blocks of sodium alginate.

Alginate hydrogels can also be used along with the traditional and usual FR materials, such as gas-generating P, O, N, and halogen-forming compounds; H_2O-eliminating hydroxides of specific metals; vitrified silica-containing materials; and expansive substances. This is because these materials retard heat transfer. Although alginate is extensively used in important applications such as food additives, mold-release agents, textile coatings, and drug delivery, its FR properties allow use in firefighting, domestic furnishings, military apparel, interior fittings, apparel, structural coatings, and interior sound and thermal insulation. Alginate is an important biopolymer, and has many unique important properties in addition to fire retardancy; it is sustainable, flexible, biodegradable, environmentally friendly, light weight, nontoxic, inexpensive to harvest, easily processed, and easy to modify. A coating of multiple layers can be created from anionic alginate (AL) and cationic chitosan (CS), deposited on the surface of PUFs layer by layer. This coating technique has little effect on the stability of PUFs under heat, but it noticeably increases the formation of char during the process of combustion. A significant reduction of fire hazards of PUFs are because of the CS–AL coatings. They produce an insulating protective layer on the surface of the material that prevents the entry of oxygen and also heat permeation, which decreases the production of flammable gases with subsequent enhancement of fire retardancy (*30–32*).

CS-Based Polymeric Material Used for Fire Retardancy of PUFs

PUFs are a flexible, three-dimensional complex form of novel material widely used in various important sectors such as furniture, decorative objects, packaging, biomedical automobiles, and so on. However, this interesting material is highly combustible because its chemical structure facilitates air permeability into the material. This means the use of these kinds of materials has posed significant fire hazards. In order to decrease the flammability property of the FPUFs, suitable FR additives, such as phosphorous- or organohalogen-based materials, must be added (33). The structures of chitin and CS are given in Figure 4.

Montmorillonite (MMT) is an important layered mineral clay material of natural origin. Because of good cation-exchange ability, it can easily exfoliate within polymer matrix. The flammability property of PUFs is drastically decreased if the PUF surface is coated with the nanoscale form of

MMT. The deposition of nanoforms of MMT and cationic form of CS on the surface of FPUFs form a condensed phase and provide increased in fire retardancy. The peak heat release rate of the new coated PUF is suddenly decreased because of the creation of a sandwichlike topology on the surface of the FPUF. The nanosized MMT and the positively charged CS coating on the surface of the FPUF greatly enhance the fire retardancy (*34*).

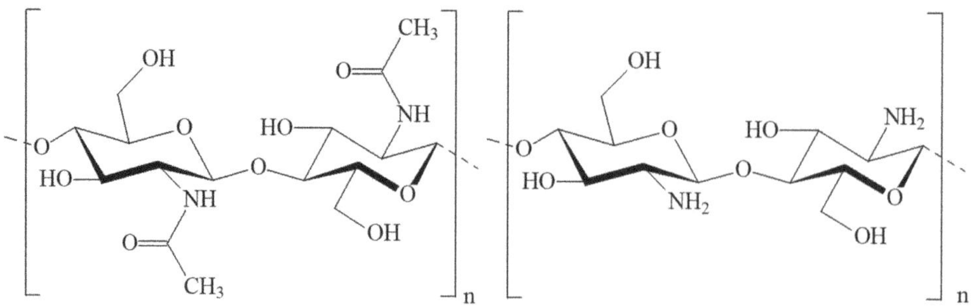

Chitin **Chitosan**

Figure 4. Structure of chitin and CS.

The fire retardancy property of the PUF by addition of FR materials is achieved by the following mechanisms:

- The FR materials initially form a condensed phase, which facilitates the creation of a highly stable carbonaceous layer residue on the surface of the PUF, popularly termed as char. This serves as a barrier in order to inhibit gaseous substances from diffusing to the flame. This protects the surface of the PUF from air and heat.
- The FR materials swell along with the degrading substance and form a porous structure, which functions as a protective barrier toward air, heat, and pyrolysis.
- The gaseous phase FR materials usually interfere with the free radicals of the reaction and retard the flammability.
- The decomposition of FR materials is an endothermic process and cools the pyrolysis area at the surface of combustion.

Among the various kinds of FR materials, halogen-based materials are more efficient, but because of their extremely toxic nature, they are strictly banned by Europe and the United States. Hence, for the past 10 years, extensive research work has been carried out to develop suitable FR materials with natural biosources having specific structural and chemical properties. Because the structure of CSs consists of *N*-acetyl-D-glucosamine haphazardly distributed over of β- (1–4)-linked D-glucosamine Figure 4, it facilitates the FR property. This is the second most widely available biopolymer, normally obtained by the deacetylation of chitin, and chitin is generally extracted naturally from shrimp and crab shells. CS can be used as FR material on its own or in combination with other FR additives. A modified form of CS with suitable FR additives is now being designed and developed as an FR material. CS, upon cross-linking with bis-(4-formylphenyl)-phenyl-phosphonate, forms a product again on combination with organo-modified MMT and ammonium polyphosphate (APP). This material, if incorporated with the PUF through melt mixing, drastically increases the fire retardancy. The mechanism of the property of fire retardancy is the development of a stable protective layer of char with simultaneous catalytic carbonization of the nanoform of MMT (clay), and the dehydration reactions stimulate the change of phosphonate into phosphonic groups.

This again cross-links CS molecules and enhances the fire retardancy. The nanosheets of modified MMT on incorporation with phosphorylated CS through melt compounding show excellent performance of fire retardancy. The formulation with modified nanoclay and AL salt with CS again show amazing performance of fire retardancy. Melamine polyphosphate may be microencapsulated through diisocyanate, and cross-linking with carboxymethyl CS exhibits good performance of FR material for the composite of thermoplastic PU. The incorporation of CS drastically decreases the peak of heat release rate and also total heat release, thereby impacting the property of fire retardancy (*35–37*).

Pectin-Based Polymeric Material Used for Fire Retardancy of PUFs

The biopolymer pectin in the form of aerogels is light weight, has low thermal conductivity and high surface area, and serves as a good FR material that can be used for thermal insulation, batteries, supercapacitors, Cherenkov detectors, and catalyst supports. The aerogels based on biopolymers are popular, cost-effective biomaterials and are attractive as FR materials. Pectin is an important natural biobased polysaccharide that has a number of unique properties such as biodegradability, nontoxicity, biocompatibility, and abundance in nature. During the past few years pectins in the form of aerogels have been used extensively as thermal insulating material because of their adequate properties of fire retardancy. Specifically, it was found that the aerogels of pectin-based materials used as super-insulating substances have thermal conductivity of nearly 0.02 W m^{-1} K^{-1}. However, these aerogels are combustible to PUFs and have less resistance toward water. The incorporation of inorganic nanoparticles, mainly $Mg(OH)_2$, $Al(OH)_3$, and clay, can form polymer-based aerogels with excellent fire retardancy. But the existence of higher percentages of inorganic nanoparticles leads to increased thermal conductivities of these hybrid aerogels. The pectin-based aerogels have increased fire resistance along with poor thermal conductivity. The low-cost, water-resistant aerogels prepared from renewable natural pectin and aniline through polymerization–coagulation, along with the method of supercritical drying, exhibit excellent FR properties. Polyaniline is normally a noncombustible polymer, which is initially utilized for modification and design of biodegradable forms of pectin-based aerogels. The FR property is because of the cross-linking between pectin and polyaniline. The aerogels of polyaniline and pectin are not only FR, but they have good water resistance and poor thermal conductivity (*38, 39*).

Cellulose-Based Composites Used as FR Material for PUFs

Cellulose is a natural biopolymer and abundantly available in nature, mostly found in cotton, ramie, linen, sisal, jute, and hessian. The biopolymer cellulose is obtained from plant tissues as microfibrils of 2–20 nm in diameter and 100–40,000 nm in length. Cellulose is mainly synthesized from wood pulp and it also is derived from an extremely hydrated state of some bacteria, like *Acetobacter xylinum*. Combustion is well defined as a fast and self-accelerating redox method with exothermic nature (*40*). Combustion takes place as a chain reaction, which is initiated and propagated through the mechanism of a free radical reaction (–OH free radical). The –OH radicals can be produced by the reaction between O_2 and the alkyl radicals. The radicals of halogen formed by the reaction of the –OH radicals with halides can function as terminator species for this kind of chain reaction. The process of burning is normally a phenomenon of the gaseous phase. After glowing, the phenomenon of burning is a nongas phase combustion, where the substrate molecules are oxidized into the condensed phase and form both gaseous and solid products. This is generally carried out at the temperature below the ignition temperature of the substance. The residue of carbon from the carbon-enriched substances is normally oxidized into the solid phase (*41*). Emman's fire triangle is usually applicable for illustrating the process of combustion, as shown in Figure 5.

In this mechanism of the process of combustion, normally three kinds of elements are required:

- Fuel: volatile combustible species form carbon-enriched materials with adequate ability of thermal degradation;
- Heat: can be provided by the oxidative destruction (exothermic reaction) of the fuel;
- Oxidizing agent: oxygen supplied from air, but in the case of polymeric materials some external source of heat is required for ignition of the substrate molecule (*42*).

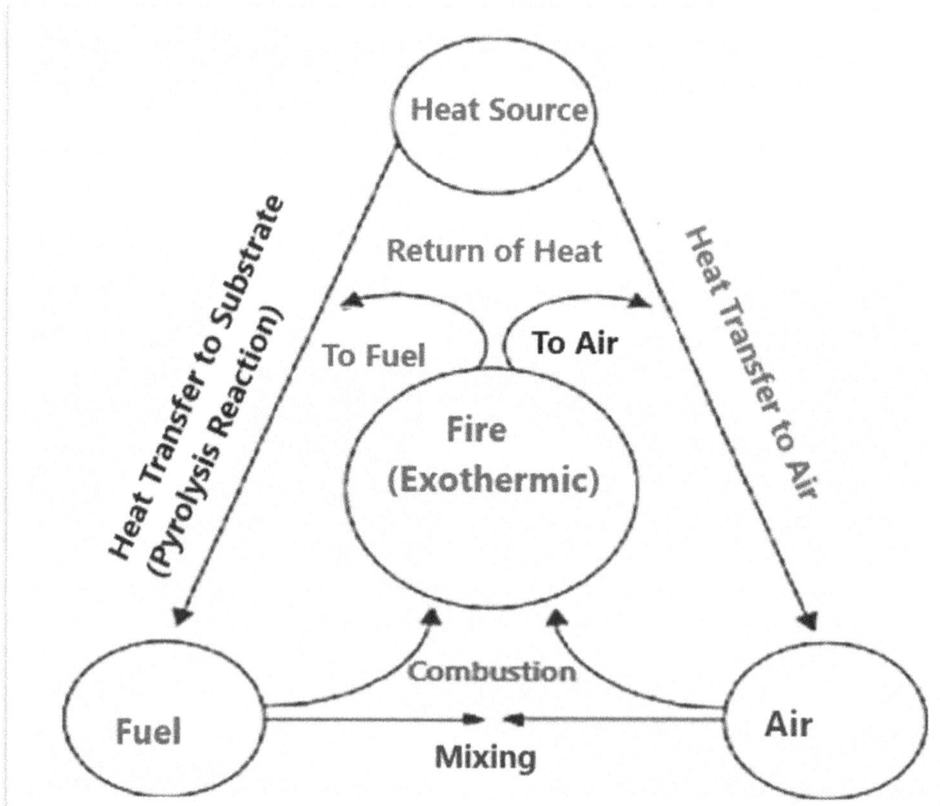

Figure 5. Emman's fire triangle explains the mechanism of the combustion process.

Cellulose is a class of organic polymeric material that can degrade into volatility at high temperature. When a sufficient quantity of these volatile species undergoes accumulation and has contact with O_2 and some ignition source, then it immediately catches fire. Here, a combustion cycle is created; substances that reduce the effectiveness of the cycle or totally stop it are termed FR. Normally, clay-based composites of cellulose such as cellulose/MMT or cellulose/kaolin are used as FR materials for PUFs. The major strategies for enhancing the property of fire retardancy of a polymeric material including PUFs are as follows:

- Complete removal or decrease in heat content;
- Decrease in the generation of volatile components through the modification of the route of thermal degradation at elevated temperature with the increase in formation of char;

- Dilution of volatile access or O_2 in order to prevent the flame;
- Retardation of combustibility of the volatile material.

FR materials are the substances that function at the gas phase or condensed phase. Normally, the compounds of Br and Cl are used as chain terminating agents in the flame. These compounds serve as deactivating materials for extremely reactive species and inhabit the thermooxidative reactions. The compounds of Cl or Br in combination with the oxides of metal release H_2O upon heating in many endothermic reactions. The cellulosic materials in combination with the halogen-, nitrogen-, and phosphorus-based FR materials develop an environmentally friendly FR novel material, which prevents combustion of PUFs because of the formation of char over its surface (*43–45*).

Starch-Based Materials Used as FRs for PUFs

Many FR materials have been developed and designed and are widely used in different polymer composite materials such as construction materials, automobile parts, and different household appliances. However, from an environmental point of view the use of phosphorus- or halogen-based FRs is not suitable, as toxic gaseous pollutants are released to the troposphere.

Microencapsulated APP incorporation with the poly(lactic acid) (PLA)/starch biocomposites enhances the property of fire retardancy of the synthesized composite materials. The PLA/starch/microencapsulated APP composite shows excellent fire retardancy, with a limiting oxygen index (LOI) value of almost 41.0.

The PLA/thermoplastic starch (TPS) composite material can also be used as excellent FR materials, where glycerol is used as plasticizer. The PLA/TPS modified starch biobased composites are also used as amazing FR materials.

The use of modified starch in PLA increases the capability of char formation and offers enhanced fire retardancy to the PLA/TPS blends (*46, 47*).

Carrageenan-Based Materials Used as FRs for PUFs

FR coatings are synthesized specifically to reduce flammability of the substances on which they are used. The coating restricts the quick spread of fire or stimulates the flame endurance over the surface of the coated materials. Among the FR coatings the intumescent FRs contain a matrix phase of polymer with a synergist and intumescent FR system. Because of extremely good fire retardancy, carrageenan-based materials are treated as a significant class of FR coating materials. Traditional FR coating systems consist of the combination of APP, melamine, and pentaerythritol; however, the conventional FR system leads to unwanted difficulties (*48, 49*).

For example, APP is less compatible with the matrix phase of the polymer and can be easily moved toward the microencapsulation of APP, which interestingly enhances the compatibility between the polymer matrix and APP (*50*). In carrageenan-based materials graphene oxide (GO) may be functioning as a synergistic agent for enhancing the effectiveness of the FR property along with decrease in emission of smoke. The addition of even a small amount of GO in the composite of carrageenan enhances the LOI value significantly. Still, because of easy agglomeration of GO within the polymer matrix, the property of fire retardancy is limited. In order to improve the property of fire retardancy functionalized GO can be used, as it exhibits better interfacial compatibility with the polymer matrix and offers good FR competence. Because of the lack of petroleum products, biobased materials such as k-carrageenan (K-CM) can be effectively used as FR materials. The microspheres of K-CM/APP, where the APP is fully covered by a wall of K-CM, function as good FR materials.

However, this property can be prevented because of less expansion of char layer and poor quality of the char. By using melamine as the source of gas, using K-CM as the source of carbon, using APP as the source of acid, and adding $C_6H5COOH$ functionalized graphene, the cause of prevention of fire retardancy is appreciably decreased (*51–53*).

Conclusion

PUFs are extremely combustible polymers, but they have a wide range of commercial significance. The ignition temperature of PUFs is 26–50 °C, whereas the decomposition temperature is 40–60 °C. The burning efficiency of urethane foams is quite a bit higher than that of other polymers. At the time of combustion, the urethane foam forms huge quantities of corrosive smoke, which impairs vision. HCN and CO are the major toxic products released during the burning of PUFs, and the level of the toxic products emitted depends upon the concentration of air or O_2 during the environmental condition of combustion. The toxicity of the products obtained during the burning of FR PUFs is comparatively more than that of the normal urethane foams because of the production of bicyclic phosphate ester within the smoke. The biological and toxicological effects mainly depend upon the amount of CO and HCN produced in the combustible gas and their ratio of formation. The phosphorus-based additives such as phosphates, phosphonitrides, phosphites, phosphoric acid, phosphonates, phosphonic acid, and phosphorous compounds containing halogen and nitrogen are effectively used as FR additives for PUFs. The addition of bromine and chlorine noticeably reduces the quantity of phosphorous in the additive materials, which subsequently decreases the synergistic effect and toxicity, but the products released during burning are still highly toxic species. Nitrogen-based additives such as melamine, dicyandiamide, and urea are also highly effective additives that prevent flammability of PUFs, but they also release highly toxic gases to the biosphere. Although modified halogen-additive-containing PUF produces less smoke with less toxicity compared with halogen-based FRs, it is still highly dangerous to our environment and biological community. Therefore, researchers are searching for new eco-friendly biobased FR additive materials for PUFs. Polysaccharide-based composite materials are found to be effective and eco-friendly additive materials for prevention of flammability of PUFs. Now different polysaccharides, such as pectin, alginate, cellulose, CS, carrageenan, and starch-based composite materials are used effectively in a sustainable way as FR additives for PUFs.

References

1. Tsuyumoto, I. Flame retardant coatings for rigid polyurethane foam based on mixtures of polysaccharides and polyborate. *J. Coat. Technol. Res.* **2020**, *2021*, 155–162.
2. Singh, H.; Jain, A. K. Ignition, combustion, toxicity, and fire retardancy of polyurethane foams: A comprehensive review. *Appl. Polym.* **2008**, *111* (2), 1115–1143.
3. Bhoyate, S.; Ionescu, M.; Kahol, P. K.; Gupta, R. K. Sustainable flame-retardant polyurethanes using renewable resources. *Industrial Crops and Products* **2018**, *123*, 480–488.
4. Michałowski, S.; Hebda, E.; Pielichowski, K. Thermal stability and flammability of polyurethane foams chemically reinforced with POSS. *J. Therm. Anal. Calorim* **2017**, *130*, 155–163.
5. Wang, C.; Chen, H.; Li, H.; Yu, J.; Wang, X.; Liu, Y. Review of emerging contaminant tris(1,3-dichloro-2-propyl)phosphate: Environmental occurrence, exposure, and risks to organisms and human health. *Environ. Int.* **2020**, *143*, 1–37.

6. Członka, S.; Strąkowska, A.; Strzelec, K.; Kairytė, A.; Kremensas, A. Melamine, silica, and ionic liquid as a novel flame retardant for rigid polyurethane foams with enhanced flame retardancy and mechanical properties. *Polym. Test.* **2020**, *87*, 1–21.
7. Levinţa, N.; Vuluga, Z.; Teodorescu, M.; Corobea, M. C. Halogen-free flame retardants for application in thermoplastics based on condensation polymers. *SN Appl. Sci.* **2019**, *1* (5), 422–440.
8. Krol, P. Polyurethanes - A review of 60 years of their syntheses and applications. *Polimery (Warsaw, Pol.)* **2009**, *54* (7), 489–500.
9. Das, A.; Mahanwar, P. A Brief Discussion on Advances in Polyurethane Applications. *Adv. Ind. Eng. Polym. Res.* **2020**, *3* (3), 93–101.
10. Król, P.; Król, B. Structures, properties and applications of the polyurethane ionomers. *J Mater. Sci.* **2020**, *55*, 73–87.
11. Kojio, K.; Nozaki, S.; Takahar, A.; Yamasaki, S. Influence of chemical structure of hard segments on physical properties of polyurethane elastomers: a review. *J. Polym. Res.* **2020**, *27* (6), 140–152.
12. Gama, N.; Ferreira, A.; Barros-Timmons, A. Polyurethane Foams: Past, Present, and Future. *Materials* **2018**, *11* (10), 1841–1875.
13. Shoaib, S.; Shahzad, M. K.; Nafisa, G.; Waqas, A.; Muhammad, S.; Tahir, J. A Comprehensive Short Review on Polyurethane Foam. *Int. J. Innovation Sci. Res.* **2014**, *12* (1), 165–169.
14. Ki-Beom, P.; Hee-Tae, K.; Nam-Yong, H.; Jae-Myung, L. Variation of Mechanical Characteristics of Polyurethane Foam: Effect of Test Method. *Materials* **2019**, *12*, 2672–2691.
15. Wei, Z.; Shu-Jie, H.; Guo-Dong, F.; Pu-You, J.; Xiao-Li, R.; Meng, Z.; Yong, H. Properties of Rigid Polyurethane Foam Modified by Tung Oil-Based Polyol and Flame-Retardant Particles. *Polymers* **2020**, *12*, 119–132.
16. Kausar, A. Polyurethane Composite Foams in High-Performance Applications: A Review. *Polym.-Plast. Technol. Eng.* **2017**, *57* (4), 346–369.
17. McKenna, S. T.; Hull, T. R. The fire toxicity of polyurethane foams. *Fire Sci. Rev.* **2016**, *5* (1), 3–30.
18. Gu, R.; Konar, S.; Sain, M. Preparation and Characterization of Sustainable Polyurethane Foams from Soybean Oils. *J. Am. Oil Chem. Soc.* **2012**, *89* (11), 2103–2111.
19. Sklenickovaa, K.; Abbrenta, S.; Haleckyc, M.; Kol, V.; Bene, H. Biodegradability and ecotoxicity of polyurethane foams: A review. *Crit. Rev. Environ. Sci. Technol.* **2020**, 1–47, DOI: 10.1080/10643389.2020.1818496.
20. Jin, F.-L.; Zhao, M.; Park, M.; Park, Soo-J Recent Trends of Foaming in Polymer Processing: A Review. *Polymers* **2019**, *11* (6), 953–975.
21. Kiss, G.; Rusu, G.; Peter, F.; Tănase, I.; Bandur, G. Recovery of Flexible Polyurethane Foam Waste for Efficient Reuse in Industrial Formulations. *Polymers* **2020**, *12* (7), 1533–1546.
22. Komurlu, E. High-density polyurethane rigid foam usability as liner support material in rock engineering. *Arabian J. Geosci.* **2020**, *13* (11), 422–434.
23. Kurańska, M.; Leszczyńska, M.; Malewska, E.; Prociak, A.; Ryszkowska, J. Implementation of Circular Economy Principles in the Synthesis of Polyurethane Foams. *Polymers* **2020**, *12* (9), 2068–2080.

24. Akindoyo, J. O.; Beg, M. D. H.; Ghazali, S.; Islam, M. R.; Jeyaratnam, N.; Yuvaraj, A. R. Polyurethane types, synthesis and applications – a review. *RSC Adv.* **2016**, *6* (115), 114453–114482.
25. Goncu, E.; Yalcin, M. E. A design of cellular automata-based PUF and its implementation on FPGA. *Int. J. Circuit Theory Appl.* **2020**, *48* (8), 1244–1255.
26. Calhoun, J.; Minwalla, C.; Helmich, C.; Saqib, F.; Che, W.; Plusquellic, J. Physical Unclonable Function (PUF)-Based e-Cash Transaction Protocol (PUF-Cash). *Cryptography* **2019**, *3*, 18–41.
27. Kim, Y.; Lee, S.; Yoon, H. Fire-Safe Polymer Composites: Flame-Retardant Effect of Nanofillers. *Polymers* **2021**, *13* (4), 540–585.
28. Illeperuma, W. R. K.; Rothemund, P.; Suo, Z.; Vlassak, J. J. Fire-Resistant Hydrogel-Fabric Laminates: A Simple Concept That May Save Lives. *ACS Appl. Mater. Interfaces* **2016**, *8* (3), 2071–2077.
29. Aarstad, O. A.; Tøndervik, A.; Sletta, H.; Skjåk-Bræk, G. Alginate Sequencing: An Analysis of Block Distribution in Alginates Using Specific Alginate Degrading Enzymes. *Biomacromolecules* **2011**, *13* (1), 106–116.
30. Imrana, I. K.; Charles, C. S.; Sajjad, S. M.; Wei, Y.; Anthony, C. Y. Y.; Muhammad, T. N.; Guan, H. Y. Alginate/Polymer-Based Materials for Fire Retardancy: Synthesis, Structure, Properties, and Applications. *Polym. Rev.* **2020**, *61* (2), 1–58.
31. Chen, H.-B.; Shen, P.; Chen, M.-J.; Zhao, H.-B.; Schiraldi, D. A. Highly Efficient Flame Retardant Polyurethane Foam with Alginate/Clay Aerogel Coating. *ACS Appl. Mater. Interfaces* **2016**, *8* (47), 32557–32564.
32. Wang, X.; Pan, Y.-T.; Wan, J.-T.; Wang, D.-Y. An eco-friendly way to fire retardant flexible polyurethane foam: layer-by-layer assembly of fully bio-based substances. *RSC Adv.* **2014**, *4* (86), 46164–46169.
33. Zheng, X.; Wang, G.; Xu, W. Roles of organically-modified montmorillonite and phosphorous flame retardant during the combustion of rigid polyurethane foam. *Polym. Degrad. Stab.* **2014**, *101*, 32–39.
34. Li, Y.-C.; Yang, Y.-H.; Shields, J. R.; Davis, R. D. Layered double hydroxide-based fire resistant coatings for flexible polyurethane foam. *Polymer* **2015**, *56*, 284–292.
35. Lin, B.; Yin Yuen, A. C.; Li, A.; Zhang, Y.; Yuan Chen, T. B.; Yu, B.; Wang, C. H. MXene/chitosan nanocoating for flexible polyurethane foam towards remarkable fire hazards reductions. *J. Hazard. Mater.* **2019**, *381*, 1–10.
36. Malucelli, G. Flame-Retardant Systems Based on Chitosan and Its Derivatives: State of the Art and Perspectives. *Molecules* **2020**, *25* (18), 4046.
37. Lin, B.; Yuen, A. C. Y.; Chen, T. B. Y. Experimental and numerical perspective on the fire performance of MXene/Chitosan/Phytic acid coated flexible polyurethane foam. *Sci. Rep.* **2021**, *11*, 4684–4696.
38. Zhao, H.-B.; Chen, M.; Chen, H.-B. Thermally Insulating and Flame-Retardant Polyaniline/Pectin Aerogels. *ACS Sustainable Chem. Eng.* **2017**, *5* (8), 7012–7019.
39. Liu, D.; Hu, A. The Influence of Environmentally Friendly Flame Retardants on the Thermal Stability of Phase Change Polyurethane Foams. *Materials* **2020**, *13* (3), 520–530.

40. Ng, H.-M.; Sin, L. T.; Tee, T.-T.; Bee, S.-T.; Hui, D.; Low, C.-Y.; Rahmat, A. R. Extraction of cellulose nanocrystals from plant sources for application as reinforcing agent in polymers. *Composites Part B* **2015**, *75*, 176–200.
41. Rogaume, T.; Valencia, L. B.; Guillaume, E.; Richard, F.; Luche, J.; Rein, G.; Torero, J. L. Development of the Thermal Decomposition Mechanism of Polyether Polyurethane Foam Using Both Condensed and Gas-Phase Release Data. *Combust. Sci. Technol.* **2011**, *183* (7), 627–644.
42. Wang, H.; Tian, Y.; Li, J. Experimental study on thermal effect and gas release laws of coal-polyurethane cooperative spontaneous combustion. *Sci. Rep.* **2021**, *11*, 1994–2006.
43. Salmeia, K. A.; Jovic, M.; Ragaisiene, A.; Rukuiziene, Z.; Milasius, R.; Mikucioniene, D.; Gaan, S. Flammability of Cellulose-Based Fibers and the Effect of Structure of Phosphorus Compounds on Their Flame Retardancy. *Polymers* **2016**, *8*, 293–307.
44. Mohamed, A. L.; Hassabo, A. G. Flame Retardant of Cellulosic Materials and Their Composites. In *Flame Retardants*; Engineering Materials Series; Springer International Publishing: Gewerbestrasse, Switzerland, 2015; Chapter 10, pp 247–314.
45. Khalili, P.; Tshai, K. Y.; Kong, I. Natural fiber reinforced expandable graphite filled composites: Evaluation of the flame retardancy, thermal and mechanical performances. *Composites Part A* **2017**, *100*, 194–205.
46. Tsuyumoto, I.; Miura, Y.; Nirei, M.; Ikurumi, S.; Kumagai, T. Highly flame retardant coating consisting of starch and amorphous sodium polyborate. *J. Mater. Sci.* **2011**, *46* (16), 5371–5377.
47. Wang, X.; Hu, Y.; Song, L.; Xuan, S.; Xing, W.; Bai, Z.; Lu, H. Flame retardancy and thermal degradation of intumescent flame retardant poly (lactic acid)/starch biocomposites. *Ind. Eng. Chem. Res.* **2011**, *50*, 713–720.
48. Liu, Q.; Wang, D.; Li, Z.; Li, Z.; Peng, X.; Liu, C.; Zheng, P. Recent Developments in the Flame-Retardant System of Epoxy Resin. *Materials* **2020**, *13* (9), 2145.
49. Corrêa-Filho, L.; Moldão-Martins, M.; Alves, V. Advances in the Application of Microcapsules as Carriers of Functional Compounds for Food Products. *Appl. Sci.* **2019**, *9* (3), 571.
50. Jena, D. K.; Sahoo, P. K. Simultaneous improvement of mechanical and fire retardant properties of synthesised biodegradable guar gum-g-poly(butyl acrylate)/montmorillonite nanocomposite. *Polym. Degrad. Stab.* **2018**, *154*, 37–45.
51. Pham, C. T.; Nguyen, B. T.; Nguyen, H. T. T.; Kang, S.-J.; Kim, J.; Lee, P.-C.; Hoang, D. Comprehensive Investigation of the Behavior of Polyurethane Foams Based on Conventional Polyol and Oligo-Ester-Ether-Diol from Waste Poly(ethylene terephthalate): Fireproof Performances, Thermal Stabilities, and Physico-mechanical Properties. *ACS Omega* **2020**, *51* (5), 33053–33063.
52. Sienkiewicz, A.; Czub, P. Flame Retardancy of Biobased Composites—Research Development. *Materials* **2020**, *13* (22), 5253–5282.
53. Wang, T.; Yang, Y.; Zhang, W. Synthesis of K-Carrageenan Flame-Retardant Microspheres and Its Application for Waterborne Epoxy Resin with Functionalized Graphene. *Polymers* **2019**, *11* (10), 1708–1720.

Chapter 3

Clay and Carbon Nanotubes as the Potential Fillers for Polyurethanes for Flame-Retardant Coatings

Anupama Mogha[*,1]

[1]Dr. S. S. Bhatnagar University Institute of Chemical Engineering & Technology, Panjab University, Chandigarh 160014, India

[*]Email: anupamapawar85@gmail.com

In the present era of synthetic products, polyurethane (PU) is one of the greatest growing industries. PUs based on novel raw material are gaining attention for having a number of commercial applications, the most significant of which is in the field of flame-retardant (FR) material. PU polymers are an extremely thermally stable FR material with potential applications under normal to insistent environments. With growing attention to environmental issues, the new idea of combining green technology with other innovative technologies is predominantly promoted. The approach results in the synthesis of PU-derived polymeric materials with a completely nontoxic and biodegradable nature. Because of the low thermal stability of pure PUs, they are reinforced using fillers, such as clay and multiwalled carbon nanotubes (MWCNTs). The MWCNTs are modified using chemical and physical functionalization methods to improve the dispersion of MWCNTs into the nanocomposite. In the present era of synthetic products, PU is one of the greatest growth industries. It makes PU polymer an extremely thermally stable FR material with potential applications under normal to insistent environments. The synthesized PU nanocomposite achieved with enhanced morphological, mechanical, and thermal properties allows researchers to explore its potential for flame retardancy.

Introduction

Polyurethanes (PUs), first developed by Otto Bayer in 1937 as an alternative to rubber, represent the most versatile, industrially important, and extensively researched group of polymers. PUs are polymeric materials composed of a backbone of urethane linkages, also called the carbamate group (−NH-C (=O) O−), along with other groups such as urea, ester, ether allophanates, hydrocarbons with aromatic and aliphatic properties, and moieties. Thus, PUs are the reaction product of the di/triisocyanate group and the hydroxyl polyol in the presence of or without a chain extender (*1*).

PUs are among the most used marketable polymers, usually manufactured using polyols from petrochemicals (2). They constitute an enormous class of polymers available as thermoplastic polymers; solid elastomers; fibers; coatings; flexible, rigid, and molded foams; fibers; sealants; and adhesives. They are known for their exceptional hardness, high mechanical strength, excellent abrasion resistance, low thermal conductivity, good chemical resistance, and low water absorption characteristics. PUs, because of their properties and segmental structure, find applications in a large range of consumer goods, appliances, footwear, furniture and bedding, insulation boards, carrier media, carpet underlayment, industrial paints and coatings, and biomedical devices such as grafts, catheters, artificial arteries, heart valves, stents, sutures, bandages, and so forth. PUs have been manufactured with considerable densities and stiffnesses for a wide range of applications. Depending on the application, their density ranges from 6 to 1220 kg/m^3. Over the last 15 years, their worldwide consumption has increased at an estimated rate of 7% annually with a massive output of 18 million tons in 2016. Thus, PUs are known as the most ubiquitous man-made multipurpose polymers on Earth (*3*, *4*).

An enormous range of PU physicochemical properties can be customized by prudently varying the composition and type of their three building blocks (i.e., polyol, isocyanate, and chain extender). Thus, it is very important to design the structure of PUs with required useful properties and end use. Most of the polyols and isocyanates are derived from petroleum sources, which lead to increasing environmental pollution and fossil fuel consumption. Various factors—like a sharp increase in oil prices, fear of exhaustion of global fuel oil, and growing environmental standards—have aggravated the problem. Thus, it is imperative and urgent to develop unconventional ways to synthesize PUs to minimize dependence on finite petroleum resources. To address the need for inspiring green technology, some novel polyols using resources such as vegetable oils have been investigated to cut down on the utilization of polyols from petroleum (*5*). There is an abundant accessibility of biobased, low-cost, and nontoxic materials, and possible functionalization routes are available to develop materials with similar properties, as petroleum-based polymers have shifted the entire focus of research to renewable resources for the synthesis of PUs.

Although PUs have remarkable mechanical and thermal properties, a chief limitation is their penetrability to various gases and vapors and poor flammability, because they decompose as soon as they are exposed to flame, releasing a large quantity of heat. PUs easily allow oxygen, carbon dioxide, helium, water, and organic vapors to pass through them. The permeability of PUs for gases like oxygen, nitrogen, carbon dioxide, and helium is in the range of 9 to 297, 3 to 1067, 175 to 2014, and 36 to 2340, respectively, which is quite lower than common polymers like low-density polyethylene and poly(tetrafluoroethylene). A pure PU has low thermal stability at elevated temperatures, and deprivation starts at temperatures of more than 180 °C (*6*). Applications such as flame-retardant (FR)-PU coatings require materials that can oppose a range of external stresses developed because of temperature changes caused in case of fire. When subject to fire or very high temperatures, pure PUs burn rapidly and degrade within 40 seconds, leaving a black char. Because of less thermal stability at elevated temperatures, the application of PU in coatings poses a threat. The biggest limitation of PU in such an application is poor fire safety, early aging, and the production of large amounts of toxic smoke and load-bearing byproducts on burning (*7*). Therefore, numerous FRs are incorporated as fillers in the PU matrix to improve its flame resistance. These include aluminum hydroxide, triphenyl phosphate, ammonium polyphosphate, dimethyl methyl phosphonate, and boric compounds. Also, numerous halogenated FRs are widely used but are prohibited because they release toxic gases while burning. Various compounds that are not halogenated—like silicon, nitrogen, and phosphorus—are used as alternatives to halogenated compounds. Metal hydroxides like aluminum trihydroxide and magnesium dihydroxide have also been explored as effective FRs. All of these FRs demonstrate a good flame retardancy but have some limitations such as easy migration, toxicity, and deterioration of polymer properties.

Various nanofillers have received high consideration. These nanofillers are comprised of nanoclays, silica nanoparticles, carbon nanotubes (CNTs), graphene, and so forth, which not only increase the flame retardancy of the polymer matrix but also increase the mechanical properties of polymers. When added to the PU matrix, layered silicates improve the barrier properties and prevent dripping while burning, thereby improving flame retardancy.

Clay as Filler

In nature, clays are present in platelet form, and they are stacked from very few to thousands of sheets in number. Clay is often used as a nanofiller because of its outstanding stacking arrangement. The structure of clay contains an octahedral layer of alumina inserted between the two layers of silicates. Aluminosilicate sheets have a width of 1 to 5 nm and a diameter of 100 to 500 nm. Platelets with a high aspect ratio (>50) result from these ideal dimensions. This causes tension to be transferred from the polymer to the clay mineral during load. As a consequence, clay is an ideal candidate for use as a filler in polymers, and the inclusion of clay results in the superior mechanical properties of the polymer (*8*).

The Cloisite 30B used in the present study is modified by using surfactant bis-2-hydroxyethyl ammonium chloride (*9*, *10*). The nanocomposite prepared using the nanoclay exhibits enhanced mechanical and solvent resistance, enhanced thermal stability, improved crystallization, and reduced penetration for fluids because of the plated structure of clays with a high aspect ratio (*11–13*). The montmorillonite (MMT) consists of multiple aluminosilicate sheets with asymmetrical polygonal shapes at the average dimension of 100 nm × 100 nm × 1 nm for single sheets. The present divalent countercations in the natural clays are interchangeable with Na+, Cu2+, Zn2+, Mg2+, and Ca2+ ions. The organic agent alters the natural clay through intercalation to become hydrophobic and harmonious with polymers (*14–17*).

The performance of clay is limited because of its close stacking arrangement, which makes it incapable of interacting with the polymer matrix. The sodium ions allow the clay mineral to swell during hydration. This increases the association of scattered clay with the polymer. Ion–dipole bonding takes place between the clay and the polymer (*18*).

Intensive work has been done on the dispersion of clay into the solvents (*19*, *20*). This method involves the sonication of clay for a predefined period. The sonication leads to the generation of mechanical energy used to resolve van der Waals forces between clay platelets, resulting in clay mineral exfoliation.

A novel solution is the integration of FRs such as organoclays and CNTs into an organic polymer to create an innovative yet nontoxic FR composite structure. The primary benefits of clay mineral and organoclay–polymer nanocomposites are their light weight, mechanical strength and durability, resistance to corrosion and solvents, and, most importantly, strong thermal efficiency and flame retardancy (*21*). A small quantity of nanoclay (<5 wt %) may significantly improve the barrier, dimensional, mechanical, and thermal characteristics of an infected polymer based on the level of inclusion between the two phases in the resulting nanocomposite (*22*).

CNTs are the only nanomaterials identified to date with the ability to outperform the FR properties of nanoclay-based polymer nanocomposites if integrated in a limited proportion alongside clay fillers in the main structure. By forming an extremely secure defensive layer, CNTs form an outstanding jammed network that can inhibit heat and shield against flames. This technique can considerably decrease the heat release rate (*23*).

Gorrasi and colleagues (*24*) used poly(caprolactone) diol, diphenylmethane diisocyanate, di(ethylene glycol), and a hydroxyl functionalized clay to study the layered silicate-based PU nanocomposites. The exfoliation occurred at 2% (w/w) of clay content, while the inorganic phase rearranged in an intercalated structure for higher concentrations. The number of silicate layers and their dispersion defined the degree of hydrogen bonding in the hard segments, which was greatly reduced. Water vapor permeability decreased linearly with inorganic material up to 4% for the organic process. The diffusion phenomenon was primarily responsible for low activity permeability behavior.

Clay-infused phenolic foam nanocomposites were studied by Rangari and colleagues (*25*). Thermogravimetric analysis/differential thermogravimetric tests showed that the formulated nanophase foams were thermally more stable than the corresponding neat system. The compressive pressure and modulus of the nanophase foams increased by 150% to 180% over the neat device. The scanning electron microscopy elaborated the existence of uniformity in size and shape of well-ordered foam cells. Layered silicates/PU nanocomposites were synthesized by Tien and colleagues (*26*). The dispersion of layered silicates in PU was observed to change from an intercalated to an exfoliated structure as the number of hydroxyl groups of the swelling agent increased. The use of reactive swelling-agent–modified silicates as pseudo chain extenders resulted in ultrahigh performance in the mechanical properties of PU.

CNTs, graphene, and clay as nanofillers were compared in polymer nanocomposites. With the addition of a single sheet of MMT, observed values of Young's modulus ranged from 178 to 265 GPa. The excellent electrical conductivity of CNTs and graphene can be used to fabricate electrically conducting materials for use in conductive adhesives, antistatic coatings, and other applications (*8*). A local triamcinolone acetonide (TA) delivery system was studied by Pinto and colleagues (*27*) based on MMT clay in a PU nanocomposite. It was observed that the rate of TA release can be decreased by adding MMT particles to the nanocomposites. Also, there was an enhancement in the mechanical characteristics of the polymer nanocomposite as compared with pure PU. This led to the conclusion that PU nanocomposite can possibly be applied as a local TA delivery system.

Solarski and colleagues (*28*) studied PU/clay nanocomposites for coating applications. The linkages among the isocyanate and clay's counterions and the prevalence on the nanocomposite morphology were discussed. Cloisite 30B led to improved thermal performance and response to fire. The main focus was to prepare PU coatings secured with Cloisite 30B clay, and the relations between the diisocyanate and Cloisite 30B and the structural properties were verified.

Composites based on PU/multiwalled CNTs (MWCNTs) were synthesized. The microstructure of the composite was studied using Fourier transform infrared analysis, and the chemical linkage of CNTs with a PU matrix was verified. Aside from a few clusters, the composition of the composite verified that CNTs were well scattered in the polymer matrix. As a result, thermal stability improved compared with pure PU. There was a rise in modulus and strength values by adding 2 wt % CNTs (*29*).

The incorporation of small quantities (0.5–10 vol %) of MWCNTs into thermoplastic elastomer was investigated. The resulting polymer nanocomposites had high electrical conductivity, small electrical percolation, and mechanical property enhancements such as increased modulus. The deformation activity of the unfilled elastomer was close to that of the nanocomposites, suggesting that the mechanisms of great distortion were mainly ruled by the matrix (*30*).

CNTs as Fillers

CNTs are one of the leading options as fillers for the polymer matrix because of their excellent properties like thermal conductivities, light weight, and high strength. These exceptional properties of pristine CNTs cannot be completely explored because of the poor dispersion of pristine CNTs in the matrix. The van der Waals interactions in CNTs result in stabilized bundles, which are particularly complicated to disperse in the matrix (*31*). The main issues in using CNTs as reinforcement for composites are the well-organized distribution of CNTs into a polymer, the approximation of the dispersal, and the organization and control of the CNTs in the matrix. Generally, CNTs are nonreactive and mostly interact through van der Waals interactions with their surrounding matrix. They are unable to offer proficient load transfer through a CNT/matrix interface without any modification (*32*). This problem can be overcome by the chemical functionalization of CNTs, which leads to enhanced dispersion and nanotubes into the matrix (*33–35*).

For high-performance polymer nanocomposites, CNTs are a crucial reinforcing agent and hold great flexibility, low mass density, excellent axial tensile strength, and high aspect ratio, and this leads to excellent electrical, mechanical, and thermal characteristics, making them a perfect material to replace the existing nanofillers in polymer nanocomposites (*36*). For the preparation of stable CNT dispersions, numerous observations have been reported by enfolding them with polymer chains using covalent as well as noncovalent functionalization (*37, 38*).

The use of MWCNTs results in a crucial reinforcing material for great performance composites. MWCNTs have high flexibility, low mass density, excellent axial tensile strength, and high aspect ratios. CNTs possess excellent mechanical, electrical, and thermal characteristics, making them the ultimate nanofillers for polymer nanocomposites (*36*). Much work has been done for achieving good dispersion of CNTs into the polymer matrix using covalent functionalization (*37–40*).

In 1994, Ajayan and colleagues first reported CNTs as filler (*41*), and ever since, CNTs have been used as a crucial reinforcing material to enhance polymer composite properties. CNTs have greater electrical properties and thermal stability up to 280 °C in a vacuum (*42, 43*), their thermal conductivity (*44*) doubles as high as diamond, and their electric current carrying capacity is 1000 times more advanced than copper wires, so they can be used to prepare anisotropic composites (*45, 46*). Elastomers using clay/MWCNTs as hybrid fillers were studied by Song (*47*). The exceptional mechanical properties, including modulus and tensile strength, were shown by the nanocomposites with the amalgamated nanofillers. In addition, hybrid nanofillers in the styrene butadiene rubber matrix have advanced thermal and electrical properties, as well as gas barrier efficiency at low filler applied loads.

The use of single-walled CNTs (SWCNTs)/MWCNTs as fillers to nanocomposites was studied by Moniruzzaman and Winey (*36*). The study indicated the challenges and opportunities for proficiently transferring the exceptional properties of CNTs to a polymer matrix. It was promising to synthesize hyperbranched poly(urea-urethane)-grafted MWCNTs based on poly(ethylene oxide-tetrahydrofuran) and aliphatic polyisocyanate resin as a preserving agent (*37*). Because of the strong dispersion of MWCNTs and improved interfacial adhesion between the PU matrix and MWCNTs, the storage modulus and glass-transition temperature increased. As composites with 0.5 to 1 wt %, MWCNTs were compared to pure PU, and the thermal conductivity increased by around 60% to 70% while retaining the high electrical resistivity.

The addition of only 2 wt % MWCNTs will lead to a rise in elastic modulus and the yield strength of the composite by 214 and 162%, respectively. Cross-linked waterborne PU (WBPU)/hexamethoxymethylmel-amine-carbon elastomer nanocomposites were manufactured by

Rahman and Shaikh (*48*). The WBPU/hexamethoxymethylmel-amine-carbon elastomer nanocomposite can take up the smallest amount of water that can be simply removed by heating without abolishing its crystalline structure. Thermoplastic polyurethane (TPU)/CNT nanocomposites were studied by Chen and colleagues (*49*). The percolated network structure in the TPU composites was observed for CNTs' content more than 2 wt %. The shape recovery behavior of the composites at room temperature was delayed by CNTs.

Carbon elastomer buckypaper/TPU elastomers composites were investigated by Han and colleagues (*50*). By incorporating CNTs into the elastomer, there was an enhancement in strength values to 120 MPa, stiffness to around 6 GPa, ductility to 30%, and toughness to 36 MJ/m^3. PU composite foams using waterblown MWCNTs were synthesized by Zhai and colleagues (*51*). The CNT/PU composite foams showed superior strength values for compression over the pure PU, and uniaxial compression led to negative pressure coefficient behavior. Fluorinated CNTs/TPU nanocomposites were synthesized by Dai and colleagues (*52*). Thermal deprivation temperature was mainly improved, and there was a rise of from 42.3 to 51.7 °C in glass-transition temperature values. There was an improvement in chemical resistance. The present literature supports the utilization of CNTs as an important reinforcing material for the polymer and polymer-based composites for numerous applications. However, because of the walled structure of the CNTs, they need modification prior to being used as reinforcing agents. Different methods have been explained as follows to enhance their interactions and interlinkages with the polymer composite, so that their superior characteristics can be utilized to enhance the polymer properties.

Functionalization Methods

The utility of CNTs as a filler for polymers depends on the dispersion of CNTs and their interaction with the matrix. Numerous methods have been developed for the surface alteration of CNTs. The surface properties for good dispersal of CNTs can be attained by adopting various functionalization methods. These methods involve chemical and physical modification of the CNTs based on their interaction with the matrix (*32*).

Chemical Functionalization Method

This method of modification involves the covalent bonding of functional groups on the carbon structure of CNTs. The group can either be attached to the sidewall or the end of the tube. There are two different methods for the chemical functionalization of CNTs.

Direct Covalent Sidewall Functionalization

There is a transformation of hybridization from sp^2 to sp^3 with loss of the π-conjugation system during the direct covalent functionalization method. This process takes place in the presence of strong chemical reactivity like fluorine (*21*, *53*), amino, alkyl, and hydroxyl groups, and so forth (*54*).

Defect Functionalization

The covalent functionalization is done using another method, known as defect functionalization. In this process, there is a chemical alteration of defect spots on CNTs. The open ends or holes in the sidewalls act as defect sites; oxygenated sites can also be considered as defects (*55*, *56*).

Physical Functionalization Method

There are two major limitations with the covalent method of functionalization (32). There is the creation of a large number of defect sites during this process, and the CNTs break into smaller pieces, leading to the poor mechanical properties of CNTs. The second limitation arises as concentrated acids are used, which are harmful to the environment. Therefore, methods are being developed for the production of less damaging and low-cost CNTs.

An alternative method for CNTs is noncovalent functionalization. In this process, the polymer is wrapped (57) using the van der Waals linkages and π–π stacking among CNTs and polymer chains containing aromatic rings.

Surfactant functionalized CNTs can also be employed, as surfactant adsorption on the surface of CNTs reduces surface stress and CNT aggregation is effectively prevented. The common surfactants studied are (1) nonionic surfactants (58), (2) anionic surfactants (59), and (3) cationic surfactants (*60*). In the surfactant method, the van der Waals forces of attraction are overcome by electrostatic repulsive forces. The effectiveness of the surfactant technique can be governed mainly by the characteristics of surfactants, intermediate chemistry, and the polymer matrix.

The endohedral method of CNT modification is a noncovalent technique. This technique uses the capillary effect to store guest atoms or molecules in the inner cavity of CNTs. Insertion also occurs at defect sites located at the ends of or on the sidewalls. The combination of these two materials, CNTs and guest molecules, is primarily useful for incorporating the properties (*61*).

Hybrid PU Nanocomposites

Polymer-based nanomaterials have arisen as a new class of hybrid materials with a wide range of interesting bioapplications, including tissue engineering, drug delivery, and a variety of biotechnological applications. Essential understanding of the structures, properties, polymer–matrix interactions, and further modifications based on fundamental specifications is needed for designing and developing them.

During combustion, a protective layer is formed by inserting a very small amount of organically modified nanoclay into the PU matrix. The use of a limited amount of nanoclay (5 wt %) improves things such as mechanical, thermal, and dimensional properties, along with barrier efficiency. In the combustion, char with increased melt viscosity against the material surface forms (*62*). The volatile matter cannot permeate the filler platelets; instead, it must go around them, leading to a tortuous path (*63*). While an MWCNT acts as a wrapping representative that can join with clay layers and reduce surface cracks of chars, it also acts as an impediment to the evolution of flammable volatile matter in the condensed phase (*64*).

Consequently, in Norouzi et al., efforts have been made to synthesize FR-PU/nanoclay/CNT nanocomposite systems that are cost-effective procedures for their fabrication. The work focused on the fabrication of an FR coating by hybridizing PU by integrating clay and CNTs (*65*). The use of these two reinforcing fillers is intended to enhance the thermal and mechanical properties of the PU matrix. The nanocomposite created in this manner has a wide range of applications, including FR coatings, foams, wires, and vehicles, among others (*66*). MWCNTs have a greater potential for enhancing foam morphology as a lower concentration of MWCNTs (0.05 wt %) will provide improved effects (higher cell density) than nanoclay (0.5 wt %). This is because of the cylindrical form of MWCNTs as opposed to nanoclay. As both ends of CNTs and their sidewalls can serve as heterogeneous nucleation sites in the MWCNT system, nucleation efficiency is increased (*67*).

Emphasis on hybrid PU nanocomposites with thermally stable PU is gaining consideration because of their potential application in areas like insulation of wire and cables, fire coatings, building materials, temperature-resistant flexible foams, and so forth. Because of this, there is a need for a system that increases safety awareness of customers, acknowledges environmental pollution generated during the fire, and recognizes economic loss that occurs during the burning of materials.

Yousefi and colleagues (*68*) studied PU nanocomposites that have highly associated graphene sheets. Fine dispersal, as well as a higher degree of orientation, were achieved using ultralarge-sized graphene oxide. A noticeable increase in the tensile modulus and strength was attained in nanocomposites. The conformity among the experimentations and academic estimates for the tensile modulus indicate that the graphene layers were spread separately on the molecular level, and a polymer with layered structure was produced. There was a regular decline with rising graphene content in moisture permeability of the nanocomposites.

Ali and colleagues (*69*) studied physicochemical properties of castor oil–based PUs as a consideration for using MWCNTs. The dissimilar weight percentages (0–1 wt %) of MWCNTs were added to castor oil–based PU nanocomposites. Good dispersion of MWCNTs in the PU matrix at different weight percentages was observed. Surface part observations exposed the existence of precise little pores, for the composition with 0.3 wt % of MWCNTs, which is in good arrangement with barrier penetrability.

Mahapatra and colleagues (*70*) studied the physical integration of MWCNTs into strongly branched polymer chains, which resulted in an innovative polymeric nanocomposite. Centered on a multihydroxy triazine nucleus, poly(e-caprolactone) diol, and diphenyl methane diisocyanate, a sequence of s-triazine–based hyperbranched PUs (HBPs) was synthesized. The potential of solubilization increased as branch density increased. Because of the cagelike superstructure of HBPs, the ability of HBPs to solubilize MWCNTs improved with increasing branch density; however, there was no further improvement for HBPs because of reduced pocket size. The crystallinity of HBPs decreased dramatically as branching increased, while it increased as MWCNT concentration increased.

Roy and colleagues (*71*) prepared TPU nanocomposites using MMT, and MWCNTs were ground in various weight ratios and characterized using a basic dry grinding process. In the creation of TPU nanocomposites, an MMT–MWCNT (1:1) hybrid was used as a reinforcing filler. Thermogravimetric analysis research revealed that a 0.25 wt % hybrid-loaded TPU nanocomposite demonstrated the greatest increase of 310 °C, corresponding to 50 wt %. Mechanical tests revealed that the tensile strength and Young's modulus of 0.50 wt %-filled TPU increased by 57 and 87.5%, respectively. Homogeneous dispersion and strong interfacial interaction were because of improvements in thermal and mechanical properties.

Clay was studied in SWCNT/epoxy composites to enhance nanotube distribution without compromising electrical conductivity or mechanical efficiency. With just 0.05 wt % SWCNT, the combined effect of nanotubes and clay improved both electrical and mechanical performance; there was also an improvement in electrical conductivity (*72*). The inclusion of clay lowered the percolation threshold of these nanocomposites from 0.05 to 0.01 wt % SWCNT. SWCNTs have a clay affinity, which allows them to become more exfoliated and networked composites.

Bahrambeygi and colleagues (*73*) used in situ polymerization for the synthesis of a PU nanocomposite. The results of nanoparticle integration on the morphological and structural development of the PU foam, as well as the relative improvement in mechanical properties, were investigated. The integration of MWCNTs and nanoclay into foams improved the morphology, resulting in increased cell density and smaller cells. The presence of nanoparticles improved the mechanical properties of foam, especially the tensile strength. Tensile strength for nanocomposite foam was increased by the nanoclay concentration, but in the case of MWCNTs, dissimilar effects were observed. A rise in the volume of nanoparticles resulted in a decrease in cell density.

Madaleno and colleagues (*74*) studied the effect of MMT–CNT hybrids on the properties and performance of PU nanocomposite foams. In situ polymerization was used to treat hybrids and rigid PU foam using the chemical vapor deposition technique. As compared to pure polymer foams, PU nanocomposite foams have smaller cell sizes and a higher cell density. Thermogravimetric experiments revealed that adding hybrid nanoparticles improved the thermal properties of the resulting nanocomposites, making them ideal for a wide range of implementations.

Kwon and Kim (*75*) studied in situ polymerization in an aqueous medium to create a collection of WBPU/MWCNT and WBPU/nitric acid–treated MWCNT (AMWCNT) composites. Tensile strength and modulus, as well as the glass-transition temperatures of the soft and hard segments, of WBPU increased as CNT and AMWCNT content increased. Alternatively, the properties of the WBPU/AMWCNT composites were higher than those of the WBPU/MWCNT composites of the same CNT material. WBPU/MWCNT and WBPU/AMWCNT composites comprising 1.5 wt % CNTs had electrical conductivities that were orders of magnitude greater than WBPU. These findings suggest that WBPU/AMWCNT composites prepared by in situ polymerization have a high potential as new materials.

Haghayegh and colleagues (*76*) used two-step in situ polymerization for the production of shape memory clay-based PU nanocomposites. The mechanical, thermal, and foam memory properties of CNTs with varying loading capacities were investigated. PU soft and hard segments were dependent on polycaprolactone diol as soft segment, 4,4′-diphenylmethane diisocyanate as hard segment, and 1,4-butanediol to preserve a molar ratio of 70:30. The glass-transition temperature rose as a result of the nanoclay filling. The incorporation of nanoclay into the PU matrix caused disturbance in the hard domains, resulting in a decrease in elongation and tensile strength. The crystallinity of soft segments and the dispersion of nanoparticles influenced the mechanical properties and foam memory conduct. The nanocomposite with 1% loading had the highest shape memory.

The enhancement of the mechanical properties and FR properties of epoxy EPIKOTE Resin 240 was studied using nanoclay I.30E and MWCNT nanocomposite. For the study, the nanoclay I.30E content (1, 2, and 3 wt %) and content of MWNT (0.01, 0.02, and 0.03 wt%) were used. The enhancements in the mechanical properties are provided in Table 1. Even the minute amount of CNTs and clay can lead to a rise in tensile strength values of the nanocomposites (*62*).

Table 1. Tensile Strength Values for Nanocomposites with Varying Clay and CNT Content[a]

S. No.	*MWCNT Content (wt %)*	*Nanoclay Content (wt %)*	*Tensile strength (MPa)*
1	0.01	1	87.56 ± 0.5
2	0.01	2	93.43 ± 0.6
3	0.01	3	93.45 ± 0.6

[a] Source: Data from reference (*62*).

Conclusion

Polymeric composites such as PU are needed in today's world, as such products have mainly been synthesized from petrochemical-based polyols, a nonrenewable and expensive source. As a result, novel polyols derived from natural resources such as vegetable oils (mostly castor oil) have been used. This polyol resource is nonhazardous and cost effective at the same time. Castor oil–based polyol is a novel polyol derived from sustainable sources that can be used to minimize use of petroleum-based polyols to satisfy the need for promoting green energy. Of the various uses of PU-based composite materials implemented in the last decade, the most prominent has been their use as an FR material. This characteristic has been mostly determined by their thermal, oxidative, and FR properties, which are now at an advanced stage of research and development. A pure PU has small thermal stability at high temperatures, and thermal degradation occurs at processing temperatures. The PUs based on castor oil are environmentally friendly and can be used for potential application in the field of flame retardancy. The reinforcement of nanofillers such as clay and nanotubes further leads to enhanced properties of the PUs. Both the nanofillers combine with PU to produce a hybrid nanocomposite. The study focuses on the synthesis of hybrid PU nanocomposites with thermally stable PUs gaining consideration because of their possible application in numerous areas like insulation of wire and cables, fire coatings, building materials, temperature-resistant flexible foams, etc. The progress in FR-PUs works to increase safety awareness, control environmental pollution, and develop cost-effective materials. It can be concluded that there is a need to develop such a novel system possessing high thermal stability, FR properties, and oxidative resistance based on renewable resources, cost effectiveness, and environmentally friendly products.

References

1. Saunders, K. J. *Organic Polymer Chemistry: An Introduction to the Organic Chemistry of Adhesives, Fibres, Paints, Plastics and Rubbers*, 2nd ed.; Springer Netherlands: Dordrecht, 1988; pp 358–387.
2. Wirpsza, Z.; Kemp, T. J. *Polyurethanes: Chemistry, Technology, and Applications*; E. Horwood: New York, 1993.
3. Woods, G. *The ICI Polyurethanes Book*, 2nd ed.; Wiley: New York, 1990.
4. Nozaki, S.; Masuda, S.; Kamitani, K.; Kojio, K.; Takahara, A.; Kuwamura, G.; Hasegawa, D.; Moorthi, K.; Mita, K.; Yamasaki, S. Superior Properties of Polyurethane Elastomers Synthesized with Aliphatic Diisocyanate Bearing a Symmetric Structure. *Macromolecules* **2017**, *50*, 1008–1015.
5. Ramadhoni, B.; Ujianto, O.; Nadapdap, M. Rigid Polyurethane Nanocomposites Prepared by Direct Incorporation: Effects of Nanoclay, Carbon Nanotubes and Mixing Speed on Physical and Morphological Properties. *IOP Conf. Ser.: Mater. Sci. Eng.* **2018**, *319*, 012041.
6. Chattopadhyay, D. K.; Webster, D. C. Thermal Stability and Flame Retardancy of Polyurethanes. *Prog. Polym. Sci.* **2009**, *34*, 1068–1133.
7. Cordero, A. I.; Amalvy, J. I.; Fortunati, E.; Kenny, J. M.; Chiacchiarelli, L. M. The Role of Nanocrystalline Cellulose on the Microstructure of Foamed Castor-Oil Polyurethane Nanocomposites. *Carbohydr. Polym.* **2015**, *134*, 110–118.
8. Bhattacharya, M. Polymer Nanocomposites: A Comparison between Carbon Nanotubes, Graphene, and Clay as Nanofillers. *Materials* **2016**, *9*, 262.
9. Filippi, S.; Paci, M.; Polacco, G.; Dintcheva, N. T.; Magagnini, P. On the Interlayer Spacing Collapse of Cloisite® 30B Organoclay. *Polym. Degrad. Stab.* **2011**, *96*, 823–832.
10. Choudalakis, G. G.; Gotsis, A. D. Permeability of Polymer/Clay Nanocomposites: A Review. *Eur. Polym. J.* **2009**, *45*, 967–984.

11. Villaluenga, J. P. G.; Khayet, M.; López-Manchado, M. A.; Valentin, J. L.; Seoane, B.; Mengual, J. I. Gas Transport Properties of Polypropylene/Clay Composite Membranes. *Eur. Polym. J.* **2007**, *43*, 1132–1143.
12. Rohlmann, C. O.; Horst, M. F.; Quinzani, L. M.; Failla, M. D. Comparative Analysis of Nanocomposites Based on Polypropylene and Different Montmorillonites. *Eur. Polym. J.* **2008**, *44*, 2749–2760.
13. Sharma, S. K.; Nayak, S. K. Surface Modified Clay/Polypropylene (PP) Nanocomposites: Effect on Physico-Mechanical, Thermal and Morphological Properties. *Polym. Degrad. Stab.* **2009**, *94*, 132–138.
14. Sparks, D. L. *Environmental Soil Chemistry*; John Wiley & Sons, Ltd.: New York, 2003.
15. Ajjou, A. N.; Harouna, D.; Detellier, C.; Alper, H. Cation-Exchanged Montmorillonite Catalyzed Hydration of Styrene Derivatives. *J. Mol. Catal. A: Chem.* **1997**, *126*, 55–60.
16. Usuki, A.; Hasegawa, N.; Kadoura, H.; Okamoto, T. Three-Dimensional Observation of Structure and Morphology in Nylon-6/Clay Nanocomposite. *Nano Lett.* **2001**, *1*, 271–272.
17. Okada, A.; Usuki, A. Twenty Years of Polymer-Clay Nanocomposites. *Macromol. Mater. Eng.* **2006**, *291*, 1449–1476.
18. Karpiński, B.; Szkodo, M. Clay Minerals—Mineralogy and Phenomenon of Clay Swelling in Oil & Gas Industry. *Adv. Mater. Sci.(Warsaw, Poland)* **2015**, *15*, 37–55.
19. Chen, B.; Evans, J. R. G.; Greenwell, H. C.; Boulet, P.; Coveney, P. V.; Bowden, A. A.; Whiting, A. A Critical Appraisal of Polymer–Clay Nanocomposites. *Chem. Soc. Rev.* **2008**, *37*, 568–594.
20. Oueslati, W.; Mefath, M.; Ben Rhaiem, H.; Ben Haj Amara, A. Cation Exchange Selectivity versus Concentration of Competing Heavy Metal Cations (Pb^{2+},Zn^{2+}): Case of Na-Montmorillonite. *Phys. Procedia* **2009**, *2*, 1059–1063.
21. Kelly, K. F.; Chiang, I. W.; Mickelson, E. T.; Hauge, R. H.; Margrave, J. L.; Wang, X.; Scuseria, G. E.; Radloff, C.; Halas, N. J. Insight into the Mechanism of Sidewall Functionalization of Single-Walled Nanotubes: An STM Study. *Chem. Phys. Lett.* **1999**, *313*, 445–450.
22. Ghadiri, M.; Chrzanowski, W.; Rohanizadeh, R. Biomedical Applications of Cationic Clay Minerals. *RSC Adv.* **2015**, *5*, 29467–29481.
23. Zhang, L.; Zhang, M.; Hu, L.; Zhou, Y. Synthesis of Rigid Polyurethane Foams with Castor Oil–Based Flame Retardant Polyols. *Ind. Crops Prod.* **2014**, *52*, 380–388.
24. Gorrasi, G.; Tortora, M.; Vittoria, V. Synthesis and Physical Properties of Layered Silicates/Polyurethane Nanocomposites. *J. Polym. Sci., Part B: Polym. Phys.* **2005**, *43*, 2454–2467.
25. Rangari, V. K.; Hassan, T. A.; Zhou, Y.; Mahfuz, H.; Jeelani, S.; Prorok, B. C. Cloisite Clay-Infused Phenolic Foam Nanocomposites. *J. Appl. Polym. Sci.* **2007**, *103*, 308–314.
26. Tien, Y. I.; Wei, K. H. High-Tensile-Property Layered Silicates/Polyurethane Nanocomposites by Using Reactive Silicates as Pseudo Chain Extenders. *Macromolecules* **2001**, *34*, 9045–9052.
27. Pinto, F. C. H.; Silva-Cunha, A.; Pianetti, G. A.; Ayres, E.; Oréfice, R. L.; Da Silva, G. R. Montmorillonite Clay-Based Polyurethane Nanocomposite as Local Triamcinolone Acetonide Delivery System. *J. Nanomater.* **2011**, *2011*, 15.
28. Solarski, S.; Benali, S.; Rochery, M.; Devaux, E.; Alexandre, M.; Monteverde, F.; Dubois, P. Synthesis of a Polyurethane/Clay Nanocomposite Used as Coating: Interactions between the

Counterions of Clay and the Isocyanate and Incidence on the Nanocomposite Structure. *J. Appl. Polym. Sci.* **2005**, *95*, 238–244.

29. Xiong, J.; Zheng, Z.; Qin, X.; Li, M.; Li, H.; Wang, X. The Thermal and Mechanical Properties of a Polyurethane/Multi-Walled Carbon Nanotube Composite. *Carbon* **2006**, *44*, 2701–2707.
30. Koerner, H.; Liu, W.; Alexander, M.; Mirau, P.; Dowty, H.; Vaia, R. A. Deformation-Morphology Correlations in Electrically Conductive Carbon Nanotube-Thermoplastic Polyurethane Nanocomposites. *Polymer* **2005**, *46*, 4405–4420.
31. RanaS.; Prasanth, R.; Tan, L. P. Functionalization of Carbon Nanotubes and Their Polyurethane Nanocomposites. In *Handbook of Polymer Nanocomposites. Processing, Performance and Application*; Kar, K. K.; Pandey, J. K.; Rana, S., Eds.; Springer-Verlag Berlin Heidelberg: Berlin, Heidelberg, 2015; Vol. B, pp 103–121.
32. Ma, P. C.; Siddiqui, N. A.; Marom, G.; Kim, J.-K. Dispersion and Functionalization of Carbon Nanotubes for Polymer-Based Nanocomposites: A Review. *Composites, Part A* **2010**, *41*, 1345–1367.
33. Shofner, M. L.; Khabashesku, V. N.; Barrera, E. V. Processing and Mechanical Properties of Fluorinated Single-Wall Carbon Nanotube Polyethylene Composites. *Chem. Mater.* **2006**, *18*, 906–913.
34. Gao, J.; Itkis, M. E.; Yu, A.; Bekyarova, E.; Zhao, B.; Haddon, R. C. Continuous Spinning of a Single-Walled Carbon Nanotube-Nylon Composite Fiber. *J. Am. Chem. Soc.* **2005**, *127*, 3847–3854.
35. Mogha, A.; Kaushik, A. Functionalized Multiwall Carbon Nanotubes to Enhance Dispersion in Castor Oil–Based Polyurethane Nanocomposites. *Fullerenes, Nanotubes, Carbon Nanostruct* **2021**, 1–8.
36. Moniruzzaman, M.; Winey, K. I. Polymer Nanocomposites Containing Carbon Nanotubes. *Macromolecules* **2006**, *39*, 5194–5205.
37. Zhao, J. C.; Du, F. P.; Zhou, X.-P.; Cui, W.; Wang, X.-M.; Zhu, H.; Xie, X.-L.; Mai, Y.-W. Thermal Conductive and Electrical Properties of Polyurethane/Hyperbranched Poly(urea-urethane)-Grafted Multi-Walled Carbon Nanotube Composites. *Composites, Part B* **2011**, *42*, 2111–2116.
38. Yadav, S. K.; Mahapatra, S. S.; Cho, J. W.; Lee, J. Y. Functionalization of Multiwalled Carbon Nanotubes with Poly(styrene-b-(ethylene-co-butylene)-b-styrene) by Click Coupling. *Journal of Physical Chemistry C* **2010**, *114*, 11395–11400.
39. Lee, J.; Park, E. J.; Choi, J.; Hong, J.; Shim, S. E. Polyurethane/PEG-Modified MWCNT Composite Film for the Chemical Vapor Sensor Application. *Synth. Met.* **2010**, *160*, 566–574.
40. Van Trinh, P.; Anh, N. N.; Tam, N. T.; Hong, N. T.; Hong, P. N.; Minh, P. N.; Thang, B. H. Influence of Defects Induced by Chemical Treatment on the Electrical and Thermal Conductivity of Nanofluids Containing Carboxyl-Functionalized Multi-Walled Carbon Nanotubes. *RSC Adv.* **2017**, 7, 49937–49946.
41. Ajayan, P. M.; Stephan, O.; Colliex, C.; Trauth, D. Aligned Carbon Nanotube Arrays Formed by Cutting a Polymer Resin—Nanotube Composite. *Science* **1994**, *265*, 1212.
42. Jin, Z.; Pramoda, K. P.; Xu, G.; ; Goh, S. H. Dynamic Mechanical Behavior of Melt-Processed Multi-Walled Carbon Nanotube/Poly(methyl methacrylate) Composites. *Chem. Phys. Lett.* **2001**, *337*, 43–47.

43. Wei, C.; Srivastava, D.; Cho, K. Thermal Expansion and Diffusion Coefficients of Carbon Nanotube-Polymer Composites. *Nano Lett.* **2002**, *2*, 647–650.
44. Park, C.; Ounaies, Z.; Watson, K. A.; Crooks, R. E.; Smith, J.; Lowther, S. E.; Connell, J. W.; Siochi, E. J.; Harrison, J. S.; Clair, T. L. S. Dispersion of Single Wall Carbon Nanotubes by in situ Polymerization under Sonication. *Chem. Phys. Lett.* **2002**, *364*, 303–308.
45. Tsui, F.; Jin, L.; Zhou, O. Anisotropic Magnetic Susceptibility of Multiwalled Carbon Nanotubes. *Appl. Phys. Lett.* **2000**, *76*, 1452–1454.
46. Zaric, S.; Ostojic, G. N.; Kono, J.; Shaver, J.; Moore, V. C.; Hauge, R. H.; Smalley, R. E.; Wei, X. Estimation of Magnetic Susceptibility Anisotropy of Carbon Nanotubes Using Magnetophotoluminescence. *Nano Lett.* **2004**, *4*, 2219–2221.
47. Song, S. H. The Effect of Clay/Multiwall Carbon Nanotube Hybrid Fillers on the Properties of Elastomer Nanocomposites. *Int. J. Polym. Sci.* **2018**, *2018*, 1–8.
48. Rahman, M. M.; Shaikh, M. N. Synthesis and Properties of Cross-Linkable Waterborne Polyurethane/HMMM-CNT Nanocomposite. *Nano Hybrids* **2014**, *7*, 87–111.
49. Chen, J.; Zhang, Z.-x.; Huang, W.-b.; Li, J.-l.; Yang, J.-h.; Wang, Y.; Zhou, Z.-w.; Zhang, J.-h. Carbon Nanotube Network Structure Induced Strain Sensitivity and Shape Memory Behavior Changes of Thermoplastic Polyurethane. *Mater. Des.* **2015**, *69*, 105–113.
50. Han, J. H.; Zhang, H.; Chen, M.-J.; Wang, G.-R.; Zhang, Z. CNT Buckypaper/Thermoplastic Polyurethane Composites with Enhanced Stiffness, Strength and Toughness. *Compos. Sci. Technol.* **2014**, *103*, 63–71.
51. Zhai, T.; Li, D.; Fei, G.; Xia, H. Piezoresistive and Compression Resistance Relaxation Behavior of Water Blown Carbon Nanotube/Polyurethane Composite Foam. *Composites, Part A* **2015**, *72*, 108–114.
52. Dai, L.; Jia, R. P.; He, X. Y.; Huang, M. S. Preparation and Thermal Performance of Fluorinated Carbon Nanotubes/Thermoplastic Polyurethane Nanocomposites. *Appl. Mech. Mater.* **2014**, *687–691*, 4273–4276.
53. Stevens, J. L.; Huang, A. Y.; Peng, H.; Chiang, I. W.; Khabashesku, V. N.; Margrave, J. L. Sidewall Amino-Functionalization of Single-Walled Carbon Nanotubes through Fluorination and Subsequent Reactions with Terminal Diamines. *Nano Lett.* **2003**, *3*, 331–336.
54. Zhang, Q. H.; Chen, D. J. Percolation Threshold and Morphology of Composites of Conducting Carbon Black/Polypropylene/EVA. *J. Mater. Sci.* **2004**, *39*, 1751–1757.
55. Karimi, M.; Solati, N.; Amiri, M.; Mirshekari, H.; Mohamed, E.; Taheri, M.; Hashemkhani, M.; Saeidi, A.; Asghari Estiar, M.; Kiani, P.; Ghasemi, A.; Moosavi Basri, S. M.; Aref, A. R.; Hamblin, M. R. Carbon Nanotubes Part I: Preparation of a Novel and Versatile Drug-Delivery Vehicle. *Expert Opin. Drug Delivery* **2015**, *12*, 1–17.
56. Hirsch, A. Functionalization of Single-Walled Carbon Nanotubes. *Angew. Chem., Int. Ed.* **2002**, *41*, 1853–1859.
57. McCarthy, B.; Coleman, J. N.; Crewz, R.; Dalton, A. B. Microscopy Studies of Nanotube-Conjugated Polymer Interactions. *Synth. Met.* **2001**, *121*, 1225–1226.
58. Gong, X.; Liu, J.; Baskaran, S.; Voise, R. D.; Young, J. S. Surfactant-Assisted Processing of Carbon Nanotube/Polymer Composites. *Chem. Mater.* **2000**, *12*, 1049–1052.

59. Strano, M. S.; Moore, V. C.; Miller, M. K.; Allen, M. J.; Haroz, E. H.; Kittrell, C.; Hauge, R. H.; Smalley, R. E. The Role of Surfactant Adsorption during Ultrasonication in the Dispersion of Single-Walled Carbon Nanotubes. *J. Nanosci. Nanotechnol.* **2003**, *3*, 81–86.
60. Whitsitt, E. A.; Barron, A. R. Silica Coated Single Walled Carbon Nanotubes. *Nano Lett.* **2003**, *3*, 775–778.
61. Georgakilas, V.; Gournis, D.; Tzitzios, V.; Pasquato, L.; Guldi, D. M.; Prato, M. Decorating Carbon Nanotubes with Metal or Semiconductor Nanoparticles. *J. Mater. Chem.* **2007**, *17*, 2679–2694.
62. Nguyen, T. A.; Nguyen, Q. T.; Bach, T. P. Mechanical Properties and Flame Retardancy of Epoxy Resin/Nanoclay/Multiwalled Carbon Nanotube Nanocomposites. *J. Chem.* **2019**, *2019*, 9.
63. Tsurko, E. *High Barrier Waterborne Polymer-Clay Nanocomposites*. Ph.D. Dissertation, University of Bayreuth, Bayreuth, Germany, Aug 2017.
64. Camargo, P. H. C.; Satyanarayana, K. G.; Wypych, F. Nanocomposites: Synthesis, Structure, Properties and New Application Opportunities. *Materials Research* **2009**, *12*, 1–39.
65. Norouzi, M.; Zare, Y.; Kiany, P. Nanoparticles as Effective Flame Retardants for Natural and Synthetic Textile Polymers: Application, Mechanism, and Optimization. *Polym. Rev.* **2015**, *55*, 531–560.
66. Wang, Z. Y.; Han, E.; Ke, W. Fire Resistant Effect of Nanoclay on Intumescent Nanocomposite Coatings. *J. Appl. Polym. Sci.* **2007**, *103*, 1681–1689.
67. Chen, L.; Ozisik, R.; Schadler, L. S. The Influence of Carbon Nanotube Aspect Ratio on the Foam Morphology of MWNT/PMMA Nanocomposite Foams. *Polymer* **2010**, *51*, 2368–2375.
68. Yousefi, N.; Gudarzi, M. M.; Zheng, Q.; Lin, X.; Shen, X.; Jia, J.; Sharif, F.; Kim, J. K. Highly Aligned, Ultralarge-Size Reduced Graphene Oxide/Polyurethane Nanocomposites: Mechanical Properties and Moisture Permeability. *Composites, Part A* **2013**, *49*, 42–50.
69. Ali, A.; Yusoh, K.; Hasany, S. F. Synthesis and Physicochemical Behaviour of Polyurethane-Multiwalled Carbon Nanotubes Nanocomposites Based on Renewable Castor Oil Polyols. *J. Nanomater.* **2014**, *2014*Article ID 564384.
70. Mahapatra, S. S.; Yadav, S. K.; Yoo, H. J.; Cho, J. W.; Park, J. S. Highly Branched Polyurethane: Synthesis, Characterization and Effects of Branching on Dispersion of Carbon Nanotubes. *Composites, Part B* **2013**, *45*, 165–171.
71. Roy, S.; Srivastava, S. K.; Pionteck, J. R.; Mittal, V. Montmorillonite and Multiwalled Carbon Nanotube Nanoarchitecture Reinforced Thermoplastic Polyurethane. *Polym. Compos.* **2014**, *37*, 1775–1785.
72. Liu, L.; Grunlan, J. Clay Assisted Dispersion of Carbon Nanotubes in Conductive Epoxy Nanocomposites. *Adv. Funct. Mater.* **2007**, *17*, 2343–2348.
73. Bahrambeygi, H.; Rabbi, A.; Nasouri, K.; Shoushtari, A. M.; Babaei, M. R. Morphological and Structural Developments in Nanoparticles Polyurethane Foam Nanocomposite's Synthesis and Their Effects on Mechanical Properties. *Adv. Polym. Technol.* **2013**, *32*, E545–E555.
74. Madaleno, L.; Pyrz, R.; Crosky, A.; Jensen, L. R.; Rauhe, J. C. M.; Dolomanova, V.; de Barros Timmons, A. M. M. V.; Cruz Pinto, J. J.; Norman, J. Processing and Characterization

of Polyurethane Nanocomposite Foam Reinforced with Montmorillonite-Carbon Nanotube Hybrids. *Composites, Part A* **2013**, *44*, 1–7.

75. Kwon, J.; Kim, H. Comparison of the Properties of Waterborne Polyurethane/Multiwalled Carbon Nanotube and Acid-Treated Multiwalled Carbon Nanotube Composites Prepared by in situ Polymerization. *J. Polym. Sci., Part A: Polym. Chem.* **2005**, *43*, 3973–3985.
76. Haghayegh, M.; Mir Mohamad Sadeghi, G. Synthesis of Shape Memory Polyurethane/Clay Nanocomposites and Analysis of Shape Memory, Thermal, and Mechanical Properties. *Polym. Compos.* **2012**, *33*, 843–849.

Chapter 4

Recent Developments in Green Flame Retardants Based on Carbon Nanotubes

Hamidreza Parsimehr,[1] Mojtaba Enayati,[2,3] and Amir Ershad Langroudi[*,1]

[1]Color and Surface Coating Group, Polymer Processing Department, Iran Polymer and Petrochemical Institute, Tehran 13115-14977, Iran

[2]Department of Chemistry and Physics, Troy University, Troy, Alabama 36082, United States

[3]Center for Materials and Manufacturing Sciences, Troy University, Troy, Alabama 36082, United States

[*]Email: a.ershad@ippi.ac.ir

Increasing attention is being paid to environmental issues, especially through developing benign and nonhazardous materials in different industries. The thermal stability and flame retardancy of nanocomposites are two significant practical parameters in developing new devices. Classic halogen-based flame-retardant (FR) materials provide an acceptable flame retardancy but lead to severe environmental contamination. Therefore, novel FR materials have been swiftly developed to reduce the environmental pollution of FR nanocomposites. Different inorganic and organic substances have been studied to find the best FR materials. Carbon-based nanomaterials, especially carbon nanotubes, provide considerable flame retardancy. The preparation of carbon-based nanomaterials via green and benign methods has also been widely developed to reduce environmental pollution. The carbon nanotubes as green FRs can be widely used in composite structures to provide flame retardancy.

Introduction

Developments in materials science have led to various useful and applicable substances that make life easier via fabrication of novel devices. Polymeric materials have a specific position in materials science for fabricating a vast variety of devices. The advanced and tailored properties of polymer-based materials are because of the extensive development of polymers in the world. Different types of thermoset (cross-linked) polymers and thermoplastic (not cross-linked) polymers—such as plastics, rubbers, foams, adhesives, and coatings—have been widely produced to respond to the increasing demands of humans' routine life. Polymeric composites have been rapidly developed because they offer better properties compared with pure polymers. Polymer-based composites are a homogeneous mixture of polymers or other materials, such as inorganic powder-like substances. Any nanosized material that is distributed in a polymer matrix can be considered a nanocomposite. Nanocomposites have great importance in modern life. Most engineering devices have been prepared from polymer-based nanocomposites. Numerous investigations have developed nanocomposites with appropriate properties. Polyurethanes (PUs) are types of polymers that are prepared through condensation polymerization of isocyanates and polyols. PU nanocomposites are considered among the most important polymer-based nanocomposites because of the versatile properties of the PUs. Several important applications of PU nanocomposites have caused extensive attention from material scientists on exploiting their possibility to be used in different applications (*1*). Durability is one of the most important features of nanocomposites. Durability has different forms such as chemical durability, mechanical durability, and thermal durability. Durability leads to reduced degradation and destruction of nanocomposites in the presence of harsh conditions such as corrosive chemical compounds, mechanical tensions, and extremely high or extremely low temperatures. These harsh conditions profoundly expedite the degradation and destruction processes of nanocomposites. Great efforts have been made to improve the durability of the nanocomposites against these destructive conditions. One of the most significant types of durability is the durability of nanocomposites against fire and high temperatures. Nanocomposites should endure in a high range of temperatures, which makes them thermally resistant. They should also extinguish themselves in the presence of the fire; this is called flame retardancy. The flame-retardant (FR) nanocomposites can resist and withstand longer in the presence of the fire compared with ordinary nanocomposites. Numerous studies have been done to optimize and improve FR and smoke-suppressant polymeric nanocomposites (*2*). The first generation of FR additives is organic-halogen material (*3*). Bromine-based FRs (BRFs) such as polybrominated diphenyl ethers (PBDEs) are considered the most common organic-halogen materials and have been used for several years because of the high atomic mass of bromine and its general versatility across a wide range of applications. There are more than 80 different types of BFRs with different properties. BRFs contribute to the fire safety of electrical and electronic equipment, along with installations where printed circuit boards are used, such as consumer electronic office supplies, automotive materials, aviation equipment, communication equipment, and entertainment equipment. Several types of FRs such as PBDEs and non-PBDEs have been used to prepare FR nanocomposites (*3*). Some of the important BRFs, including PBDEs and non-PBDEs, are illustrated in Figure 1 and Figure 2, respectively. The BRFs in Figure 1 have the PBDE structure, but the BRFs in Figure 2 do not have it (non-PBDE). The nanocomposites containing BFRs and other halogen-based FRs have created several dangerous problems in the environment. Several case studies in the United States (*4, 5*), China (*6, 7*), and North African coastal cities (*8*) have determined the significant hazards of BRFs. BRFs have been detected in the atmosphere (*6*) and different living organisms such as human blood (serum) (*9*), trees (*10*), fish (*4*), chicken eggs (*7*), and bobcats (*5*). The presence of BRFs inside a living organism is extremely dangerous and may cause several diseases (*6, 9*). Usually, there are two main contingent pervasiveness processes for halogen-based FRs: dispersion after burning the nanocomposite and percolation after discarding a nanocomposite. Halogen-based FRs reach the environment in both scenarios. Therefore, a large number of halogen-based FRs are released in the environment.

Several investigations have examined different methods for the remediation of halogenated organic materials because of the severe impact of these materials on the environment (*11–13*). The absorption and remediation of the released halogen-based FRs are the first solutions to reducing environmental pollution. The next solution is developing green FRs. Green FR nanocomposites do not contaminate and damage the environment, because these novel nanocomposites do not contain hazardous materials.

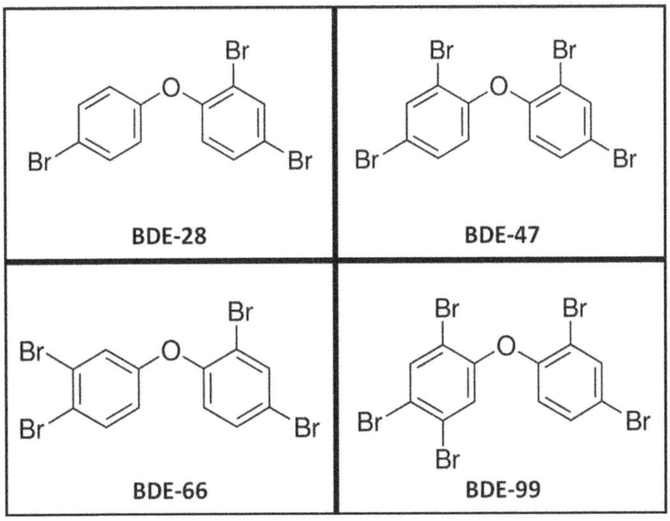

Figure 1. Some of the most important PBDEs of BRFs.

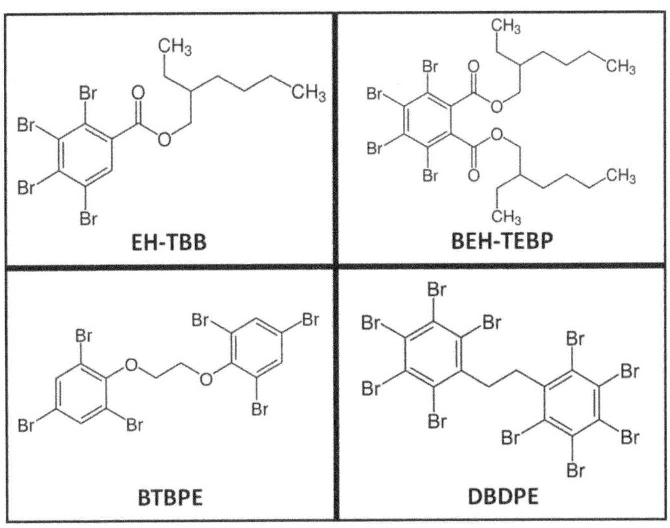

Figure 2. Some of the most important non-PBDEs of BRFs. EH-TBB = 2-ethylhexyl-2,3,4,5-tetrabrombenzoat, BEH-TEBP = bis(2-ethylhexyl) tetrabromophthalate, BTBPE = 1,2-bis(2,4,6-tribromphenoxy)ethan, DBDPE = decabromodiphenyl ethane.

Tremendous efforts have been made to find out the next generation of FR materials. The next generation must have two important advantages. First, the flame retardancy should be equal to or better than that of common BRF materials. Second, the novel FR materials must not contaminate the environment. Many investigations have examined different inorganic, organic, and organic–inorganic hybrid materials to find the best benign FR materials. Using nanomaterials is a suitable approach for the FR issue. Various one-dimensional (1D), two-dimensional (2D), and three-dimensional (3D) nanomaterials in different shapes—such as nanofibers (*14*), nanoparticles (*15*), nanoplates (*16*), nanorods (*17*), nanosheets (*18*), and nanospheres (*19*)—have been widely investigated as FR agents in nanocomposites. These nanomaterials have shown considerable flame retardancy, which is even better than that of the ordinary halogen-based FRs. Therefore, the halogen-based FRs can be completely replaced by novel FR nanomaterials.

FR Nanocomposites

FR nanocomposites have been developed via preparing a homogenous mixture of polymers and nanomaterials. The nanomaterials must be distributed and dispersed homogeneously to reach their best performances, because the accumulation or aggregation of them in the structure of the nanocomposites reduces the nanocomposite quality (*20*). The appropriate loading of FR nanomaterials in the polymer matrix leads to the fabrication of the FR nanocomposites. The structure and chemical composition of the FR nanomaterials are the most important factors in improving flame retardancy. Many different chemical substances have been investigated to obtain the best possible FR efficiency.

FR Nanomaterials

The most critical parameters of appropriate FR materials include high flame retardancy, environmental friendliness, green preparation processes, and cost effectiveness. According to these crucial parameters, organic and inorganic nanomaterials have been developed to obtain flame retardancy.

Inorganic Nanomaterials as FRs

Inorganic nanomaterials have been developed as the second generation of FR materials to improve the disadvantages of BRFs (*21*). Different types of inorganic materials are available, such as metal oxide–based FRs (*22, 23*), boron-based FRs (*24*), phosphorus-based FRs (*25*), nitrogen-based FRs (*26*), silicon-based FRs (*27*), or a hybrid of these inorganic substances (*28*). The inorganic materials have provide the flame retardancy in nanocomposites. These 1D, 2D, and 3D inorganic nanomaterials (such as graphene) are developed rapidly because they do not contaminate the environment.

The considerable ablation resistance, thermal resistance, and flame retardancy of different carbon-based nanomaterials have led to an increased interest in developing carbon-based nanomaterials such as graphene, graphite, multiwalled carbon nanotubes (MWCNTs), and single-walled carbon nanotubes (SWCNTs).

Organic Nanomaterials as FRs

Organic and biomass materials play a critical role in materials science to reduce environmental pollution and production costs (*29–34*). Organic FR materials are considered as the FRs in this case. Different organic materials, such as polydopamine (*35*), have been examined to prepare FR nanocomposites. State-of-the-art research has focused on developing low-cost FR materials from disposable biomass. Generally, the thermal durability and chemical stability parameters of the biomass and organic materials are much fewer compared with those of the inorganic materials. The novel modification of the biomass and organic materials has led to enhanced thermal durability of these materials. Especially low-cost, disposable biomass materials have been rapidly developed as novel types of FR materials. For example, chitosan (*36*), modified lignin (*37*), cellulose nanofibers (*38*, *39*), and wood nanofibers (*40*) have been used as novel biomass-based green FRs. However, biomass-based FRs have not yet been developed appropriately and need further research. Biochar-based FRs have also not yet been completely developed (*41*, *42*). Biomass-based and biochar-based FRs should be widely investigated in future studies. An example in this area is the modification of wood nanofiber (in the form of sawdust) to convert it to an effective FR (*40*). State-of-the-art investigations in the preparation of green FRs have focused on the use of renewable and disposable biomass materials to reduce production costs and environmental pollution.

Organic–Inorganic Hybrid Nanomaterials as FRs

Organic–inorganic hybrid nanomaterials are a combination of organic and inorganic materials and contain most properties of both the organic segment and inorganic segment. The combination of the organic part and the inorganic part provides a unique structure and performance in materials (*43*). Organic–inorganic hybrid materials—such as metal–organic frameworks (*44*), modified MXenes (*45*) (a class of 2D inorganic compounds), and cyclo-triphosphazene-bridged periodic mesoporous organosilica (*46*)—have been used to prepare FR nanocomposites. The properties' consolidation in organic–inorganic hybrid nanomaterials is responsible for emerging novel materials with new properties. Preparing FR nanocomposites with organic–inorganic hybrid nanomaterials is still considered a novel approach in FR development. It is expected that investigations into this subject will increase in future studies. The preparation of FR from 2,2,6,6-tetramethyl-1-piperidinyloxy, free radical–oxidized softwood pulp cyclo-triphosphazene-bridged periodic mesoporous organosilica hybrid nanospheres has determined the significant importance of hybrid FRs (*46*). Hybrid FRs should be widely developed in future studies.

Green Synthesis Methods of Nanomaterials

Emerging serious concerns about environmental contaminations have been the main reason for the swift development of green chemistry (*47*, *48*). Today, green chemistry priorities have been considered by most scientists and even politicians. Various pieces of legislation will guarantee the acceptance of these novel criteria by industries. Ignorance toward green chemistry has resulted in the spread of the perilous halogen-based FRs in the environment. Therefore, green chemistry is considered a key function in all industrial procedures. Use of benign and nonhazardous materials to fabricate different chemical substances must be carried out in all chemical synthesis methods (*48*, *49*). The performance of the final products produced through green synthesis routes depends on the chemical composition, morphology, and structure of the products (*50*). Benign materials should achieve equal or better performances than previously created products. Preparing different chemical substances via benign methods that do not contaminate the environment is considered one of the most important principles in green chemistry (*47*). Nanomaterials are considered an appropriate solution to addressing some of the issues related to the contamination associated with FRs. Halogen-based FRs have been replaced by nanomaterials as novel green FRs (*15*). Organic and inorganic nanomaterials as novel FRs contaminate the environment much less compared with halogen-based FRs. This is a considerable achievement, because the contamination of FRs is dramatically decreased. There is another important aspect of FR production based on green chemistry: the preparation of these nanomaterials must not contaminate the environment in any of the synthesis process steps.

The replacement of these novel nanomaterials helps to decrease environmental contamination, but it is not enough. The synthesis process must be benign and cannot produce hazardous byproducts. Tremendous efforts have been made in the last 2 decades to prepare nanomaterials via completely green methods (*51*). For example, nanomaterials synthesis should be performed by using benign solvents (like water) instead of some hazardous and common solvents (like carbon tetrachloride, *N*-methylpyrrolidone, and toluene). Numerous investigations have been conducted to develop the green synthesis of nanomaterials such as green nanobelts (*52*), green nanodiscs (*53*), green nanofibers (*54*), green nanoparticles (*55*), green nanoplates (*56*), green nanorods (*57*), green nanosheets (*58*), green nanospheres (*59*), and green nanowires (*60*). Nanoscale materials have better performance in composites because of high surface-to-volume ratio of nanomaterials. State-of-the-art publications on nanocomposites have focused on the preparation of nanocomposite ingredients based on green chemistry, including green synthesis of polymers, green synthesis of nanomaterials, and green preparation methods for nanocomposites. Therefore, FR nanomaterials must follow green chemistry principles to be considered appropriate FR nanocomposites for the future. Several techniques have been used to reduce environmental pollution of FRs. Some of these methods have related to their preparation processes, such as using nonhazardous materials and solvents in the synthesis procedure.

Carbon-Based FRs

Carbon-based materials have earned great importance in chemistry and materials science (*61*). Various types of carbon-based materials—such as graphene, graphene oxide (GO), graphite, MWCNTs, SWCNTs, and fullerenes—have been widely developed (*62*). The extraordinarily wide applications of carbon-based materials are the main motivation for the development of these materials. Carbon-based materials have developed widely since Hummers started their modification and functionalization (*63*). The modification of carbon-based materials that is called the Hummers method, and also updates of this method (called the modified Hummers methods), cause fast development of those carbon-based materials. Numerous investigations have developed several applications of modified and functionalized carbon-based materials, such as functionalized GO (*61*).

The considerable chemical durability and thermal stability of carbon-based materials have been obtained from the strong covalent bonds (C–C) in these carbon-based materials. Carbon-based FRs have been developed to fabricating FR nanocomposites (*62*). Various types of carbon-based materials—such as graphite (*64*), graphite oxide (*65*), graphene (*66*), GO (*67*), SWCNTs (*68*), MWCNTs (*69*), and fullerenes (*70*)—have been extensively used to prepare FR nanocomposites. Carbon-based materials are considered to be the appropriate alternatives for common halogen-based FRs. Carbon-based materials provide better flame retardancy and do not contaminate the environment. Because of their very effective performances, all of the halogen-based FRs can be replaced by carbon-based FRs. The considerable thermal durability of carbon nanotubes (CNTs) has led to the development of CNT-based FRs. Numerous investigations into CNTs and CNT nanocomposites have indicated the importance of CNTs in materials science.

CNT-Based FRs

The considerable flame retardancy of CNT-based nanocomposites has led to the immediate development of FRs based on CNTs (*71*). Various polymers have gained considerable flame retardancy through compounding with CNTs (*72*), such as polypropylene (*73*), poly(methyl methacrylate) (*74*), epoxy resins (*75*), poly(1,4-butylene terephthalate) (*76*), and even cotton (*77*) as a natural polymer. The modification and functionalization of CNTs have a significant influence on improving the flame retardancy of those CNTs (*76*). Future studies will likely be focused on CNT modification and functionalization to improve the flame retardancy of CNT-based nanocomposites.

2D Carbon-Based FRs

Numerous investigations have used different types of carbon-based nanomaterials to provide flame retardancy and smoke suppression in composites. Biochar materials are considered the most novel carbon-based FRs (*42*). The 2D carbon-based nanomaterials—such as graphene (*66*), GO (*67*), graphite (*64*), and graphite oxide (*65*)—have high thermal resistance and are widely used to provide flame retardancy. These 2D carbon-nanomaterials have some other advantages, such as appropriate corrosion resistance (*78*) and mechanical durability (*79*). Therefore, 2D carbon-based nanomaterials have been widely used to improve the performance and applications of nanocomposites (*80*).

FR-PU Nanocomposites

PU nanocomposites are a class of very important materials because of their various useful applications. Several investigations report the FR-PUs that use different types of FRs (*81–84*). The flame retardancy has an important influence on the development of PUs. For example, FR-PU foams (PUFs) have earned many more applications than ordinary PUFs (*85*). The large number of FR-PUs produced annually is a measure of the importance of finding novel FRs for the PUs, because ignoring this issue can bring catastrophic consequences. The investigations into improving flame retardancy of PUs are still ongoing. Different soft (flexible) PUs (FPUFs) (*86*) and hard (rigid) Pus (*87*) have obtained appropriate flame retardancy through nanomaterials. Novel investigations have focused on providing flame retardancy in PUs through green FRs, including organic and inorganic nanomaterials (*88*). Also, low-cost and disposable biomass materials can be considered ideal options for providing flame retardancy in Pus (*89–91*).

Biomass-based FRs have three valuable advantages, including: they do not contaminate the environment after burning; they do not contaminate the environment in the preparation process; and they have a low cost (or sometimes they are even free).

The first and second advantages are common for all green nanomaterials, but the third advantage only applies to disposable biomass materials. Perhaps some industries resist accepting novel FRs, but the minimum costs of the biomass materials should easily convince them to modify their production procedures. The economic issues have a huge importance in mass production, and they must be considered along with environmental issues to provide sustainable and reliable production. The low-cost biomass-based FR-PUs are expected to be quickly developed in future investigations.

CNT-Based FR-PU Nanocomposites

Preparing CNT-based FR-PUs is considered a novel objective in the FR-PU industry. Few investigations have determined the considerable potential of CNTs to provide appropriate FR-Pus (*92–100*). CNT-based FR-PUs should be developed to yield better results. However, the current results of CNT-based FR-PUs have determined their considerable potential. The research into CNT-based FR-PUs is listed in Table 1, indicating the structure of these nanocomposites. According to Table 1, the investigations into CNT-based FR-PUs could be widely developed. The presence of

CNTs can improve the flame retardancy of PUs and other important parameters such as mechanical properties, electromagnetic shielding properties, and other thermal properties. The investigations in Table 1 focus on FPUFs and thermoplastic PUs (TPUs). The flame retardancy of the other important types of PUs—such as rigid PUFs, cross-linked PUs, PU elastomers, PU adhesives, and PU coatings—should be carefully investigated by using CNTs as the novel FR agent.

Table 1. CNT-Based FR-PU Nanocomposites

Article	*Main Issue*	*Year*	*Ref.*
1	Layer-by-layer assembled thin films based on clay and CNTs to reduce the flammability of FPUFs	2014	(92)
2	CNT multilayer nanocoatings for preventing flame spread on FPUFs	2016	(93)
3	CNTs/ZnAl-layered double hydroxide composite as FR for FPUFs		(94)
4	Improving mechanical, thermal, and flammability properties of TPU/CNT composites	2018	(95)
5	Synergistic effect between CNTs and intumescent FRs on flammability and smoke suppression of copolymer TPU		(96)
6	Synergistic effect of FRs and CNTs on flame-retarding and electromagnetic shielding properties of TPU		(97)
7	Flexible and flame-retarding TPU-based electromagnetic interference shielding composite	2019	(98)
8	Constructing electromagnetic interference shielding and flame retarding in poly(butylene succinate) and TPU multilayered composite	2020	(100)
9	Simultaneously improved electromagnetic interference shielding and flame retarding of poly(butylene succinate)/TPU blends by constructing segregated FRs and MWCNT double network		(99)

An appropriate preparation method for CNT-based FR FPUF is illustrated in Figure 3. This benign method is environmentally friendly and provides green CNT-based flame retardancy for PUFs.

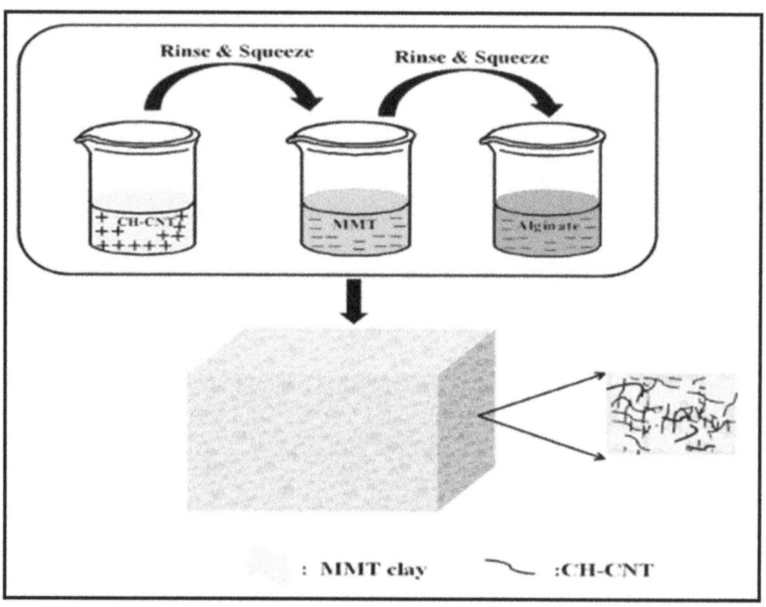

Figure 3. Preparation of montmorillonite (MMT in figure)–CNT-coated FPUF via the layer-by-layer assembly method. Reproduced with permission from reference (92). Copyright 2014 American Chemical Society.

Conclusion

The various and progressive applications of PUs and PU composites have caused the wide development of these polymers. Environmental issues are the most important driving force for finding novel FR materials and eliminating the production of halogen-based FRs. The severe damages and contamination to the environment caused by halogen-based FRs are the main reasons for the development of novel FR materials. Different organic, inorganic, and organic–inorganic hybrid materials have been used as FR materials. Nanomaterials are recognized as the best alternative to halogen-based FRs. FR nanomaterials can be designed to not contaminate the environment and therefore can be green FRs. The next environmentally friendly step is the preparation of these nanomaterials via green methods to produce even greener FRs. Based on green chemistry principles, green methods of nanomaterials synthesis have been widely developed. CNTs have the significant potential to be used as FR agents. According to certain investigations, CNT-based FPUFs and CNT-based TPUs provide appropriate flame retardancy. However, the flame retardancy of the other types of PUs should be investigated. Halogen-based FRs should be replaced with CNTs as green and efficient FR materials. The future of FR-PUs belongs to some types of FRs with the advantages of having a low cost, being benign, and having high performance in flame retardancy.

References

1. Moradi, G.; Monazzam, M.; Ershad-Langroudi, A.; Parsimehr, H.; Keshavarz, S. T. Organoclay Nanoparticles Interaction in PU:PMMA IPN Foams: Relationship between the Cellular Structure and Damping-Acoustical Properties. *Appl. Acoust.* **2020**, *164*, 107295. https://doi.org/10.1016/j.apacoust.2020.107295.
2. Lu, S.-Y.; Hamerton, I. Recent Developments in the Chemistry of Halogen-Free Flame Retardant Polymers. *Prog. Polym. Sci.* **2002**, *27* (8), 1661–1712. https://doi.org/10.1016/S0079-6700(02)00018-7.
3. Allgood, J. M.; Jimah, T.; McClaskey, C. M.; La Guardia, M. J.; Hammel, S. C.; Zeineddine, M. M.; Tang, I. W.; Runnerstrom, M. G.; Ogunseitan, O. A. Potential Human Exposure to Halogenated Flame-Retardants in Elevated Surface Dust and Floor Dust in an Academic Environment. *Environ. Res.* **2017**, *153*, 55–62. https://doi.org/10.1016/j.envres.2016.11.010.
4. Widelka, M.; Lydy, M. J.; Wu, Y.; Chen, D. Statewide Surveillance of Halogenated Flame Retardants in Fish in Illinois, USA. *Environ. Pollut.* **2016**, *214*, 627–634. https://doi.org/10.1016/j.envpol.2016.04.063.
5. Boyles, E.; Tan, H.; Wu, Y.; Nielsen, C. K.; Shen, L.; Reiner, E. J.; Chen, D. Halogenated Flame Retardants in Bobcats from the Midwestern United States. *Environ. Pollut.* **2017**, *221*, 191–198. https://doi.org/10.1016/j.envpol.2016.11.063.
6. Li, Q.; Yang, K.; Li, K.; Liu, X.; Chen, D.; Li, J.; Zhang, G. New Halogenated Flame Retardants in the Atmosphere of Nine Urban Areas in China: Pollution Characteristics, Source Analysis and Variation Trends. *Environ. Pollut.* **2017**, *224*, 679–688. https://doi.org/10.1016/j.envpol.2017.02.052.
7. Huang, C.-C.; Zeng, Y.-H.; Luo, X.-J.; Tang, B.; Liu, Y.-E.; Ren, Z.-H.; Mai, B.-X. Level Changes and Human Dietary Exposure Assessment of Halogenated Flame Retardant Levels in Free-Range Chicken Eggs: A Case Study of a Former e-Waste Recycling Site, South China. *Sci. Total Environ.* **2018**, *634*, 509–515. https://doi.org/10.1016/j.scitotenv.2018.03.386.

8. Barhoumi, B.; Tedetti, M.; Peris, A.; Guigue, C.; Aznar-Alemany, Ò.; Touil, S.; Driss, M. R.; Eljarrat, E. Halogenated Flame Retardants in Atmospheric Particles from a North African Coastal City (Bizerte, Tunisia): Pollution Characteristics and Human Exposure. *Atmos. Pollut. Res.* **2020**, *11* (4), 831–840. https://doi.org/10.1016/j.apr.2020.01.011.
9. Ma, Y.; Li, P.; Jin, J.; Wang, Y.; Wang, Q. Current Halogenated Flame Retardant Concentrations in Serum from Residents of Shandong Province, China, and Temporal Changes in the Concentrations. *Environ. Res.* **2017**, *155*, 116–122. https://doi.org/10.1016/j.envres.2017.02.010.
10. Dreyer, A.; Neugebauer, F.; Rüdel, H.; Klein, R.; Lohmann, N.; Rauert, C.; Koschorreck, J. Halogenated Flame Retardants in Tree Samples Applied as Bioindicators for Atmospheric Pollution. *Chemosphere* **2018**, *208*, 233–240. https://doi.org/10.1016/j.chemosphere.2018.05.033.
11. Martin, E. T.; McGuire, C. M.; Mubarak, M. S.; Peters, D. G. Electroreductive Remediation of Halogenated Environmental Pollutants. *Chem. Rev.* **2016**, *116* (24), 15198–15234. https://doi.org/10.1021/acs.chemrev.6b00531.
12. Huang, D.; Xu, Q.; Cheng, J.; Lu, X.; Zhang, H. Electrokinetic Remediation and Its Combined Technologies for Removal of Organic Pollutants from Contaminated Soils. *Int. J. Electrochem. Sci.* **2012**, *7* (5), 4528–4544.
13. Tratnyek, P. G.; Edwards, E.; Carpenter, L.; Blossom, S. Environmental Occurrence, Fate, Effects, and Remediation of Halogenated (Semi)Volatile Organic Compounds. *Environ. Sci. Process. Impacts* **2020**, *22* (3), 465–471. https://doi.org/10.1039/D0EM90008G.
14. Wang, D.; Peng, H.; Yu, B.; Zhou, K.; Pan, H.; Zhang, L.; Li, M.; Liu, M.; Tian, A.; Fu, S. Biomimetic Structural Cellulose Nanofiber Aerogels with Exceptional Mechanical, Flame-Retardant and Thermal-Insulating Properties. *Chem. Eng. J.* **2020**, *389*, 124449. https://doi.org/10.1016/j.cej.2020.124449.
15. Zhang, Z.; Li, X.; Yuan, Y.; Pan, Y.-T.; Wang, D.-Y.; Yang, R. Confined Dispersion of Zinc Hydroxystannate Nanoparticles into Layered Bimetallic Hydroxide Nanocapsules and Its Application in Flame-Retardant Epoxy Nanocomposites. *ACS Appl. Mater. Interfaces* **2019**, *11* (43), 40951–40960. https://doi.org/10.1021/acsami.9b15393.
16. Zhang, H.; Mao, J.; Li, M.; Cai, Q.; Li, W.; Huang, C.; Yuan, C.; Xu, Y.; Zeng, B.; Dai, L. Design of H-BN@boronate Polymer Core-Shell Nanoplates to Simultaneously Enhance the Flame Retardancy and Mechanical Properties of Epoxy Resin through the Interficial Regulation. *Compos. Part A Appl. Sci. Manuf.* **2020**, *130*, 105751. https://doi.org/10.1016/j.compositesa.2019.105751.
17. Wei, Y.-X.; Deng, C.; Wei, W.-C.; Chen, H.; Wang, Y.-Z. Hybrid Nanorods Composed of Titanium, Silicon, and Organophosphorus as Additives for Flame-Retardant Polycarbonate. *ACS Appl. Nano Mater.* **2019**, *2* (8), 4859–4868. https://doi.org/10.1021/acsanm.9b00814.
18. Li, X.; Feng, Y.; Chen, C.; Ye, Y.; Zeng, H.; Qu, H.; Liu, J.; Zhou, X.; Long, S.; Xie, X. Highly Thermally Conductive Flame Retardant Epoxy Nanocomposites with Multifunctional Ionic Liquid Flame Retardant-Functionalized Boron Nitride Nanosheets. *J. Mater. Chem. A* **2018**, *6* (41), 20500–20512. https://doi.org/10.1039/C8TA08008A.
19. Zhang, L.; Wang, Q.; Jian, R.-K.; Wang, D.-Y. Bioinspired Iron-Loaded Polydopamine Nanospheres as Green Flame Retardants for Epoxy Resin via Free Radical Scavenging and

Catalytic Charring. *J. Mater. Chem. A* **2020**, *8* (5), 2529–2538. https://doi.org/10.1039/C9TA11021F.

20. Azadi, N.; Parsimehr, H.; Ershad-Langroudi, A. Cultural Heritage Protection via Hybrid Nanocomposite Coating. *Plast. Rubber Compos.* **2020**, *49* (9), 414–424. https://doi.org/10.1080/14658011.2020.1784589.
21. Laoutid, F.; Bonnaud, L.; Alexandre, M.; Lopez-Cuesta, J.-M.; Dubois, P. New Prospects in Flame Retardant Polymer Materials: From Fundamentals to Nanocomposites. *Mater. Sci. Eng. R Reports* **2009**, *63* (3), 100–125. https://doi.org/10.1016/j.mser.2008.09.002.
22. Gao, Y.; Wu, J.; Wang, Q.; Wilkie, C. A.; O'Hare, D. Flame Retardant Polymer/Layered Double Hydroxide Nanocomposites. *J. Mater. Chem. A* **2014**, *2* (29), 10996–11016. https://doi.org/10.1039/c4ta01030b.
23. Morgan, A. B. A Review of Transition Metal-Based Flame Retardants: Transition Metal Oxide/Salts, and Complexes. In *Fire and Polymers V: Materials and Concepts for Fire Retardancy*; ACS Symposium Series; American Chemical Society, Washington, DC, 2009; Vol. 1013, pp 312–328. https://doi.org/10.1021/bk-2009-1013.ch019.
24. Cheng, X.-W.; Wu, Y.-X.; Huang, Y.-T.; Jiang, J.-R.; Xu, J.-T.; Guan, J.-P. Synthesis of a Reactive Boron-Based Flame Retardant to Enhance the Flame Retardancy of Silk. *React. Funct. Polym.* **2020**, *156*, 104731. https://doi.org/10.1016/j.reactfunctpolym.2020.104731.
25. Levchik, S. V.; Weil, E. D. A Review of Recent Progress in Phosphorus-Based Flame Retardants. *J. Fire Sci.* **2006**, *24* (5), 345–364. https://doi.org/10.1177/0734904106068426.
26. Ma, H.; Fang, Z. Synthesis and Carbonization Chemistry of a Phosphorous–Nitrogen Based Intumescent Flame Retardant. *Thermochim. Acta* **2012**, *543*, 130–136. https://doi.org/10.1016/j.tca.2012.05.021.
27. Li, J.; Gao, M.; Zheng, Y.; Guan, Y.; Yi, D. Effects of Low-Load Boron/Silicon-Based Graphene Oxide on Combustion and Thermal Degradation of Flame-Retardant Unsaturated Polyester Resin. *Macromol. Mater. Eng.* **2020**, 2000454. https://doi.org/10.1002/mame.202000454.
28. Li, S.; Lin, X.; Liu, Y.; Li, R.; Ren, X.; Huang, T.-S. Phosphorus-Nitrogen-Silicon-Based Assembly Multilayer Coating for the Preparation of Flame Retardant and Antimicrobial Cotton Fabric. *Cellulose* **2019**, *26* (6), 4213–4223. https://doi.org/10.1007/s10570-019-02373-5.
29. Ehsani, A.; Parsimehr, H.; Nourmohammadi, H.; Safari, R.; Doostikhah, S. Environment-friendly Electrodes Using Biopolymer Chitosan/Poly Ortho Aminophenol with Enhanced Electrochemical Behavior for Use in Energy Storage Devices. *Polym. Compos.* **2019**, *40* (12), 4629–4637. https://doi.org/10.1002/pc.25330.
30. Ehsani, A.; Parsimehr, H. Electrochemical Energy Storage Electrodes via Citrus Fruits Derived Carbon: A Minireview. *Chem. Rec.* **2020**, *20* (8), 820–830. https://doi.org/10.1002/tcr.202000003.
31. Parsimehr, H.; Ehsani, A. Corn-based Electrochemical Energy Storage Devices. *Chem. Rec.* **2020**, *20* (10), 1163–1180. https://doi.org/10.1002/tcr.202000058.
32. Ehsani, A.; Parsimehr, H. Electrochemical Energy Storage Electrodes from Fruit Biochar. *Adv. Colloid Interface Sci.* **2020**, *284*, 102263. https://doi.org/10.1016/j.cis.2020.102263.
33. Parsimehr, H.; Ehsani, A. Algae-Based Electrochemical Energy Storage Devices. *Green Chem.* **2020**, *22* (23), 8062–8096. https://doi.org/10.1039/D0GC02246B.

34. Parsimehr, H.; Ehsani, A.; Reyes, L.; Nikitine, C.; Vilcocq, L.; Fongarland, P. Cutting-Edge Research for a Greener Sustainable Future. *Green Chem.* **2020**, *22* (23), 8051–8061. https://doi.org/10.1039/D0GC90131H.
35. Zhang, L.; Li, Z.; Pan, Y.-T.; Yáñez, A. P.; Hu, S.; Zhang, X.-Q.; Wang, R.; Wang, D.-Y. Polydopamine Induced Natural Fiber Surface Functionalization: A Way towards Flame Retardancy of Flax/Poly(Lactic Acid) Biocomposites. *Compos. Part B Eng.* **2018**, *154*, 56–63. https://doi.org/10.1016/j.compositesb.2018.07.037.
36. Zhang, T.; Yan, H.; Shen, L.; Fang, Z.; Zhang, X.; Wang, J.; Zhang, B. Chitosan/Phytic Acid Polyelectrolyte Complex: A Green and Renewable Intumescent Flame Retardant System for Ethylene–Vinyl Acetate Copolymer. *Ind. Eng. Chem. Res.* **2014**, 53 (49), 19199–19207. https://doi.org/10.1021/ie503421f.
37. Costes, L.; Laoutid, F.; Aguedo, M.; Richel, A.; Brohez, S.; Delvosalle, C.; Dubois, P. Phosphorus and Nitrogen Derivatization as Efficient Route for Improvement of Lignin Flame Retardant Action in PLA. *Eur. Polym. J.* **2016**, *84*, 652–667. https://doi.org/10.1016/j.eurpolymj.2016.10.003.
38. Huang, Y.; Zhou, T.; He, S.; Xiao, H.; Dai, H.; Yuan, B.; Chen, X.; Yang, X. Flame-Retardant Polyvinyl Alcohol/Cellulose Nanofibers Hybrid Carbon Aerogel by Freeze Drying with Ultra-Low Phosphorus. *Appl. Surf. Sci.* **2019**, *497*, 143775. https://doi.org/10.1016/j.apsusc.2019.143775.
39. Nechyporchuk, O.; Bordes, R.; Köhnke, T. Wet Spinning of Flame-Retardant Cellulosic Fibers Supported by Interfacial Complexation of Cellulose Nanofibrils with Silica Nanoparticles. *ACS Appl. Mater. Interfaces* **2017**, *9* (44), 39069–39077. https://doi.org/10.1021/acsami.7b13466.
40. Li, P.; Sirviö, J. A.; Hong, S.; Ämmälä, A.; Liimatainen, H. Preparation of Flame-Retardant Lignin-Containing Wood Nanofibers Using a High-Consistency Mechano-Chemical Pretreatment. *Chem. Eng. J.* **2019**, *375*, 122050. https://doi.org/10.1016/j.cej.2019.122050.
41. Zhang, Q.; Cai, H.; Yang, K.; Yi, W. Effect of Biochar on Mechanical and Flame Retardant Properties of Wood – Plastic Composites. *Results Phys.* **2017**, 7, 2391–2395. https://doi.org/10.1016/j.rinp.2017.04.025.
42. Barbalini, M.; Bartoli, M.; Tagliaferro, A.; Malucelli, G. Phytic Acid and Biochar: An Effective All Bio-Sourced Flame Retardant Formulation for Cotton Fabrics. *Polymers.* **2020**, *12* (4), 811. https://doi.org/10.3390/polym12040811.
43. Ehsani, A.; Mirtamizdoust, B.; Karimi, F.; Bigdeloo, M.; Parsimehr, H. Influence of Nanostructured VO-Acetylacetonate Coordination System with 2-(Pyridin-4-Ylmethylene) Hydrazine-1-Carbothioamide in Pseudocapacitance Performance of p-Type Conductive Polymer Composite Film. *Plast. Rubber Compos.* **2020**. https://doi.org/10.1080/14658011.2020.1859875.
44. Wang, H.; Qiao, H.; Guo, J.; Sun, J.; Li, H.; Zhang, S.; Gu, X. Preparation of Cobalt-Based Metal Organic Framework and Its Application as Synergistic Flame Retardant in Thermoplastic Polyurethane (TPU). *Compos. Part B Eng.* **2020**, *182*, 107498. https://doi.org/10.1016/j.compositesb.2019.107498.
45. Si, J.-Y.; Tawiah, B.; Sun, W.-L.; Lin, B.; Wang, C.; Yuen, A.; Yu, B.; Li, A.; Yang, W.; Lu, H.-D.; Chan, Q.-N.; Yeoh, G.-H. Functionalization of MXene Nanosheets for Polystyrene

towards High Thermal Stability and Flame Retardant Properties. *Polymers.* **2019**, *11* (6), 976. https://doi.org/10.3390/polym11060976.

46. Wang, D.; Feng, X.; Zhang, L.; Li, M.; Liu, M.; Tian, A.; Fu, S. Cyclotriphosphazene-Bridged Periodic Mesoporous Organosilica-Integrated Cellulose Nanofiber Anisotropic Foam with Highly Flame-Retardant and Thermally Insulating Properties. *Chem. Eng. J.* **2019**, *375*, 121933. https://doi.org/10.1016/j.cej.2019.121933.
47. Anastas, P.; Eghbali, N. Green Chemistry: Principles and Practice. *Chem. Soc. Rev.* **2010**, *39* (1), 301–312. https://doi.org/10.1039/b918763b.
48. Erythropel, H. C.; Zimmerman, J. B.; de Winter, T. M.; Petitjean, L.; Melnikov, F.; Lam, C. H.; Lounsbury, A. W.; Mellor, K. E.; Janković, N. Z.; Tu, Q.; Pincus, L. N.; Falinski, M. M.; Shi, W.; Coish, P.; Plata, D. L.; Anastas, P. T. The Green ChemisTREE: 20 Years after Taking Root with the 12 Principles. *Green Chem.* **2018**, *20* (9), 1929–1961. https://doi.org/10.1039/C8GC00482J.
49. Goharshenas Moghadam, S.; Pazokifard, S.; Parsimehr, H. Performance of Room Temperature Cross-Linkable Acrylic Latexes Synthesized Using DAAM and AAEMA Functional Monomers. In *Eco-friendly and Smart Polymer Systems*; Springer International Publishing: Cham, Tehran, 2020; pp 125–128. https://doi.org/10.1007/978-3-030-45085-4_30.
50. Parsimehr, H.; Ershad Langroudi, A. Superhydrophobic Coatings Applications. In *Polymer Coatings*; Wiley, Beverly, MA, 2020; pp 95&120. https://doi.org/10.1002/9781119655145.ch5.
51. Lu, Y.; Ozcan, S. Green, Nanomaterials: On Track for a Sustainable Future. *Nano Today* **2015**, *10* (4), 417–420. https://doi.org/10.1016/j.nantod.2015.04.010.
52. Xie, Z.; Lai, J.; Zhu, X.; Wang, Y. Green Synthesis of Vanadate Nanobelts at Room Temperature for Superior Aqueous Rechargeable Zinc-Ion Batteries. *ACS Appl. Energy Mater.* **2018**, *1* (11), 6401–6408. https://doi.org/10.1021/acsaem.8b01378.
53. Zhou, L.; Zou, J.; Yu, M.; Lu, P.; Wei, J.; Qian, Y.; Wang, Y.; Yu, C. Green Synthesis of Hexagonal-Shaped WO3·0.33H2O Nanodiscs Composed of Nanosheets. *Cryst. Growth Des.* **2008**, *8* (11), 3993–3998. https://doi.org/10.1021/cg800609n.
54. Yassin, M. A.; Gad, A. A. M.; Ghanem, A. F.; Abdel Rehim, M. H. Green Synthesis of Cellulose Nanofibers Using Immobilized Cellulase. *Carbohydr. Polym.* **2019**, *205*, 255–260. https://doi.org/10.1016/j.carbpol.2018.10.040.
55. Duan, H.; Wang, D.; Li, Y. Green Chemistry for Nanoparticle Synthesis. *Chem. Soc. Rev.* **2015**, *44* (16), 5778–5792. https://doi.org/10.1039/C4CS00363B.
56. Wijaya, Y. N.; Kim, J.; Choi, W. M.; Park, S. H.; Kim, M. H. A Systematic Study of Triangular Silver Nanoplates: One-Pot Green Synthesis, Chemical Stability, and Sensing Application. *Nanoscale* **2017**, *9* (32), 11705–11712. https://doi.org/10.1039/C7NR03077K.
57. Purbia, R.; Paria, S. Green Synthesis of Single-Crystalline Akaganeite Nanorods for Peroxidase Mimic Colorimetric Sensing of Ultralow-Level Vitamin B1 and Sulfide Ions. *ACS Appl. Nano Mater.* **2018**, *1* (3), 1236–1246. https://doi.org/10.1021/acsanm.7b00390.
58. Guo, H.; Wang, X.-F.; Qian, Q.; Wang, F.; Xia, X. A Green Approach to the Synthesis of Graphene Nanosheets. *ACS Nano* **2009**, *3* (9), 2653–2659. https://doi.org/10.1021/nn900227d.

59. Ruan, M. M.; Song, L. X.; Yang, Z.; Teng, Y.; Wang, Q. S.; Wang, Y. Q. Novel Green Synthesis and Improved Solar-Blind Detection Performance of Hierarchical γ-Ga2O3 Nanospheres. *J. Mater. Chem. C* **2017**, *5* (29), 7161–7166. https://doi.org/10.1039/C7TC02615C.
60. Maji, N. C.; Chakraborty, J. Gram-Scale Green Synthesis of Copper Nanowire Powder for Nanofluid Applications. *ACS Sustain. Chem. Eng.* **2019**, *7* (14), 12376–12388. https://doi.org/10.1021/acssuschemeng.9b01814.
61. Ehsani, A.; Esfahaniha, M.; Khodaei kahriz, P.; Safari, R.; Parsimehr, H. Functionalized Graphene Oxide Aerogel as a High Efficient Material for Electrochemical Sensing of Organic Pollutant. *Surfaces and Interfaces* **2021**, *22*, 100817. https://doi.org/10.1016/j.surfin.2020.100817.
62. Wang, X.; Kalali, E. N.; Wan, J.-T.; Wang, D.-Y. Carbon-Family Materials for Flame Retardant Polymeric Materials. *Prog. Polym. Sci.* **2017**, *69*, 22–46. https://doi.org/10.1016/j.progpolymsci.2017.02.001.
63. Hummers, W. S.; Offeman, R. E. Preparation of Graphitic Oxide. *J. Am. Chem. Soc.* **1958**, *80* (6), 1339–1339. https://doi.org/10.1021/ja01539a017.
64. Modesti, M.; Lorenzetti, A.; Simioni, F.; Camino, G. Expandable Graphite as an Intumescent Flame Retardant in Polyisocyanurate–Polyurethane Foams. *Polym. Degrad. Stab.* **2002**, *77* (2), 195–202. https://doi.org/10.1016/S0141-3910(02)00034-4.
65. Higginbotham, A. L.; Lomeda, J. R.; Morgan, A. B.; Tour, J. M. Graphite Oxide Flame-Retardant Polymer Nanocomposites. *ACS Appl. Mater. Interfaces* **2009**, *1* (10), 2256–2261. https://doi.org/10.1021/am900419m.
66. Yu, B.; Shi, Y.; Yuan, B.; Qiu, S.; Xing, W.; Hu, W.; Song, L.; Lo, S.; Hu, Y. Enhanced Thermal and Flame Retardant Properties of Flame-Retardant-Wrapped Graphene/Epoxy Resin Nanocomposites. *J. Mater. Chem. A* **2015**, *3* (15), 8034–8044. https://doi.org/10.1039/C4TA06613H.
67. Bao, C.; Guo, Y.; Yuan, B.; Hu, Y.; Song, L. Functionalized Graphene Oxide for Fire Safety Applications of Polymers: A Combination of Condensed Phase Flame Retardant Strategies. *J. Mater. Chem.* **2012**, *22* (43), 23057–23063. https://doi.org/10.1039/c2jm35001g.
68. Beyer, G. Filler Blend of Carbon Nanotubes and Organoclays with Improved Char as a New Flame Retardant System for Polymers and Cable Applications. *Fire Mater.* **2005**, *29* (2), 61–69. https://doi.org/10.1002/fam.866.
69. Gu, L.; Qiu, J.; Yao, Y.; Sakai, E.; Yang, L. Functionalized MWCNTs Modified Flame Retardant PLA Nanocomposites and Cold Rolling Process for Improving Mechanical Properties. *Compos. Sci. Technol.* **2018**, *161*, 39–49. https://doi.org/10.1016/j.compscitech.2018.03.033.
70. Song, P.; Liu, H.; Shen, Y.; Du, B.; Fang, Z.; Wu, Y. Fabrication of Dendrimer-like Fullerene (C60)-Decorated Oligomeric Intumescent Flame Retardant for Reducing the Thermal Oxidation and Flammability of Polypropylene Nanocomposites. *J. Mater. Chem.* **2009**, *19* (9), 1305–1313. https://doi.org/10.1039/b815610g.
71. Peeterbroeck, S.; Laoutid, F.; Swoboda, B.; Lopez-Cuesta, J.-M.; Moreau, N.; Nagy, J. B.; Alexandre, M.; Dubois, P. How Carbon Nanotube Crushing Can Improve Flame Retardant Behaviour in Polymer Nanocomposites? *Macromol. Rapid Commun.* **2007**, *28* (3), 260–264. https://doi.org/10.1002/marc.200600614.

72. Kausar, A.; Rafique, I.; Muhammad, B. Significance of Carbon Nanotube in Flame-Retardant Polymer/CNT Composite: A Review. *Polym. Plast. Technol. Eng.* **2017**, *56* (5), 470–487. https://doi.org/10.1080/03602559.2016.1233267.
73. Song, P.; Xu, L.; Guo, Z.; Zhang, Y.; Fang, Z. Flame-Retardant-Wrapped Carbon Nanotubes for Simultaneously Improving the Flame Retardancy and Mechanical Properties of Polypropylene. *J. Mater. Chem.* **2008**, *18* (42), 5083–5091. https://doi.org/10.1039/b808309f.
74. Isitman, N. A.; Kaynak, C. Nanoclay and Carbon Nanotubes as Potential Synergists of an Organophosphorus Flame-Retardant in Poly(Methyl Methacrylate). *Polym. Degrad. Stab.* **2010**, *95* (9), 1523–1532. https://doi.org/10.1016/j.polymdegradstab.2010.06.013.
75. Wang, S.; Xin, F.; Chen, Y.; Qian, L.; Chen, Y. Phosphorus-Nitrogen Containing Polymer Wrapped Carbon Nanotubes and Their Flame-Retardant Effect on Epoxy Resin. *Polym. Degrad. Stab.* **2016**, *129*, 133–141. https://doi.org/10.1016/j.polymdegradstab.2016.04.011.
76. Zhu, S. E.; Wang, L. L.; Chen, H.; Yang, W.; Yuen, A. C. Y.; Chen, T. B. Y.; Luo, C.; Bi, W. M.; Hu, E. Z.; Zhang, J.; Si, J. H.; Lu, H. D.; Hu, K. H.; Chan, Q. N.; Yeoh, G. H. Comparative Studies on Thermal, Mechanical, and Flame Retardant Properties of PBT Nanocomposites via Different Oxidation State Phosphorus-Containing Agents Modified Amino-CNTs. *Nanomaterials* **2018**, *8* (2), 1–17. https://doi.org/10.3390/nano8020070.
77. Bharathi Yazhini, K.; Gurumallesh Prabu, H. Study on Flame-Retardant and UV-Protection Properties of Cotton Fabric Functionalized with Ppy–ZnO–CNT Nanocomposite. *RSC Adv.* **2015**, *5* (61), 49062–49069. https://doi.org/10.1039/C5RA07487H.
78. Hikku, G. S.; Jeyasubramanian, K.; Venugopal, A.; Ghosh, R. Corrosion Resistance Behaviour of Graphene/Polyvinyl Alcohol Nanocomposite Coating for Aluminium-2219 Alloy. *J. Alloys Compd.* **2017**, *716*, 259–269. https://doi.org/10.1016/j.jallcom.2017.04.324.
79. Young, R.; Kinloch, I.; Gong, L.; Novoselov, K. The Mechanics of Graphene Nanocomposites: A Review. *Compos. Sci. Technol.* **2012**, *72* (12), 1459–1476. https://doi.org/10.1016/j.compscitech.2012.05.005.
80. Saleem, H.; Edathil, A.; Ncube, T.; Pokhrel, J.; Khoori, S.; Abraham, A.; Mittal, V. Mechanical and Thermal Properties of Thermoset-Graphene Nanocomposites. *Macromol. Mater. Eng.* **2016**, *301* (3), 231–259. https://doi.org/10.1002/mame.201500335.
81. Visakh, P. M.; Semkin, A. O.; Rezaev, I. A.; Fateev, A. V. Review on Soft Polyurethane Flame Retardant. *Constr. Build. Mater.* **2019**, *227*, 116673. https://doi.org/10.1016/j.conbuildmat.2019.116673.
82. Yang, H.; Yu, B.; Song, P.; Maluk, C.; Wang, H. Surface-Coating Engineering for Flame Retardant Flexible Polyurethane Foams: A Critical Review. *Compos. Part B Eng.* **2019**, *176*, 107185. https://doi.org/10.1016/j.compositesb.2019.107185.
83. Cho, J. H.; Vasagar, V.; Shanmuganathan, K.; Jones, A.; Nazarenko, S.; Ellison, C. Bioinspired Catecholic Flame Retardant Nanocoating for Flexible Polyurethane Foams. *Chem. Mater.* **2015**, *27* (19), 6784–6790. https://doi.org/10.1021/acs.chemmater.5b03013.
84. Lin, B.; Yuen, A. C. Y.; Li, A.; Zhang, Y.; Chen, T. B. Y.; Yu, B.; Lee, E. W. M.; Peng, S.; Yang, W.; Lu, H.-D.; Chan, Q. N.; Yeoh, G. H.; Wang, C. H. MXene/Chitosan Nanocoating for Flexible Polyurethane Foam towards Remarkable Fire Hazards Reductions. *J. Hazard. Mater.* **2020**, *381*, 120952. https://doi.org/10.1016/j.jhazmat.2019.120952.

85. Chen, H.-B.; Shen, P.; Chen, M.-J.; Zhao, H.-B.; Schiraldi, D. A. Highly Efficient Flame Retardant Polyurethane Foam with Alginate/Clay Aerogel Coating. *ACS Appl. Mater. Interfaces* **2016**, *8* (47), 32557–32564. https://doi.org/10.1021/acsami.6b11659.
86. Xu, W.; Li, A.; Liu, Y.; Chen, R.; Li, W. CuMoO4@hexagonal Boron Nitride Hybrid: An Ecofriendly Flame Retardant for Polyurethane Elastomer. *J. Mater. Sci.* **2018**, *53* (16), 11265–11279. https://doi.org/10.1007/s10853-018-2390-5.
87. Yang, R.; Hu, W.; Xu, L.; Song, Y.; Li, J. Synthesis, Mechanical Properties and Fire Behaviors of Rigid Polyurethane Foam with a Reactive Flame Retardant Containing Phosphazene and Phosphate. *Polym. Degrad. Stab.* **2015**, *122*, 102–109. https://doi.org/10.1016/j.polymdegradstab.2015.10.007.
88. Zeng, S.-L.; Xing, C.-Y.; Chen, L.; Xu, L.; Li, B.-J.; Zhang, S. Green Flame-Retardant Flexible Polyurethane Foam Based on Cyclodextrin. *Polym. Degrad. Stab.* **2020**, *178*, 109171. https://doi.org/10.1016/j.polymdegradstab.2020.109171.
89. Yue, D.; Oribayo, O.; Rempel, G. L.; Pan, Q. Liquefaction of Waste Pine Wood and Its Application in the Synthesis of a Flame Retardant Polyurethane Foam. *RSC Adv.* **2017**, *7* (48), 30334–30344. https://doi.org/10.1039/C7RA03546B.
90. Nabipour, H.; Wang, X.; Song, L.; Hu, Y. A Fully Bio-Based Coating Made from Alginate, Chitosan and Hydroxyapatite for Protecting Flexible Polyurethane Foam from Fire. *Carbohydr. Polym.* **2020**, *246*, 116641. https://doi.org/10.1016/j.carbpol.2020.116641.
91. Bhoyate, S.; Ionescu, M.; Kahol, P. K.; Gupta, R. K. Sustainable Flame-Retardant Polyurethanes Using Renewable Resources. *Ind. Crops Prod.* **2018**, *123*, 480–488. https://doi.org/10.1016/j.indcrop.2018.07.025.
92. Pan, H.; Pan, Y.; Wang, W.; Song, L.; Hu, Y.; Liew, K. M. Synergistic Effect of Layer-by-Layer Assembled Thin Films Based on Clay and Carbon Nanotubes To Reduce the Flammability of Flexible Polyurethane Foam. *Ind. Eng. Chem. Res.* **2014**, *53* (37), 14315–14321. https://doi.org/10.1021/ie502215p.
93. Holder, K. M.; Cain, A. A.; Plummer, M. G.; Stevens, B. E.; Odenborg, P. K.; Morgan, A. B.; Grunlan, J. C. Carbon Nanotube Multilayer Nanocoatings Prevent Flame Spread on Flexible Polyurethane Foam. *Macromol. Mater. Eng.* **2016**, *301* (6), 665–673. https://doi.org/10.1002/mame.201500327.
94. Xie, H.; Ye, Q.; Si, J.; Yang, W.; Lu, H.; Zhang, Q. Synthesis of a Carbon Nanotubes/ZnAl-Layered Double Hydroxide Composite as a Novel Flame Retardant for Flexible Polyurethane Foams. *Polym. Adv. Technol.* **2016**, *27* (5), 651–656. https://doi.org/10.1002/pat.3735.
95. Kanbur, Y.; Tayfun, U. Investigating Mechanical, Thermal, and Flammability Properties of Thermoplastic Polyurethane/Carbon Nanotube Composites. *J. Thermoplast. Compos. Mater.* **2018**, *31* (12), 1661–1675. https://doi.org/10.1177/0892705717743292.
96. Chen, C.; Zhao, X.; Shi, C.; Chen, J. Synergistic Effect between Carbon Nanoparticle and Intumescent Flame Retardant on Flammability and Smoke Suppression of Copolymer Thermoplastic Polyurethane. *J. Mater. Sci.* **2018**, *53* (8), 6053–6064. https://doi.org/10.1007/s10853-017-1970-0.
97. Ji, X.; Chen, D.; Wang, Q.; Shen, J.; Guo, S. Synergistic Effect of Flame Retardants and Carbon Nanotubes on Flame Retarding and Electromagnetic Shielding Properties of Thermoplastic Polyurethane. *Compos. Sci. Technol.* **2018**, *163*, 49–55. https://doi.org/10.1016/j.compscitech.2018.05.007.

98. Ji, X.; Chen, D.; Shen, J.; Guo, S. Flexible and Flame-Retarding Thermoplastic Polyurethane-Based Electromagnetic Interference Shielding Composites. *Chem. Eng. J.* **2019**, *370*, 1341–1349. https://doi.org/10.1016/j.cej.2019.03.293.
99. Shi, Y.; He, L.; Chen, D.; Wang, Q.; Shen, J.; Guo, S. Simultaneously Improved Electromagnetic Interference Shielding and Flame Retarding Properties of Poly (Butylene Succinate)/Thermoplastic Polyurethane Blends by Constructing Segregated Flame Retardants and Multi-Walled Carbon Nanotubes Double Network. *Compos. Part A Appl. Sci. Manuf.* **2020**, *137*, 106037. https://doi.org/10.1016/j.compositesa.2020.106037.
100. He, L.; Shi, Y.; Wang, Q.; Chen, D.; Shen, J.; Guo, S. Strategy for Constructing Electromagnetic Interference Shielding and Flame Retarding Synergistic Network in Poly (Butylene Succinate) and Thermoplastic Polyurethane Multilayered Composites. *Compos. Sci. Technol.* **2020**, *199*, 108324. https://doi.org/10.1016/j.compscitech.2020.108324.

Chapter 5

Expandable Graphite for Flame-Retardant Polyurethane Foams

Ashesh Mahto[1] and Mahima Khandelwal[*,2]

[1]Centre for Nano and Material Science, Jain University, Jain Global Campus, Bengaluru, Karnataka 562112, India

[2]School of Mechanical Engineering, Chung-Ang University, Seoul 06974, Republic of Korea

[*]Email: mahimaiitr@gmail.com

Intumescent materials, like expandable graphite (EG), inorganic acids, polyols, and their combinations, have gained considerable attention because of their exceptional flame-retardant properties and are being extensively used in various polymeric systems, including coatings and blends. The foamed multicellular char generated by heating the polymer–EG system acts as a physical barrier on the surface of the polymer and reduces the heat and mass transfer rates between the system and its surroundings, thereby acting as a flame-retardant coating. EG is a graphite intercalation compound that is generally produced via oxidation or electrochemical treatment of graphite flakes in an intercalating acid. The present chapter sheds light on the mechanism by which EG induces flame retardancy in polyurethane-based materials like foams and coatings. An in-depth discussion on the mode of action, the chemical and physical changes in the polymer matrix because of EG incorporation, and the major bottlenecks has been provided.

Introduction

Polyurethane (PU), as a class of alternating copolymers, has found extensive applications in industries as well as in our day-to-day lives. PUs are available in a variety of forms, such as foams (rigid or semirigid), coatings, adhesives, and so forth, and like any other material built from organic building blocks, they are flammable. Because of the rapid increase in the usage of polymeric materials in various sectors (like manufacturing, packaging, etc.), regulations on the fire safety of these materials have become more and more stringent. Apart from decreasing feasibility of ignition and rate of flame spread, current flame retardants (FRs) are also expected to reduce generation of smoke, thereby mitigating obscuration. The nature of the smoke generated during a fire hazard also plays a crucial role and must be taken into account to limit overall damage. Environmental concerns regarding toxic byproducts (smoke/gases) released during the decomposition of halogen-based FRs in the case of a fire have led to a comprehensive assessment of overall fire hazards. The corrosive nature of hydrogen halides and other toxic combustion products emanating from these FRs supersedes the risk posed by smoke/fumes generated from pristine polymeric systems (*1*). This has led to a systematic phasing out of widely used phosphohalogen compounds and replacing them with environmentally benign halogen-free FRs.

Intumescent systems are known to act as a shield against fire propagation in flammable materials. Many research groups have proven the efficiency of various intumescent additives in diverse polymer matrices to induce FR properties (*2–5*). Among these, expandable graphite (EG) has received considerable attention because of the low cost of the raw material (graphite flakes), widespread availability, low chemical footprint, and nontoxicity. EG systems are ionic graphite acid salts; the graphite sheets are oxidized in the presence of an intercalating inorganic acid (H_2SO_4, HNO_3, etc.), resulting in a positively charged graphitic network with acid anions as the counterions. Some acid molecules are also intercalated in the resultant graphitic matrix as shown in eq 1 and depicted schematically in Figure 1. Sulfuric acid is an example. Although acid-intercalated graphite was discovered approximately 2 centuries ago by Dr C. Schafhaeutl during analyzation of graphite flakes in aqueous sulfuric acid (*6*), its application as an FR additive in polymer matrices has gained traction since the late 1990s and early 2000s (*7*):

$$24n\mathrm{C} + m\mathrm{H_2SO_4} + \frac{1}{2}\mathrm{O_2} \xrightarrow{yields} \mathrm{C}_{24n}^{+}(\mathrm{HSO_4^-})(m-1)\mathrm{H_2SO_4} + \frac{1}{2}\mathrm{H_2O} \quad (1)$$

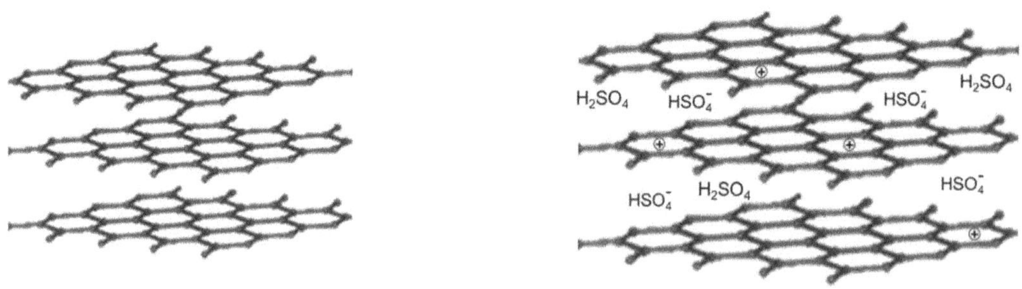

Figure 1. Chemical equation and schematic for the formation of graphite–H_2SO_4 salt. The resultant graphite intercalation compound (GIC) structure shows the presence of bisulfate ions and sulfuric acid intercalated in the graphitic network (7).

Many studies have ascertained the efficiency of EG, both standing alone and in combination with other halogen-free FRs, as an intumescent additive that acts both as a carbonization compound and a blowing agent (*8–10*). Heating up the polymer–EG blend leads to the formation of a low-density, wormlike charred layer, which acts as a physical barrier between the polymeric material and the flame. This hinders the heat and mass transfer rates. This property of EG has been attributed to the expansion/exfoliation of graphitic sheets along the c-axis when subjected to heat (*9*). The carbon atoms of EG are bonded in layers (hexagon arrangement), whereas van der Waals interactions keep the layers together. Rapid thermal treatment of EG leads to volatilization of intercalated compounds, which increases the layer spacing and volume expansion ratio. Together with the degrading polymer layer, a boundary of charred material is produced on the polymer surface exposed to the flame. The formation of char effectively reduces the rate of heat release (RHR) and the effective heat of combustion values, thereby decreasing the chances of "flashover" phenomenon in real fire conditions (*11, 12*). The final result is a low-risk, slow, and smooth burning polymer blend.

A cursory glance over the scanning electron microscopy (SEM) images in Figures 2a and b shows a clear difference in the morphologies of natural graphite flakes and after treating them with a 4:1 v/v mixture of sulfuric and nitric acid for 24 h (*13*). The resultant graphite intercalation compound (GIC) is an acid-intercalated graphite material, a type of EG. Lu et al. prepared CrO_3-intercalated graphite particles (Figure 2c) to obtain expanded graphite by both thermal expansion and chemical (H_2O_2) treatment (Figure 2d) (*14*). The authors observed "balloonlike" structures comprised of bundles of graphitic layers, which in turn are made up of multiple individual graphene layers. As the gases leave the graphitic layers of EG, they exert an outward-directed force, which leads to the exfoliation of those layers.

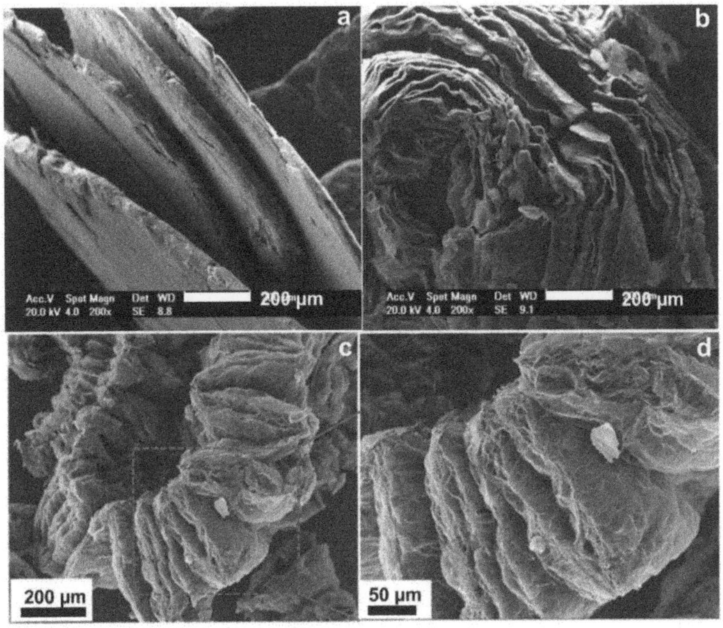

Figure 2. SEM micrographs of (a) natural graphite flakes and (b) intercalated graphite after acid treatment (parts a and b: Reproduced with permission from reference (13). Copyright 2011 John Wiley & Sons); (c) thermally expanded GIC and (d) expanded graphite at higher magnification. (parts c and d: Reproduced with permission from reference(14). Copyright 2016 American Chemical Society).

The following sections focus on PU–EG composite synthesis and its characterization, thermal FR properties, and the degradation mechanism. A correlation is drawn to account for the enhancement in fire retardancy of different types of EG-incorporated PUs. Current research trends are compared, and the bottlenecks are assessed. Finally, a brief perspective for future research is provided.

Materials Synthesis and Characterization

PUs are alternate block-copolymers ($[A\text{-}B\text{-}A\text{-}B\text{-}]_n$) with alternating hard and soft segments along the polymeric chain and are synthesized via addition polymerization of polyols (polyester, polyether, polycarbonate, etc.) and isocyanates. First synthesized and reported by Otto Bayer (*15*) using hexane diisocyanate and 1,4-butanediol monomers, the underlying isocyanate reaction and carbamate chemistry have been extensively studied and well documented because of their significant commercial importance (*16–18*). Depending on the desired properties of the final PU material, different types of polyols, isocyanates and chain extenders can be used as raw materials. Cross-linking agents, fillers, blowing agents/surfactants, and plasticizers are usually incorporated to impart different properties like rigidity, flexibility, tensile strength, and so forth (*19*). The commonly used technique to prepare PUs is the "one-shot" process (*20*), which has been employed by most research groups to study the effects of EG on PU's FR properties. Although the prepolymer method (two steps) provides better control over the packing of soft and hard segments, the one-step (one-shot) technique is preferred commercially because of ease of operation (*21*). A detailed analysis of synthetic methods and different types of PUs can be found in many articles (*19, 21*). Industrially relevant processing technologies, which are generally employed for large-scale production, are molding, slabstock, and spraying. The choice of technology depends on the final material requirement and the type of FR filler in the polyol solution. Additive manufacturing, like three-dimensional printing technology, has gained widespread attention because of the need for new raw materials and fillers (*22*).

Incorporation of EG is generally achieved by physically mixing (stirring) it in polyol-containing solution before the addition of polyisocyanate (*23–25*). In a few cases, EG is premixed with polyisocyanate before it is added to the polyol (*26*), but the synthetic method remains one shot and free rise. It should be noted that the polyol solution is comprised of different components, such as the catalyst, blowing agent, surfactant, and so forth. Distilled water has been used as the blowing agent in most of the studies discussed in later sections. Along with the polyester/ether polyol starting material, these substances together are usually designated as component A. The individual components are stirred together to obtain a uniform mixture, followed by the addition of EG as the FR additive and further stirring. The isocyanate (component B) is added to component A, and the mixture is stirred for a few seconds (5–30 s). It is then transferred into an open mold for free rise. The polymerizing mixture is cured either at room temperature or around 50–100 °C for few hours to complete the foaming process. Depending on the final product requirement (rigid/semirigid), the curing temperature and time will vary. A high curing temperature of around 100 °C will increase the cross-linking density of the resultant rigid PU foam (RPUF), thereby increasing the mechanical strength. The effects of various parameters—like particle size of EG, its concentration in the polymer blend, and density—have been studied by varying the synthetic parameters. Tabulated as follows is the formulation used in a study to prepare semirigid PU foams (SPFs) with EG and ammonium polyphosphate (APP) as synergistic FRs (Table 1). Most of the studies have employed a similar ratio of the components mentioned in Table 1 while using different curing processes. All of the components are commercially available, including a premixed polyol solution, and they come with a set of guidelines for mixing and foam preparation (*7, 27*).

The viscosity of the reaction mixture plays an important role in the processability and subsequent formation of the PU foam (PUF). According to Hu et al., the industrially accepted upper limit for polyols viscosity is <6000 mPa·s (6 Pa·s). The addition of the lower viscosity isocyanate component further makes the reaction mixture easily processable (*29*). Addition of EG into the PU formulation increases the viscosity to some extent, but it remains less than the permissible limit for further processing and foam formation. As can be observed from the values in Table 2, the viscosity increases gradually from 3.1 to 5.5 Pa·s for different concentrations of EG.

Table 1. Example of a Typical Formulation Used for the Preparation of SPFs: EG and Polyphosphate Used as FRs[a]

Composition	*Parts per Hundred of Polyol (w/w) (pphp)*
Polyol 3050A	100.00
Chain extender	6.00
Organic amine catalyst	2.00
Organic tin catalyst	0.25
Silicone surfactant	2.10
Distilled water	2.40
Polymethylene polyphenyl isocyanate	3.10
FR: EG, APP	30.00

[a] Adapted with permission from reference (*28*). Copyright 2018 Springer Nature.

Table 2. Effect of EG Concentration on the Viscosity of the Polyol Mixture[a]

Sample	*Amount of EG (pphp)*	*Viscosity at 25 °C (Pa·s)*
1	0	3.1
2	5	3.2
3	15	4.8
4	25	4.9
5	50	5.5

[a] Adapted with permission from reference (*24*). Copyright 2008 John Wiley & Sons.

Morphology and Physical–Mechanical Properties

A high magnification SEM image of the cross-section of an SPF (Figure 3a) shows an open-cell structure with a roughly spherical cell shape. Incorporating EG of smaller particle size compared with the cell dimensions in the SPF results in the encapsulation of graphite particles in the polymeric matrix (Figure 3b). Meanwhile, larger particles, with a similar mean particle size as that of the cell dimensions, tend to stay between the cell walls (Figure 3c and d) (*30*). These results are in agreement with previous findings of Modesti et al., using rigid PU–EG foams (*11*), which also recorded an increase in the cell size of the PU–EG-blended foams, as the bigger EG particles do not fit in the struts. A slight increase in the density of foams with higher EG loadings (72.5 pphp and greater) was observed, which was anticipated for the solid-fillers incorporated polymeric matrix. Maximum reductions of ~20% in parallel compression strength and ~50% in perpendicular compression strength were observed for a high EG loading of 140 pphp. In the same study, the introduction of triethyl phosphate (TEP) in the PU–EG composite foams preserved some of the loss in compression strength. The authors attributed this to the plasticizing effect of TEP, which hindered the increase in cell dimensions because of EG incorporation and tried to retain the original foam structure. Another study carried out by Li et al. during the early 2000s reported an opposite trend. The average cell size of the RPUF decreased with the incorporation of EG particles (*31*). The cell diameter distribution of 67–267 μm for pure RPUFs reduced to 47–207 μm for EG-blended RPUFs with EG loading of ~100 pphp. The authors reasoned that some EG particles with smaller sizes have nucleation ability, which increases the concentration of cells produced during the foam-blowing step. As the density of cells increased, the size decreased. The EG appeared as agglomerate sheets, which were distributed randomly in the PU matrix. A consequence was the decrease in compressive strength and modulus because of poor compatibility of EG carbon with RPUF's polymeric matrix. The spheroid cell structure of the pure RPUFs was altered in the presence of EG, which caused inhomogeneity in the cellular assembly of the resultant composite foam. The authors emphasized that the reduction in mechanical properties was nominal, even in the presence of high EG loading (100 pphp). Similar results were observed by Thirumal et al. while working with low-density PUFs (*24*). The compressive strength decreased as the EG concentration in the blend was increased. A drastic decrease in the density of PU–EG foams compared to the pure ones was also reported by the authors (see Table 3). The density of PU–EG foams, even with 25 pphp EG loading, was roughly 40% lower than that of the pure foam. This was attributed to the increase in the cell volume of the resultant foam because of the volatilization of intercalated acid in EG during the highly exothermic reaction between polyol and isocyanate moieties. The release of gases led to the expansion of cell walls and in some cases, rupture. The density of the foams increased only when the concentration of EG reached 50 pphp, which the authors attributed to the high density of EG particles (1250 kg/m^3). The resultant foam also exhibited higher water absorption capacity than the control. The bulk electrical resistivity, as reported by Li et al., remained almost unaltered even after varying the amount of EG in the RPUF's matrix. Although EG was electrically conductive, an uneven and agglomerated distribution in an insulating polymeric matrix made the overall composite an insulator (*31*).

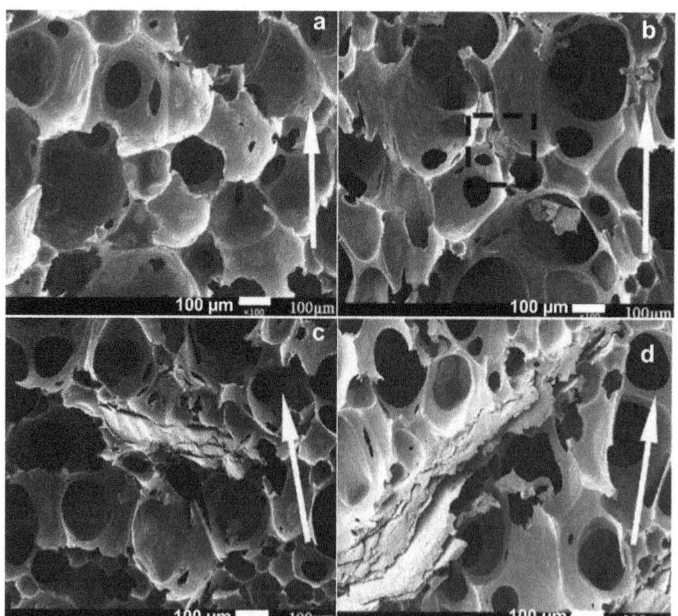

Figure 3. Cross-section SEM images of (a) pristine SPF, (b-d) EG-blended SPF foams with increasing EG mean particle size from b to d. Reproduced with permission from reference (30). Copyright 2013 John Wiley & Sons.

Table 3. Effect of EG Concentration on the Density and Compressive Strength of RPUFs[a]

Sample	*EG Content (pphp)*	*Density (kg/m³)*		*Compressive Strength at 10% Strain (kg/cm²)*	
		EG: 180 μm	*EG: 300 μm*	*EG: 180 μm*	*EG: 300 μm*
1	0	103	103	8.1	8.1
2	5	58	67	3.5	3.9
3	15	54	61	2.4	3.1
4	25	63	83	2.9	3.9
5	50	143	103	1.0	4.1

[a] Adapted with permission from reference (*24*). Copyright 2008 John Wiley & Sons.

The inhomogeneous distribution of EG flakes in the PU matrix is responsible for the diminishing compression strength of the composite foams (Table 3). Some studies argue that the viscosity of the polyol solution increases when dispersing EG flakes into them, which leads to agglomeration (*24, 32*). Poor dispersion and incompatibility between the EG sheets and PU polymeric matrix of the foam results not only in the loss of the compression modulus but also the elastic modulus (*32, 33*). Weak interfacial adhesion between the EG and PU surfaces leads to small gaps in the composite foam, which can result in slippage between the graphite layer and polymer matrix (*30*). A study by Liang and Zou reported that EG particles with a mean size of ~70 μm exhibited a positive impact on the compression properties of SPF–EG foams (*27*). Around a 50% increase in the compressive strength at a 25% strain of SPF–EG foams was observed for high loading of EG flakes (50 pphp). A corresponding increase in the density of the foams was also recorded with EG loading. With bigger EG particles (430 and 960 μm), although a slight increase in foam density was observed, the specific compressive strength decreased. The smaller EG particles could easily be encapsulated in the PU polymeric matrix, thereby reducing the chances of foam deformation because of cell wall rupture and EG agglomeration. Small particles also act as nucleation spots, leading to the formation of smaller, closely packed cells in foam with a higher number of cells per unit volume. These factors account for the high increase in density and compressive strength of SPFs. Meanwhile, bigger particles, as discussed earlier, lie between cell walls and create defects in the resultant PU–EG foam. A study conducted by Kurańska et al. using rapeseed oil–based polyols (bio-polyols) for PU–EG foams suggested that incorporation of EG had a beneficial effect on the mechanical properties of the final foam. Another notable finding was that the introduction of EG into the reaction formulation reduced the polymerization reaction rate, which might be responsible for the formation of denser cell structure without disruption (*34*).

Dynamic mechanical analysis is a widely used technique to study the viscoelastic properties of different materials and has been employed by a few research groups to examine the effects of EG on PUFs (*30, 32, 35*). Tang and Li reported that the addition of EG led to a decrease in the glass-transition temperature (T_g) of the PU–EG composite foam calculated via loss factor (tan δ) analysis. With an increase in temperature, the rate of increase in storage modulus was lower than that for loss modulus. This trend continued until the T_g was reached; afterward, the reverse was true. Lower tan δ maxima for PU–EG composites compared with those of pure RPUFs suggested a more elastic (springlike) nature of the composite formulation. In another study by Zou and Liang, the authors compared the area under the curve for tan δ versus temperature (Figures 4a, b, and c) and found that the incorporation of EG with a larger particle size led to better damping properties compared with pure SPFs and smaller particle size–incorporated SPF–EGs (27). A larger peak area was indicative of better damping properties, as suggested by the authors. This was attributed to frictional energy loss between the interfaces of EG and PU polymer matrix and also to the larger cell volume of the foam because of bigger EG particle size. An increase in the EG concentration restricted foam deformation, which was reflected in the decrease of peak area (Figure 4d). On the other hand, smaller particle size (EG with 70 μm average size) did not show any significant impact on the energy dissipation, and the lowest peak area was observed for composite foam with 50 pphp concentration of 70 μm EG particles. The incorporation of EG with 960 μm in the SPF yielded better damping properties, which the authors attributed to the larger surface area and cell size of the resultant foam. They correlated the decrease in T_g to an increase in the free volume of the PU–EG composites. Because of the larger particle size, the incorporation of EG led to an inhomogeneous packing of the PU matrix, along with the expansion in the cell dimensions of the foam. This resulted in an additional free volume that could accommodate bigger polymeric chain segments before deformation.

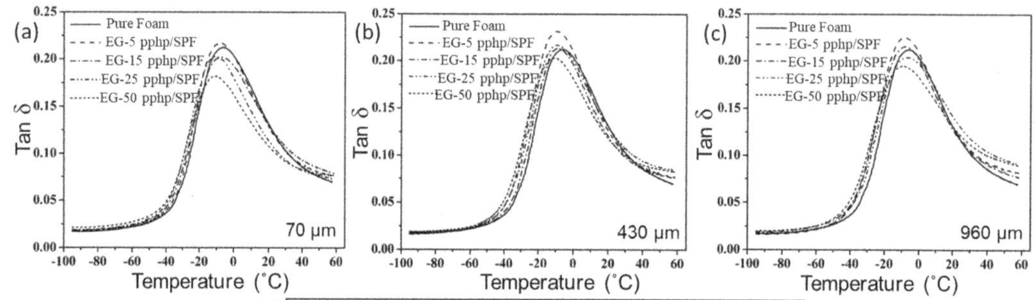

(d)

Contents	Peak Area				
	Pure foam	5 pphp	15 pphp	25 pphp	50 pphp
EG-70 μm per SPF	13.512	13.865	13.917	13.430	12.920
EG-430 μm per SPF	13.512	14.701	13.538	14.430	14.098
EG-960 μm per SPF	13.512	14.656	14.444	14.571	14.490

Figure 4. Change in tan δ with temperature for pure SPF and EG-incorporated SPF of different particle sizes: (a) 70 μm SPF–EG, (b) 430 μm SPF–EG, and (c) 960 μm SPF–EG. (d) Corresponding peak areas of tan δ curve obtained from dynamic mechanical analysis of different particle sizes of EG-incorporated PUFs at different concentrations. Reproduced with permission from reference (27). Copyright 2014 Royal Society of Chemistry.

Thermal Properties and Degradation Mechanism

Analyzing thermal properties of PU composites using different characterization techniques and the study of degradation products are vital to thoroughly assess the filler's (EG's) efficiency. Various tools and tests such as thermogravimetric analysis (TGA), in conjunction with IR spectroscopy and pyrolysis gas chromatography/mass spectroscopy (GC-MS), help in understanding the degradation mechanism and the role of filler during thermal degradation. On the other hand, thermal conductivity measurements help to generate valuable information about the insulation behavior of the composite material, which is a valuable property of PUFs.

The thermal conductivity of PUFs shows a slight increase with the addition of EG in the polymer matrix (*11*, *36*). The overall thermal conductivity of foam (λ_f) is shown in eq 2:

$$\lambda_f = \lambda_g + \lambda_s + \lambda_r + \lambda_c \quad (2)$$

In eq 2, g, s, r and c subscripts are the gaseous, solid phase, radiative, and convective thermal conductivity components, respectively. EG, as a solid material, increases the λ_s contribution toward the overall conductivity of the foam. The incorporation of larger EG particles leads to an increase in cell dimensions of the resultant foam, which invariably increases the radiation contribution toward overall conductivity (*11*). Similarly, higher concentrations of EG in the reaction formulation increase the thermal conductivity of the resultant PUF (Figure 5). Loss in thermal resistance of PU–EG foams hampers their efficient applicability in fields requiring good insulation properties. The addition of TEP as a cofire retardant with EG helps in reducing thermal conductivity values. This is attributed to the plasticizing effect of TEP, which assists in keeping the cell sizes of PU–EG foams to a minimum, thereby mitigating radiative thermal conductivity.

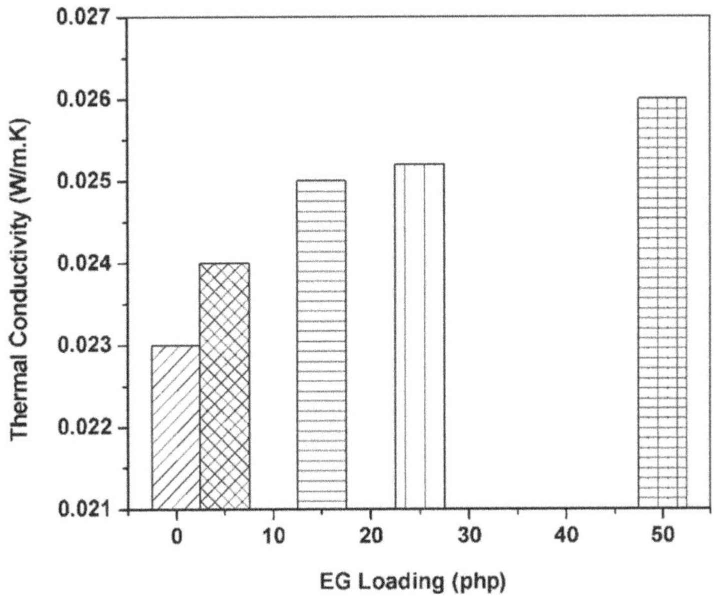

Figure 5. Effect of EG (300 μm) on the thermal conductivity of RPUFs. Reproduced with permission from reference (24). Copyright 2008 John Wiley & Sons.

TGA and its first derivative thermogravimetric (DTG) curves provide valuable insight into the thermal stability of various materials. TGA of PU and PU–EG composites has been widely reported in literature. Some studies have combined different analytical tools, like IR spectroscopy and GC-MS techniques, with TGA to study degradation products as well (*7, 23, 27*). An important contribution toward understanding the thermal degradation mechanism of PU–EG coatings was made by Camino and coworkers.[7] Apart from TGA under a nitrogen atmosphere, the authors also carried out isothermal heating studies at different temperatures for the completion of every degradation step. The degradation products—like high boiling products (HBPs), gases, liquids, and solid residues—were characterized using a custom pyrolysis Fourier transform infrared GC-MS unit. The thermogram of EG was recorded under a nitrogen atmosphere at a heating rate of 10 °C/min and showed that the major weight loss occurred between 200 and 350 °C. Concomitant blowing measurements performed using a transducer probe corroborated the weight loss with the expansion of EG particles. To understand the expansion mechanism, the authors collected the gases formed during EG decomposition under isothermal conditions (250 °C, 30 min) and characterized them using IR spectroscopy. It was concluded that the irreversible expansion of EG was caused by CO_2, SO_2, and H_2O vapors escaping through the edges of the graphitic sheets, leading to a weight loss of ~25 wt %. The EG consisted of roughly 8.5 wt % H_2SO_4, and the acid had a decomposition temperature of 340 °C. A redox reaction between the intercalated acid and graphitic carbon was proposed to account for the aforementioned observations, as depicted in eq 3:

$$C + 2H_2SO_4 \xrightarrow{yields} CO_2 + 2H_2O + 2SO_2 \quad (3)$$

Partial oxidation of carbon-to-carbon monoxide was ruled out following a thermal volatilization experiment. Only a fraction of intercalated H_2SO_4 participated in the aforementioned redox reaction with graphitic carbon, while further volatilization/decomposition at elevated temperatures (>340 °C) did not contribute to the expansion process. The gases released between 200 and 350 °C led to partial exfoliation of the EG particles, making the cell structure more open. A further release of gases did not generate enough internal push/pressure, which could lead to further expansion. The final result was a wormlike, low-density multicellular structure.

Neat PUF degrades via a two-step process, with major weight loss occurring between ~230 and 400 °C. The residue from the first step further volatilizes at a much slower rate between 400 and 600 °C. The first degradation step may be envisaged as a reverse polycondensation reaction resulting in the formation of hydroxyl and isocyanate groups containing fragments of the polymer chain. The residual (surplus in foam), as well as newly formed isocyanates, dimerize in this temperature range to form carbodiimide and release carbon dioxide gas. Carbodiimide reacts with now free alcohol groups to produce substituted urea. There is a possibility of trimerization of the isocyanate groups, resulting in the formation of stable isocyanurate rings. The formation of bifunctional polyols by the depolymerization reaction may also lead to cross-linking via free OH groups in the first degradation step. This might explain the slow decomposition rate after 400 °C. As the temperature is increased to 500 °C, a complete disintegration of the carbonyl-containing structure is achieved, along with dehydration of alcohols, releasing more CO_2 and H_2O. HBP analysis shows the presence of aromatic amines. At more than 500 °C, the urea bonds start to decompose, leading to the final charring of the foam. All of the probable reactions for PU degradation are depicted in Figure 6 (7).

The introduction of EG slightly modifies the degradation profile of the PUF. The first degradation step is almost the same as that of pure PU, with an additional release of SO_2 because of EG decomposition. Some compounds containing free carbonyl groups have also been observed in isothermal IR experiments. They might have been formed because of the oxidation of the polymer chain by H_2SO_4. A peculiar feature of the second degradation step is the formation of thiol groups by the reaction of PU polymeric chains and H_2SO_4. The HBP analysis using GC-MS also reveals the presence of substituted quinolines and aniline, which might form by acid-catalyzed degradation of PU (Figure 7).

Figure 8a depicts the TGA curves for pure SPF and SPFs filled with EG of different particle sizes at different concentrations (pphp). The two-step decomposition process can be seen from a cursory glance. The initial decomposition temperature ($T_{5\%}$) was defined as the temperature at which 5% weight loss was completed. Meanwhile, the temperature at which maximum weight loss was observed was designated as T_{max}. Larger particle size and high EG loading in PUFs led to an increase in $T_{5\%}$ values. The T_{max} values obtained from the DTG curves (Figure 8b) were slightly altered in the presence of EG, while a shoulder peak between 260 and 320 °C could also be observed for SPF–EG composites. This additional peak might be because of acid- (HNO_3 in this case) catalyzed oxidation of polymeric chains leading to fragmentation and weight loss (27).

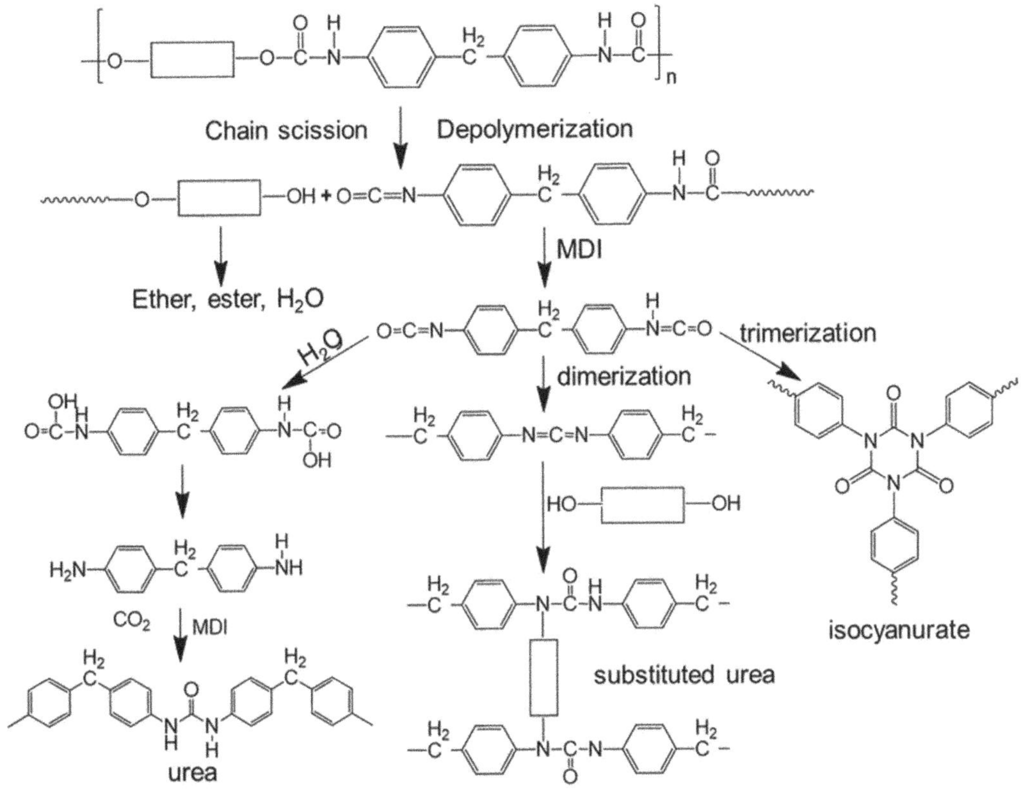

Figure 6. Mechanism of degradation of PU (MDI stands for methylene diphenyl diisocyanate) (7).

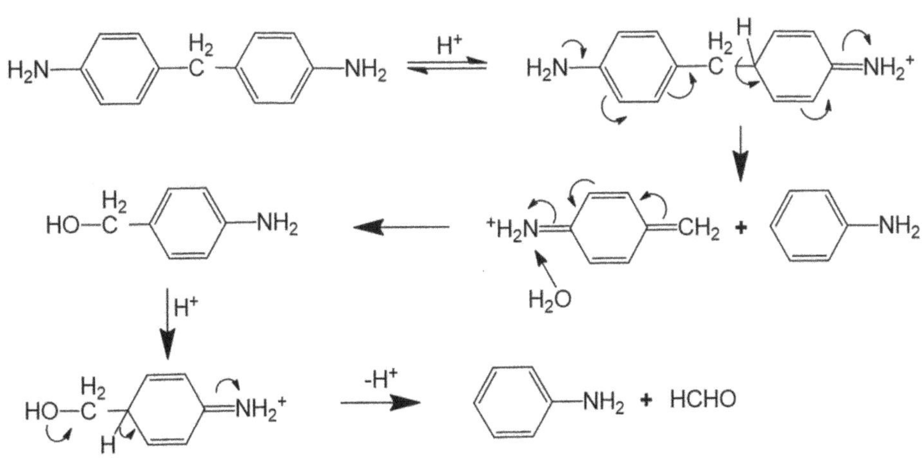

Figure 7. Formation of aniline via acid catalysis during thermal degradation of PU (7).

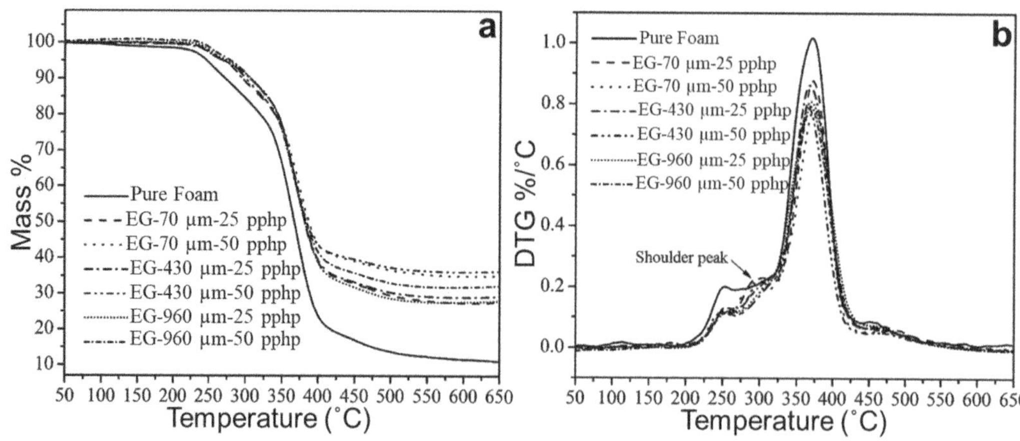

Figure 8. TGA (a) and DTG (b) curves for neat and EG-filled SPFs. Reproduced with permission from reference (27). Copyright 2014 Royal Society of Chemistry.

FR Properties

Different standardized tests have been proposed by the International Organization for Standardization to check the burning behavior of materials and to provide ratings for their classification according to test results. A cone calorimeter, along with different types of burn tests (vertical and horizontal), are employed to check for the fire safety of the test material (*37, 38*). Another important parameter is the limiting oxygen index (LOI), which is the measure of the tendency of a material to sustain the flame. Introduced by Fenimore and Martin in 1966, this method involves ignition of a polymeric material (generally at the top), followed by finding out the lowest O_2 concentration, in an upward flowing O_2/N_2 mixed gas stream (~4 ± 1 cm/sec), which supports combustion (*39*). Lower LOI values signify that the material will support combustion, even at low O_2 concentrations.

Figure 9a depicts the variation in the LOI values of the pristine PUFs and those incorporated with EG of different particle sizes, measured using the standard International Organization for Standardization 4589-2:1996 technique. It is evident that the LOI value remains the same for pristine PUFs and those incorporated with EG of particle size70 μm, signifying the negligible effect of small particle size EG on the FR properties of the foam. The trend changes to a positively sloped line as one moves from 150 to 430 μm size EG particles. It is worth mentioning here that the normal O_2 concentration in the atmosphere is ~21%, and any material with an LOI value less than that will probably support open-air burning. It can be seen from the graph that the PU–EG composite foams cross this LOI mark at EG particle size more than ~275 μm. At more than 430 μm EG particle size, the increase in LOI values for the PU–EG composites is less pronounced. As the mass loading of EG in different composites remains constant for all particle sizes, the lowering of the number of particles per unit volume of PU–EG composites for bigger sizes might justify the observed trend (*30*). Increasing the mass loading of EG has a similar effect of raising the LOI values as has been reported by several research groups (*7, 11, 23, 40*). Small EG particles (70 μm) are unable to form a comprehensive barrier layer, as the expanded graphite formed during the heat treatment does not provide an adequate hindrance to heat and mass transfer. However, as the particle size increases, the expanded volume of EG becomes larger to the point that it can effectively shield the inner layer of the foam by forming a dense isolation layer. This is evident from the digital images (Figure 9b) of the burned samples after LOI test, wherein the expanded graphite in the form of protruding wormlike structures can be seen by naked eyes only after the particle size reaches 430 μm. Layer segregation (burned outside and pristine inside) by the expanded graphite and degraded polymeric layer at the surface of the PUF are responsible for cutting off the O_2 supply and preserving the integrity of the inner layer of the foam.

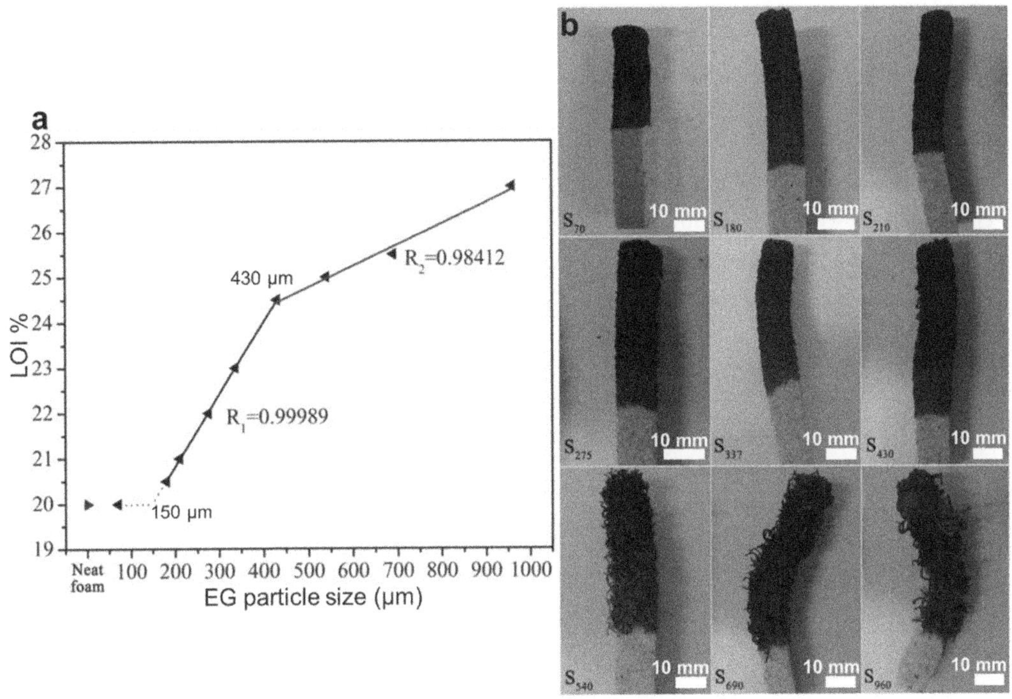

Figure 9. (a) LOI results of SPF in the presence of EG with different particle sizes (20 pphp), (b) digital photographs of SPF–EG composites after LOI test. Reproduced with permission from reference (30). Copyright 2013 John Wiley & Sons.

Horizontal burning (HB) and vertical burning tests are also important measurements and can corroborate with the LOI data while at the same time provide FR ratings. Table 4 provides a detailed burn rate analysis for EG-incorporated PUFs with different particle sizes along with the material ratings. For example, a material can be assigned the "HB" rating provided that the burn rate is lower than 75 mm/min and the flame does not cross the length mark of 100 mm. One peculiar feature observed with the incorporation of EG particles that were 70 μm in the SPF matrix was the increase in burn rate compared with neat SPF. For a sample containing EG particles that were 430 μm, an HB-3 rating was achieved at the EG loading of 15 pphp, while the same rating was achieved at 5 pphp for EG particles of average size ~960 μm. Also, for all of the samples tested, only the SPF–EG composite with 960 μm particle size EG loaded at a concentration of 50 pphp passed the vertical burning test and achieved a V-0 rating.

RHR is another important parameter that reflects the FR behavior of materials and is obtained through a cone calorimeter. The CO/CO_2 weight ratio can also be obtained to verify the extent of complete combustion. Modesti et al. studied these parameters for n-pentane-blown PU–EG composite foams and reported that the addition of EG significantly reduced the peak and mean RHR values (*11*). While a maximum RHR value of ~180 kW/m^2 was reported for neat PUF, it was reduced to roughly one-third of the original value (~65–70 kW/m^2) for 140 pphp EG loaded PUF. The authors also reported that a significant increase in the CO/CO_2 ratio was observed for very high EG loading (140 pphp) only. A higher value of this ratio suggested an increase in monoxide levels, which pointed toward lower combustion completeness. Other parameters—like total heat evolved (THE), total mass loss (TML), and effective heat of combustion, defined in terms of the ratio THE/TML—can also be obtained through cone calorimetry studies. In another study by Modesti et al., the researchers compared the FR efficiency of sulfur-intercalated EG versus phosphorus-intercalated EG (S-EG vs P-EG) and found that the peak RHR was reduced by almost 70% for S-EG, while for P-EG, the reduction was around 60%. S-EG performed better than P-EG in terms of THE as well the amount of residue left after burning. While 34 MJ/m^2 of total heat was evolved for pure PU, S-EG incorporation led to a sharp decrease in the THE value (12–14 MJ/m^2) as well as 60–65% of weight retention after burning (*41*). The better performance of S-EG was attributed to the dilution of fuel gas (O_2) by nonflammable gases (namely H_2O, CO_2, and SO_2) released during the expansion of S-EG. P-EG upon heat treatment retained a large part of phosphorous in the condensed phase, while releasing some nitrogen dioxide.

Table 4. Results of Horizontal and Vertical Burning Tests for EG-Incorporated SPF[a]

Sample	*Neat*	*5 pphp EG*	*15 pphp EG*	*25 pphp EG*	*50 pphp EG*
EG-70 µm	72.86 mm/min	72.2 mm/min	70.2 mm/min	86.7 mm/min	73.3 mm/min
EG-430 µm		56.6 mm/min	HB-3-38.7[b] mm/min	HB-2-9.2[c] mm	HB-1[d]
EG-960 µm		HB-3-39.8 mm/min	HB-1	HB-1	HB-1 V-0

[a] Adapted with permission from reference (*27*). Copyright 2014 Royal Society of Chemistry. [b] HB-3, the burning rate is less than 40 mm/min. [c] HB-2, the flame front passes the 25 mm reference mark and is self-extinguishing. [d] HB-1, the flame front does not pass the 25 mm reference mark and quenches after the ignition source is removed.

The FR effect of EG particles on RPUFs of different densities was studied and reported by Bian et al. (*26*) However, the authors did not mention the reaction mixture composition in terms of pphp, which slightly hinders the quantitative comparison with existing reports. Nevertheless, the LOI tests revealed that increasing the weight percentage of EG from 5 to 20% in dense PUF (0.510 g/cm^3) led to a drastic increase in the LOI values (27–41%). While under the same conditions, a nominal increase from 21.5 to 27% was recorded for low-density foam (0.065 g/cm^3). The authors observed that the high-density foams performed better in terms of FR properties, especially for higher EG loadings. To further study the effect of density on burning and understand the better performance of high-density foams, the authors recorded the cross-sectional SEM images of the burned PU–EG foams (10 wt % EG). Figures 10a and b depict the fractured surface SEM micrographs after burning the low- and high-density PU–EG composite foams, respectively. As can be observed from these images, the burned PU–EG composite has been divided into three layers: the outer one, which is the burned layer; an interface with cells exhibiting plastic deformation because of heat; and the inner pristine layer, which is unharmed by the fire. The interface layer is more compact for the denser foam (Figure 10b), suggesting a low degree of plastic deformation because of the heating effect. The corresponding high magnification SEM images of the interface layer are represented in

Figures 10a′ and b′. For low-density foam (Figure 10a′), some expanded graphite particles can be observed, along with a largely destroyed cell structure. Meanwhile, for the high-density foam (Figure 10b′), anamorphic cells can be observed, along with some plastic deformation on the edge of the burned layer (near fire spots). Although the weight ratio of the EG/PU polymer matrix is the same in both cases, the real EG loading per unit volume increases with the density of the foam (*26*). Upon exposure to fire, the outer layer of the denser foam produces more expanded graphite intertwined in a degrading polymeric matrix (wormlike structures), which effectively reduces the heat flux toward inner sections of the foam. A direct implication of this study is that the amount (and probably the size as well) of EG required to induce efficient flame retardancy to PU depends on the density of the foam.

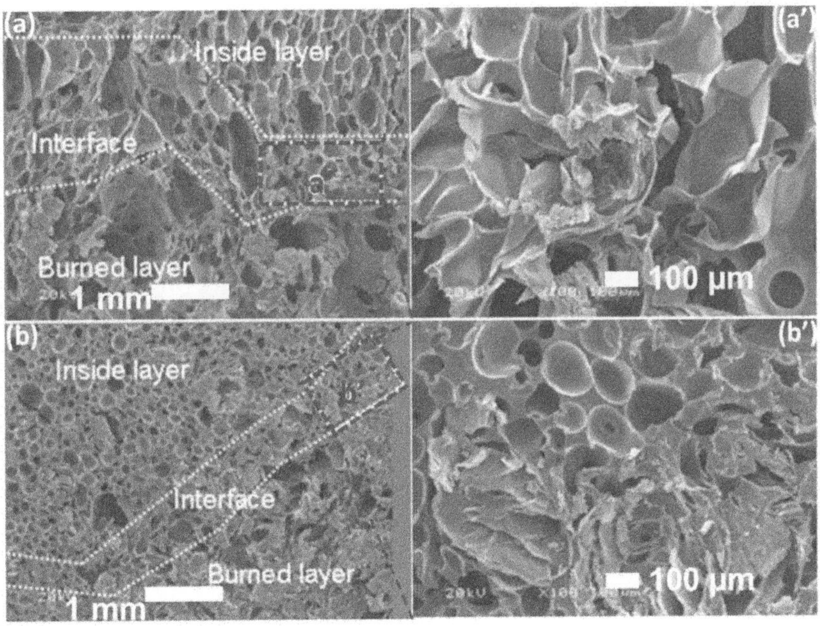

Figure 10. SEM micrographs of the fractured surfaces of low- and high-density EG-RPUF composite foams after exposure to fire: (a) 0.065 g/cm³ 10 wt % EG/RPUF; (b) 0.510 g/cm³ 10 wt % EG/RPUF; (a') high magnification of the highlighted area in figure (a); (b') high magnification of the highlighted area in figure (b). Reproduced with permission from reference (26). Copyright 2007 John Wiley & Sons.

Figure 11 depicts the overall schematic of the FR action of EG in PUF (SPF in this case). When subjected to fire, the EG particles contained in the outer polymeric layer get exposed to heat and start expanding. This expanded graphite starts forming a wormlike, low-density boundary layer, which physically separates the flame from the inner polymer layer, thereby providing protection. In the presence of fire, the outer PU polymer matrix starts degrading as well, and in combination with the expanded graphite, it forms the charred outer layer. Upon close inspection via high magnification electron microscope imaging, it is apparent that the wormlike structures of expanded graphite are glued together with the burned PU residue. This layer is responsible for limiting the heat and mass transfer rates, along with hindering the fuel (O_2) supply.

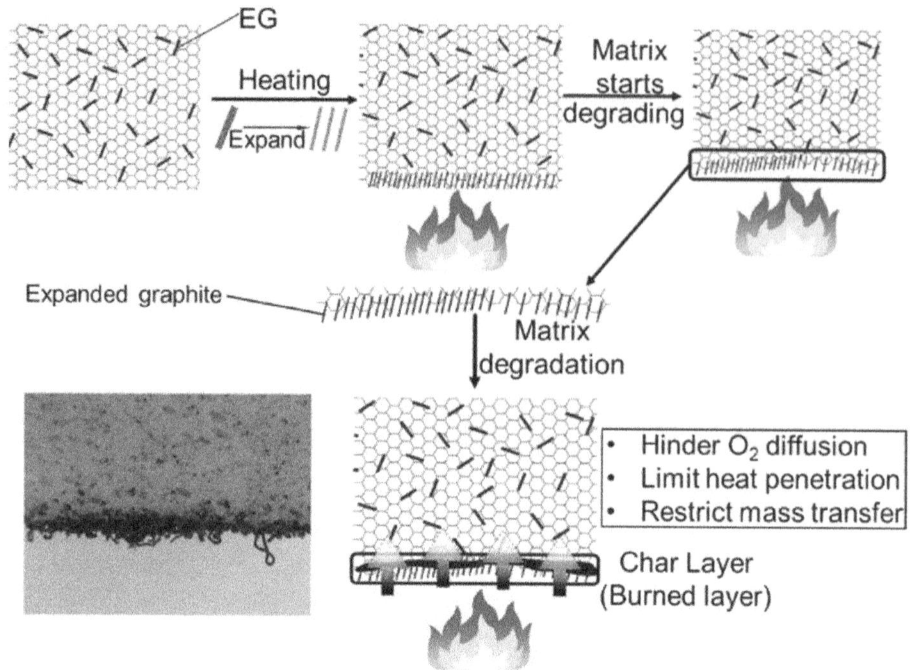

Figure 11. FR mechanism of SPF–EG composites: formation of expanded graphite inside the polymer matrix, resulting in the formation of a charred barrier layer. Reproduced with permission from reference (27). Copyright 2014 Royal Society of Chemistry.

Research Trends and Perspectives

As discussed in the previous sections, EG has proven to be an effective FR additive in PUFs, with many independent studies providing detailed material characterization and corroborating the FR properties. Still, with all of the advantages that EG offers as a greener and cleaner alternative to halogen-based FRs, a few bottlenecks persist. Researchers are now trying to improve the compatibility of EG in the polymeric matrix of PUFs using different methodologies. Surface modification of EG, encapsulation of EG particles in another material (mostly polymeric) before incorporation in PU, and addition of cofire retardants that can act synergistically with EG are some of the commonly reported methods.

For instance, poly(methylmethacrylate) (PMMA)-encapsulated EG particles have been reported and studied for their effects on the mechanical and FR properties of RPUFs (*42*). It has been found that compared to the unmodified EG particles, PMMA-EG showed better compatibility with the polymeric matrix of PU, resulting in better interfacial stability. This was reflected in the increase in the mechanical and damping properties of the resultant foam. The FR properties were also substantially improved because of the increase in the expanded volume of the PMMA-EG particles compared to EG. This was attributed to the inability of the blowing gases to escape through the graphitic edges of the EG particles because of the presence of a PMMA layer. Following a similar approach, EG-based core–shell type structures in a melamine-formaldehyde resin was synthesized and studied for its effect on the FR properties of PU (*43*). This, again, led to a significant increase in the expanded volume of the modified EG particles upon burning, thereby providing better fire resistance. The thermal conductivity of the pristine EG was also shielded by incorporation in the melamine-formaldehyde resin, leading to the preservation of the original thermal insulation of the PUFs. EG encapsulated in magnesium hydroxide (MH) nanosheets has also been used as an FR additive in RPUFs (*44*), wherein the reaction between the isocyanate functional groups of PU and MH led to better interfacial adhesion between the EG particles and the PU matrix. Better FR properties were observed for these particles when compared with pristine EG or physically mixed EG-MH particles.

Unlike in other studies, Wang et al. employed a liquid phosphorus-containing polyol as one of the starting materials for flexible PUF preparation, along with EG particles. The resultant foams exhibited FR as well as smoke-suppressant properties. Cone calorimetry data pointed toward an effective reduction in RHR and THE values, along with a reduction in smoke production release and total smoke production. The phosphorus-containing polyol acted as a reactive FR and worked in conjunction with the expanded graphite to mitigate severe burns (*45*). APP is another reactive FR additive that has been used in combination with EG to enhance the FR behavior of different types of PUFs (*28, 40*). Addition of dimethyl methylphosphonate along with EG has also been reported to improve the FR properties of PUFs (*35, 36*). In the case of incorporation in flexible PUF, dimethyl methylphosphonate led to an increase in carbon monoxide emission because of incomplete burning. Ionic liquid–modified EG, along with silanes/silica particle fillers, have also been tested as a novel FR system (*10, 46*). It was concluded that these additives act synergistically toward achieving better FR behavior and stronger mechanical properties.

Current research on improving the performance of EG-based FR additives in PUFs is focused not only on fire behavior but also on other physicochemical properties such as thermal conductivity, mechanical strength, and damping nature. The root cause behind the negative impact of EG addition on certain properties of foam is the incompatibility between the EG particles and the PU polymeric matrix. Changing the surface properties of the EG particles by coating/encapsulating them with different materials, which can interact with the PU polymer chains or free isocyanate functionalities, has proven to be successful. The addition of cofire retardants containing phosphorous reactive groups has also yielded good results. Modifying the synthetic process (raw materials etc.) to obtain PUFs is another approach gaining interest among the research community. One study reported the synthesis of aerogel–graphite PU spray-foam composites with enhanced mechanical properties, thermal insulation, and, of course, FR properties (*47*). Suppressing the release of toxic smoke is another area of interest that is currently in focus.

Summary

FR-PUs are in exceptionally high demand because of their widespread application in diverse fields. Neat PUFs are highly flammable with LOIs of less than 20%, making them vulnerable to fire hazards without having self-extinguishing properties. EG as an intumescent FR additive in PUs has been extensively studied because of its ease of availability at low costs. Various research groups have ascertained the efficacy of EG as a competent FR additive in different types of PUFs, such as RPUFs, SPFs, flexible PU coatings, and so forth. The mode of action involves the expansion of graphitic layers because of the escaping gases (intercalated substances), which leads to the formation of a low-density, wormlike char layer acting as a boundary between the condensed and gas phases. These wormlike structures embedded in the degraded outer PU layer in contact with the fire help to cut the fuel (O_2) supply and restrict heat infusion into the inner layers of the foam. A trade-off exists between the particle size/FR activity and the mechanical properties of the foams, in the sense that bigger EG particles have shown better fire behavior but have also led to worsening mechanical and compression strength. The density of the foam also plays a crucial role when deciding on the size and amount of EG to be incorporated. For high-density foams, smaller EG particles at an optimized concentration acts as a better FR. However, for low-density foams, bigger EG particles must be used to obtain similar flame retardancy behavior, but this comes at the cost of reduced mechanical and compression strength. To circumvent issues arising because of incompatibility between EG and PU surfaces, modified EG particles have also been studied. Encapsulation of EG particles in a polymer matrix before incorporation in PU has proven to be effective in enhancing FR behavior without hampering mechanical properties and thermal conductivity of the foam. Finally, cofire retardants containing phosphorous active groups have been used in conjunction with EG particles to enhance fire retardancy, suppress smoke production, and preserve the mechanical properties of the resultant PUF.

Table 5. List of Abbreviations

Abbreviations	*Full Name*
APP	Ammonium polyphosphate
DTG	Derivative thermogravimetric
EG	Expandable graphite
FR	Flame retardant
GC-MS	Gas chromatography/mass spectroscopy
GIC	Graphite intercalation compound
HB	Horizontal burning
HBP	High boiling products
LOI	Limiting oxygen index
MH	Magnesium hydroxide
P-EG	Phosphorus-intercalated expandable graphite
PMMA	Poly(methylmethacrylate)
pphp	Parts per hundred of polyol
PU	Polyurethane
PUF	Polyurethane foam
RHR	Rate of heat release
RPUF	Rigid polyurethane foam
S-EG	Sulfur-intercalated expandable graphite
SEM	Scanning electron microscopy
SPF	Semirigid polyurethane foam
tan δ	Loss factor
TEP	Triethyl phosphate
Tg	Glass-transition temperature
TGA	Thermogravimetric analysis
THE	Total heat evolved
TML	Total mass loss

References

1. Casal, J., Ed.; *Chemical Industry and Environment: General Aspects. Risk Analysis*; Barcelona: Universitat Politècnica de Catalunya. 2-4 June 1993. http://hdl.handle.net/10068/293552, ISBN 8460084876 (accessed July 20, 2021).
2. Camino, G.; Costa, L.; Trossarelli, L. Study of the mechanism of intumescence in fire retardant polymers: Part I—Thermal degradation of ammonium polyphosphate-pentaerythritol mixtures. *Polym. Degrad. Stab.* **1984**, *6* (4), 243–252. https://doi.org/10.1016/0141-3910(84)90004-1.
3. Camino, G.; Costa, L.; Trossarelli, L. Study of the mechanism of intumescence in fire retardant polymers: Part II—Mechanism of action in polypropylene-ammonium polyphosphate-pentaerythritol mixtures. *Polym. Degrad. Stab.* **1984**, 7 (1), 25–31. https://doi.org/10.1016/0141-3910(84)90027-2.
4. Camino, G.; Costa, L.; Martinasso, G. Intumescent fire-retardant systems. *Polym. Degrad. Stab.* **1989**, *23* (4), 359–376. https://doi.org/10.1016/0141-3910(89)90058-X.
5. Xie, R.; Qu, B. Synergistic effects of expandable graphite with some halogen-free flame retardants in polyolefin blends. *Polym. Degrad. Stab.* **2001**, *71* (3), 375–380. https://doi.org/10.1016/S0141-3910(00)00188-9.
6. Schafhaeutl, C. Ueber die Verbindungen des Kohlenstoffes mit Silicium, Eisen und anderen Metallen, welche die verschiedenen Gallungen von Roheisen, Stahl und Schmiedeeisen bilden. *J. Prakt. Chem.* **1840**, *21* (1), 129–157. https://doi.org/10.1002/prac.18400210117.
7. Camino, G.; Duquesne, S.; Delobel, R.; Eling, B.; Lindsay, C.; Roels, T. Mechanism of Expandable Graphite Fire Retardant Action in Polyurethanes. *Fire and Polymers, American Chemical Society* **2001**, 797, 90–109, DOI: 10.1021/bk-2001-0797.ch008.
8. Xie, R.; Qu, B. Expandable graphite systems for halogen-free flame-retarding of polyolefins. II. Structures of intumescent char and flame-retardant mechanism. *J. Appl. Polym. Sci.* **2001**, *80* (8), 1190–1197. https://doi.org/10.1002/app.1203.
9. Duquesne, S.; Le Bras, M.; Bourbigot, S.; Delobel, R.; Camino, G.; Eling, B.; Lindsay, C.; Roels, T. Thermal degradation of polyurethane and polyurethane/expandable graphite coatings. *Polym. Degrad. Stab.* **2001**, *74* (3), 493–499. https://doi.org/10.1016/S0141-3910(01)00177-X.
10. Chen, Y.; Luo, Y.; Guo, X.; Chen, L.; Jia, D. The Synergistic Effect of Ionic Liquid-Modified Expandable Graphite and Intumescent Flame-Retardant on Flame-Retardant Rigid Polyurethane Foams. *Materials* **2020**, *13* (14), 3095. https://doi.org/10.3390/ma13143095.
11. Modesti, M.; Lorenzetti, A.; Simioni, F.; Camino, G. Expandable graphite as an intumescent flame retardant in polyisocyanurate–polyurethane foams. *Polym. Degrad. Stab.* **2002**, 77 (2), 195–202. https://doi.org/10.1016/S0141-3910(02)00034-4.
12. Vanspeybroeck, R.; Van Hees, P.; Vandevelde, P. Combustion behaviour of polyurethane flexible foams under cone calorimetry test conditions. *Fire Mater.* **1992**, *11* (2), 96–117. https://doi.org/10.1002/fam.810170403.
13. Montagna, L. S.; Fim, F. d. C.; Galland, G. B.; Basso, N. R. d. S. Synthesis of Poly(propylene)/Graphite Nanocomposites by in Situ Polymerization. *Macromol. Symp.* **2011**, *299-300* (1), 48–56. https://doi.org/10.1002/masy.200900133.

14. Lin, S.; Dong, L.; Zhang, J.; Lu, H. Room-Temperature Intercalation and ~1000-Fold Chemical Expansion for Scalable Preparation of High-Quality Graphene. *Chem. Mater.* **2016,** *28* (7), 2138–2146. https://doi.org/10.1021/acs.chemmater.5b05043.
15. Bayer, O. Das di-isocyanat-polyadditionsverfahren (polyurethane). *Angew. Chem.* **1947**, *59* (9), 257–272. https://doi.org/10.1002/ange.19470590901.
16. Melville, H. W. Advances in polyurethane technology. Buist, J. M., Gudgeon, H., Eds.; pp. xii + 311, 1968. London: Maclaren and Sons Ltd., 90s. *Br. Polym. J.* **1969**, *1* (4), 191–191. https://doi.org/10.1002/pi.4980010410.
17. Ionescu, M. *Chemistry and Technology of Polyols for Polyurethanes*, 2nd ed.; iSmithers Rapra Publishing Shawbury: Shrewsbury, Shropshire, UK 2016; Volume 1. ISBN 978-1-91024-215-5 (ebook).
18. Gaylord, N. G.; Seeger, N. V. Polyurethanes: chemistry and technology, Part I. (High Polymers, Vol. XVI). Saunders, J. H., Frisch, K. C., Eds.; Interscience, New York, 1962. xi + 368 pp. *J. Appl. Polym. Sci.* **1964**, *8* (3), 1498–1499. https://doi.org/10.1002/app.1964.070080342.
19. Akindoyo, J. O.; Beg, M. D. H.; Ghazali, S.; Islam, M. R.; Jeyaratnam, N.; Yuvaraj, A. R. Polyurethane types, synthesis and applications – a review. *RSC Adv.* **2016**, *6* (115), 114453–114482. https://doi.org/10.1039/C6RA14525F.
20. Shanks, R.; Kong, I. Thermoplastic elastomers. El-Sonbati, A. , Ed.; *Thermoplastic Elastomers*; IntechOpen, 2012; Ch 8, pp 137—154, DOI: 10.5772/36807, ISBN 978-953-51-0346-2.
21. Petrović, Z. S.; Ferguson, J. Polyurethane elastomers. *Prog. Polym. Sci.* **1991**, *16* (5), 695–836. https://doi.org/10.1016/0079-6700(91)90011-9.
22. Gama, N. V.; Ferreira, A.; Barros-Timmons, A. Polyurethane Foams: Past, Present, and Future. *Materials* **2018**, *11* (10), 1841. https://doi.org/10.3390/ma11101841.
23. Shi, L.; Li, Z.-M.; Xie, B.-H.; Wang, J.-H.; Tian, C.-R.; Yang, M.-B. Flame retardancy of different-sized expandable graphite particles for high-density rigid polyurethane foams. *Polym. Int.* **2006**, *55* (8), 862–871. https://doi.org/10.1002/pi.2021.
24. Thirumal, M.; Khastgir, D.; Singha, N. K.; Manjunath, B. S.; Naik, Y. P. Effect of expandable graphite on the properties of intumescent flame-retardant polyurethane foam. *J. Appl. Polym. Sci.* **2008**, *110* (5), 2586–2594. https://doi.org/10.1002/app.28763.
25. Kurańska, M.; Cabulis, U.; Auguścik, M.; Prociak, A.; Ryszkowska, J.; Kirpluks, M. Bio-based polyurethane-polyisocyanurate composites with an intumescent flame retardant. *Polym. Degrad. Stab.* **2016**, *127*, 11–19. https://doi.org/10.1016/j.polymdegradstab.2016.02.005.
26. Bian, X.-C.; Tang, J.-H.; Li, Z.-M.; Lu, Z.-Y.; Lu, A. Dependence of flame-retardant properties on density of expandable graphite filled rigid polyurethane foam. *J. Appl. Polym. Sci.* **2007**, *104* (5), 3347–3355. https://doi.org/10.1002/app.25933.
27. Luo, W.; Li, Y.; Zou, H.; Liang, M. Study of different-sized sulfur-free expandable graphite on morphology and properties of water-blown semi-rigid polyurethane foams. *RSC Adv.* **2014**, *4* (70), 37302–37310. https://doi.org/10.1039/C4RA05559D.
28. Li, J.; Mo, X.; Li, Y.; Zou, H.; Liang, M.; Chen, Y. Influence of expandable graphite particle size on the synergy flame retardant property between expandable graphite and ammonium polyphosphate in semi-rigid polyurethane foam. *Polym. Bull.* **2018**, *75* (11), 5287–5304. https://doi.org/10.1007/s00289-018-2309-y.

29. Jin, J. F.; Chen, Y. L.; Wang, D. N.; Hu, C. P.; Zhu, S.; Vanoverloop, L.; Randall, D. Structures and physical properties of rigid polyurethane foam prepared with rosin-based polyol. *J. Appl. Polym. Sci.* **2002,** *84* (3), 598–604. https://doi.org/10.1002/app.10312.
30. Li, Y.; Zou, J.; Zhou, S.; Chen, Y.; Zou, H.; Liang, M.; Luo, W. Effect of expandable graphite particle size on the flame retardant, mechanical, and thermal properties of water-blown semi-rigid polyurethane foam. *J. Appl. Polym. Sci.* **2013,** *131* (3), 39885. https://doi.org/10.1002/app.39885.
31. Shi, L.; Li, Z.-M.; Yang, M.-B.; Yin, B.; Zhou, Q.-M.; Tian, C.-R.; Wang, J.-H. Expandable graphite for halogen-free flame-retardant of high-density rigid polyurethane foams. *Polym. Plast. Technol. Eng.* **2005,** *44* (7), 1323–1337. https://doi.org/10.1080/03602550500208145.
32. Meng, X.-Y.; Ye, L.; Zhang, X.-G.; Tang, P.-M.; Tang, J.-H.; Ji, X.; Li, Z.-M. Effects of expandable graphite and ammonium polyphosphate on the flame-retardant and mechanical properties of rigid polyurethane foams. *J. Appl. Polym. Sci.* **2009,** *114* (2), 853–863. https://doi.org/10.1002/app.30485.
33. Kirpluks, M.; Cabulis, U.; Zeltins, V.; Stiebra, L.; Avots, A. Rigid polyurethane foam thermal insulation protected with mineral intumescent mat. *Autex Res. J.* **2014,** *14* (4), 259–269. https://doi.org/10.2478/aut-2014-0026.
34. Kurańska, M.; Prociak, A.; Cabulis, U.; Kirpluks, M.; Ryszkowska, J.; Auguścik, M. Innovative porous polyurethane-polyisocyanurate foams based on rapeseed oil and modified with expandable graphite. *Ind. Crops Prod.* **2017,** *95*, 316–323. https://doi.org/10.1016/j.indcrop.2016.10.039.
35. Wang, C.-Q.; Ge, F.-Y.; Sun, J.; Cai, Z.-S. Effects of expandable graphite and dimethyl methylphosphonate on mechanical, thermal, and flame-retardant properties of flexible polyurethane foams. *J. Appl. Polym. Sci.* **2013,** *130* (2), 916–926. https://doi.org/10.1002/app.39252.
36. Feng, F.; Qian, L. The flame retardant behaviors and synergistic effect of expandable graphite and dimethyl methylphosphonate in rigid polyurethane foams. *Polym. Compos.* **2014,** *35* (2), 301–309. https://doi.org/10.1002/pc.22662.
37. Bakhtiyari, S.; Taghi-Akbari, L.; Ashtiani, M. J. Evaluation of thermal fire hazard of 10 polymeric building materials and proposing a classification method based on cone calorimeter results. *Fire Mater.* **2015,** *39* (1), 1–13. https://doi.org/10.1002/fam.2219.
38. Schartel, B.; Hull, T. R. Development of fire-retarded materials—Interpretation of cone calorimeter data. *Fire Mater.* **2007,** *31* (5), 327–354. https://doi.org/10.1002/fam.949.
39. Fenimore, C. P.; Martin, F. J. Flammability of polymers. *Combust. Flame* **1966,** *10* (2), 135–139. https://doi.org/10.1016/0010-2180(66)90059-9.
40. Hu, X.-M.; Wang, D.-M. Enhanced fire behavior of rigid polyurethane foam by intumescent flame retardants. *J. Appl. Polym. Sci.* **2013,** *129* (1), 238–246. https://doi.org/10.1002/app.38722.
41. Lorenzetti, A.; Dittrich, B.; Schartel, B.; Roso, M.; Modesti, M. Expandable graphite in polyurethane foams: The effect of expansion volume and intercalants on flame retardancy. *J. Appl. Polym. Sci.* **2017,** *134* (31), 45173. https://doi.org/10.1002/app.45173.
42. Ye, L.; Meng, X.-Y.; Ji, X.; Li, Z.-M.; Tang, J.-H. Synthesis and characterization of expandable graphite–poly(methyl methacrylate) composite particles and their application to flame

retardation of rigid polyurethane foams. *Polym. Degrad. Stab.* **2009**, *94* (6), 971–979. https://doi.org/10.1016/j.polymdegradstab.2009.03.016.

43. Duan, H.-J.; Kang, H.-Q.; Zhang, W.-Q.; Ji, X.; Li, Z.-M.; Tang, J.-H. Core–shell structure design of pulverized expandable graphite particles and their application in flame-retardant rigid polyurethane foams. *Polym. Int.* **2014**, *63* (1), 72–83. https://doi.org/10.1002/pi.4489.
44. Wang, Y.; Wang, F.; Dong, Q.; Yuan, W.; Liu, P.; Ding, Y.; Zhang, S.; Yang, M.; Zheng, G. Expandable graphite encapsulated by magnesium hydroxide nanosheets as an intumescent flame retardant for rigid polyurethane foams. *J. Appl. Polym. Sci.* **2018**, *135* (39), 46749. https://doi.org/10.1002/app.46749.
45. Rao, W.-H.; Liao, W.; Wang, H.; Zhao, H.-B.; Wang, Y.-Z. Flame-retardant and smoke-suppressant flexible polyurethane foams based on reactive phosphorus-containing polyol and expandable graphite. *J. Hazard. Mater.* **2018**, *360*, 651–660. https://doi.org/10.1016/j.jhazmat.2018.08.053.
46. Strąkowska, A.; Członka, S.; Konca, P.; Strzelec, K. New flame retardant systems based on expanded graphite for rigid polyurethane foams. *Appl. Sci.* **2020**, *10* (17), 5817. https://doi.org/10.3390/app10175817.
47. Wi, S.; Berardi, U.; Loreto, S. D.; Kim, S. Microstructure and thermal characterization of aerogel–graphite polyurethane spray-foam composite for high efficiency thermal energy utilization. *J. Hazard. Mater.* **2020**, *397*, 122656. https://doi.org/10.1016/j.jhazmat.2020.122656.

Chapter 6

Minerals as Flame-Retardant Fillers in Polyurethanes

Przemysław Bartczak[1] and Aleksandra Grząbka-Zasadzińska[*,1]

[1]Faculty of Chemical Technology, Division of Polymers, Poznań University of Technology, 60-965 Poznań, Poland

[*]Email: aleksandra.grzabka-zasadzinska@put.poznan.pl

Polyurethanes (PUs), like all organic materials, are flammable, and in order to meet stringent conditions regarding the performance of polymeric materials and their safety, they should be flame retardant (FR). Limiting the flammability of PU materials is an extremely important problem that poses a great challenge for the industry. Finding an effective solution can be a decisive factor for further development of this group of polymers. The primary aim of FRs is to stop or slow down the burning process. The broad group of these compounds includes the following minerals: aluminum hydroxide, calcium hydroxide, magnesium hydroxide, boehmite, huntite, and hydromagnesite. They are often used as FRs for economic and ecological reasons. During endothermic decomposition, mineral fillers release an inert gas (water vapor and carbon dioxide) or vapors, and a number of occurring physical phenomena make it possible to slow down or even stop the burning process. This chapter deals with characteristics and flame retardancy mechanisms of some mineral fillers that have a potential for application in PU materials. Also, much focus is put on industrial aspects of using minerals as FRs.

Introduction

A broad spectrum of designable properties of polyurethane (PU) products makes them almost irreplaceable. As the analysis of a report on PUs consumed by major end-use market shows, in only the United States, in 2018, 7.314 million lb of PU were manufactured (*1*).

High and versatile performance properties make PUs essential to numerous applications. As in the case of all construction polymers, requirements for PU refer not only to their physicochemical and mechanical properties but flammability as well. The presence of carbon, hydrogen, and other atoms prone to decomposition and production of flammable volatile materials at elevated temperature makes PU easy to burn. Apart from obvious loss of some specific properties of materials, during the combustion process, harmful smoke and unwanted products are produced (2). So to overcome these difficulties, flame retardants (FRs) are used.

In today's world, industry offers numerous types of FR; however, halogen-containing FRs are still the most common ones. FRs based on halogens are widely used because of their high susceptibility to decrease flammability, low price, good compatibility with polymers, and satisfactory effectiveness even at low loadings. On the other hand, during combustion, they may emit toxic and corrosive gaseous products (3). This, in conjunction with growing awareness regarding the ozone layer and increasing understanding of the ecological impact of plastics on the planet, has resulted in some serious limitations placed on use of halogen FR (e.g., agreements under the Stockholm Convention on Persistent Organic Pollutants) (*4*). Bans on brominated products are being implemented at a time when fire-protection standards on performance of polymeric materials used in everyday products are continuously increasing. This raises the demand for new, ecofriendly FRs, and mineral fillers seem to be a promising option.

Mineral fillers—such as metal hydroxides, calcium carbonate, talc, and clays—have already been successfully used for lowering the flammability of numerous polymers: polypropylene, polyethylene, poly(vinyl chloride), polyamides, and so forth (*5, 6*). Also, chemical modification of other fillers has been found to increase flame retardancy of composite materials, but this approach is less common (*7–10*). Surprisingly, minerals that are readily used for thermoplastics have not yet been explored to enhance the functional properties PU. In fact, in industrial practice, aluminum hydroxide (ATH) is still in the lead.

Therefore, the purpose of this chapter is to show that other mineral fillers also have a potential for being used as FR for PU.

Minerals as FRs for PUs

Apart from a few exceptions, minerals have a specific crystalline structure, with the molecules arranged according to some internal order. Minerals occur in nature as physically and chemically homogeneous solids. Their chemical composition is uniform and constant, although it can vary within a predetermined range. Therefore, they offer a great variety of properties, and a number of them have been found to act like FRs.

There are several mineral materials that can be used as FRs for polymers. Chemical formulas and decomposition temperatures for chosen substances are presented in Table 1. The most commonly used mineral fillers are the following: ATH and magnesium hydroxide (MDH), carbonates (nesquehonite, huntite, hydromagnesite), and clays (mainly montmorillonites [MMTs]).

The aforementioned minerals have already been proven to effectively lower the flame retardancy of polymers. They can limit or delay the spread of fire because of two possible mechanisms: by forming a protective layer or by suppressing chemical reactions occurring on the surface of a material. In general, all mineral fillers work in a condensed phase; they dilute the combustible polymer, creating barriers to heat and mass transfer and subsequently reducing the rate of mass loss and heat generation (*14*).

FRs can be introduced into a polymer matrix either as nonreactive additives (*15, 16*) or reactive additives (*17, 18*). It is also possible to coat the product with some nonflammable layer (*19*). In most cases, mineral fillers are added to a polymer matrix as nonreactive additives, and their performance is based on lowering the fuel content. Application of mineral fillers may be slightly more problematic than use of halogen-based FR. This is because addition of mineral filler will most likely affect rheology of polymer melt and phase separation, and loss of homogeneity may be observed as well. Also, the content of this FR will be higher than in the case of classical ones. It is also possible that some synergistic or antagonistic effects will occur (*20*). Therefore, understanding of mechanisms leading to decreased flame retardancy is of high importance. One should also remember that the decomposition temperature of the mineral fillers should be higher than processing parameters of the polymer matrix so they do not undergo decomposition too early.

Table 1. Chemical Formula and Decomposition Temperature for Some Mineral Fillers[a]

Filler	*Chemical Formula*	*Decomposition Temp. (°C)*
ATH	$Al(OH)_3$	180–200
Calcium hydroxide	$Ca(OH)_2$	430–450
MDH	$Mg(OH)_2$	300–320
Beohmite	$AlO(OH)$	340–350
Huntite	$Mg_3Ca(CO_3)_4$	400
Hydromagnesite	$Mg_5Ca(CO_3)_4(OH)_2 \cdot 4H_2O$	220–240
Nesquehonite	$MgCO_3 \cdot 3H_2O$	70–100
Na-MMT	$(Na,Ca)_{0.3}(Al,Mg)_2Si_4O_{10}(OH)_{2\backslash'b7}n(H_2O)$	271
Organomontmorillonite[b]		297–475

[a] Source: Data from references (*11–13*). [b] Chemical formula depends on organic modifier used.

Characteristics of Some Minerals with FR Properties

Minerals applied as particulate reinforcement for polymer can influence various properties of composites, such as mechanical properties, dimensional stability, crystallinity, melt flow index, and, of course, flammability. However, obtaining good effects requires careful selection of materials based on knowledge of the properties of each mineral. Therefore, basic characteristics of the most popular minerals used as FRs will be discussed.

Metal Hydroxides

Relatively cheap and environmentally friendly metal hydroxides are the most popular mineral FRs in the PU industry. There are three types of hydroxides exhibiting flame retardancy potential: calcium, aluminum, and magnesium. Of these hydroxides, it is $Al(OH)_3$ and $Mg(OH)_2$ that constitute the vast majority. In the case of each, the mineral mechanism of flame retardancy is based on endothermic dehydration.

ATH is a side stream of aluminum production in the so-called Bayer process (*21*). The starting material of this process is bauxite ore, containing approximately 50% alumina. After reaction with caustic soda, filtering, and precipitation at 60–70 °C, ATH or gibbsite ($Al_2O_3 \cdot 3H_2O$) is obtained. Then material undergoes calcination and as the final product, alumina, is sold to smelters (*22*). If ATH is meant to be applied as a filler for polymer composites, especially as FR, the material has to exhibit some specific properties—a large surface area and the appropriate grain size (*21*).

Figure 1 and Figure 2 present the Bayer process and preparation of fine ATH, respectively (*21*).

The decomposition of ATH occurs at less than 200 °C, and thus it is used for polymers processed at relatively low temperatures (up to 200–220 °C)—that is, elastomers, thermosetting resins, and thermoplastics. If higher processing temperatures are required—as in poly(phenylene oxide) and polyamides—then MDH is preferred, as its thermal decomposition takes place at about 300 °C.

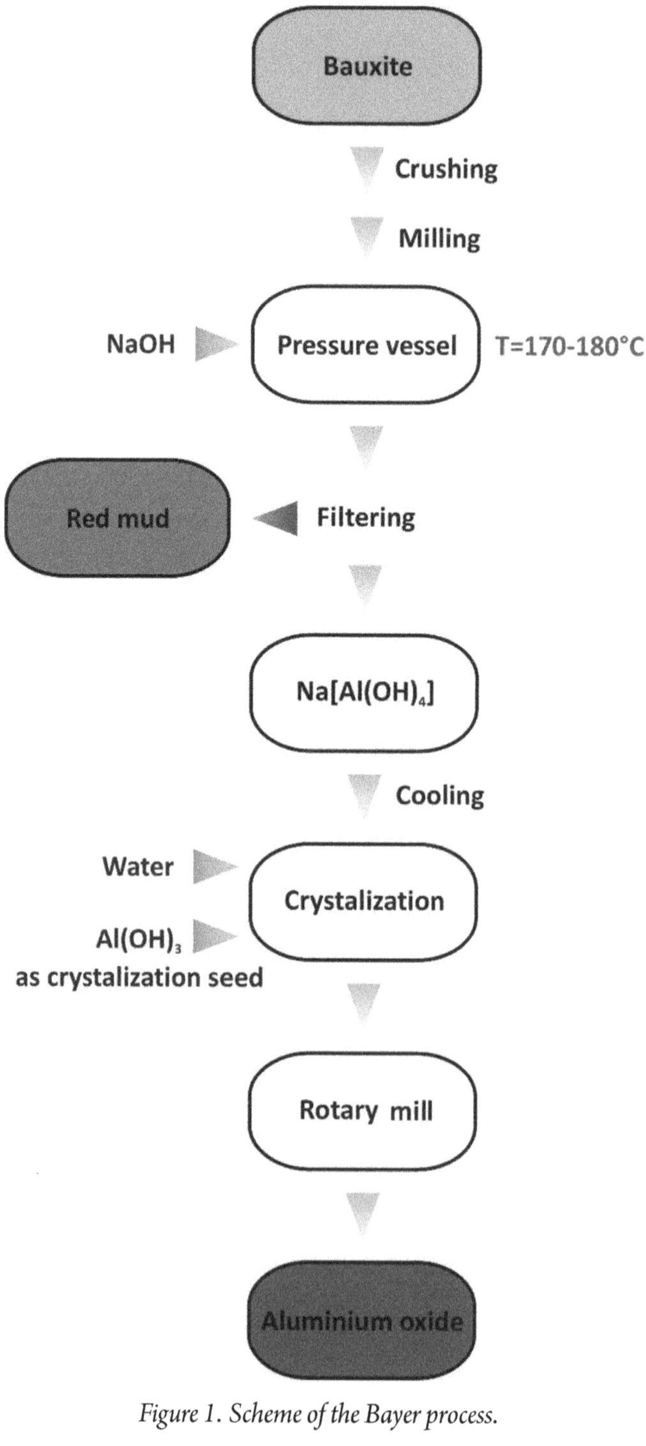

Figure 1. Scheme of the Bayer process.

In nature, MDH is naturally found as the mineral brucite. Industrially, MDH is produced mainly as a result of precipitation from solutions of magnesium salts—MgSO4, MgCl2, and Mg(NO3) (*23*). Similarly to ATH, process parameters, including precursors and chemicals used, affect morphological–dispersive properties of mineral filler (*24*). A simplified scheme of the precipitation process of MDH is presented in Figure 3. There are also other techniques like magnesite calcination, the hydration method, the pyrolysis method, and the seawater bittern magnesium–deposition method (*23*). Increasing the flame retardancy of polymers with MDH may be problematic because of two aspects: high filler content needed to achieve desired flame retardancy and resultant impairment of mechanical properties of the composite.

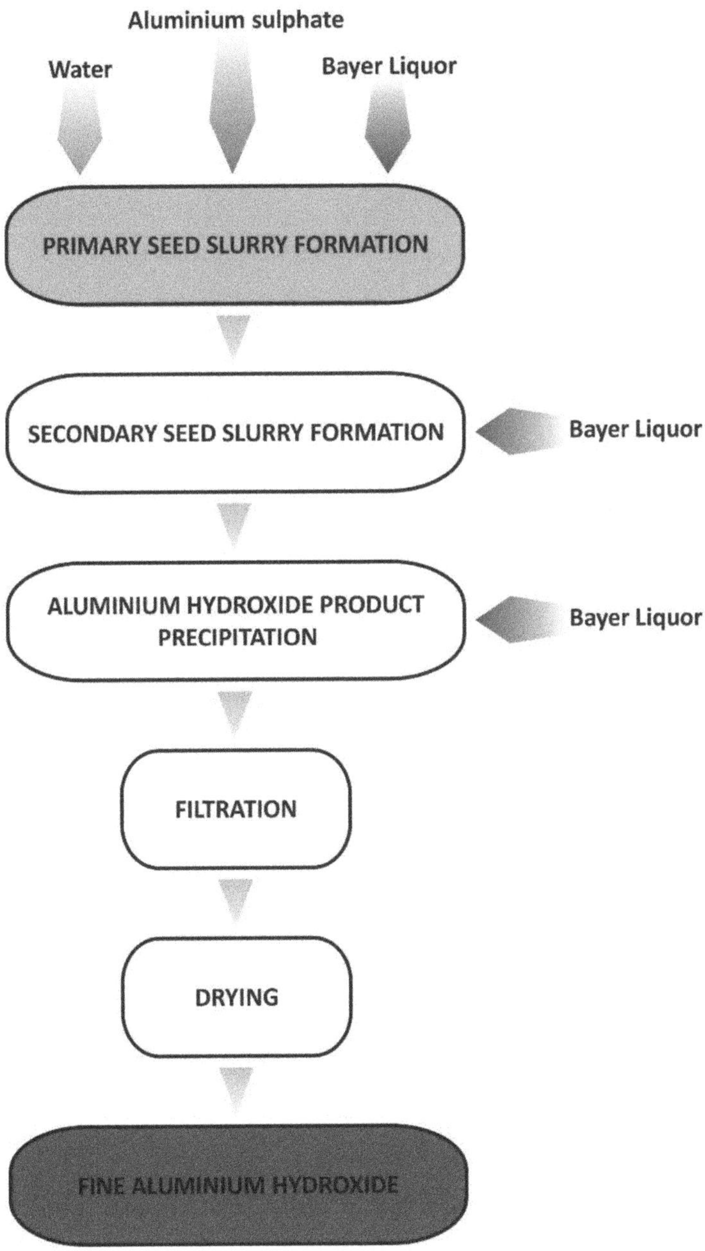

Figure 2. Schematic production process of fine ATH.

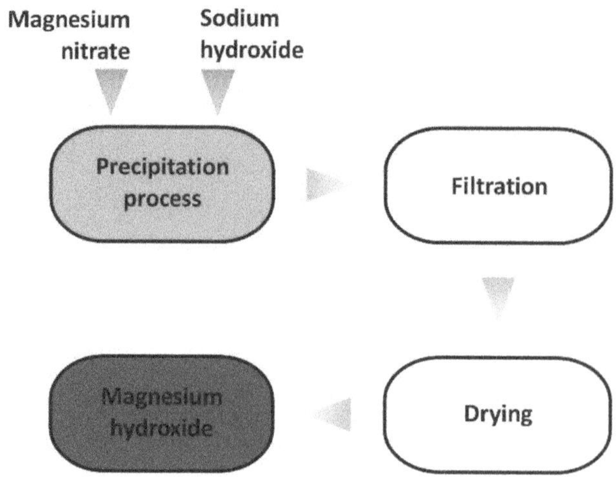

Figure 3. Scheme of precipitation process of MDH.

Carbonates

The group of carbonate minerals used as FR for polymers includes nesquehonite, hydromagnesite, and huntite.

Natural occurrence of the first mentioned mineral, nesquehonite ($MgCO_3·3H_2O$), is rather rare, especially when compared to other carbonates. Nonetheless, it is believed that nesquehonite is a precursor of the more abundant hydromagnesite or magnesite (*25*).

Among all discussed minerals, nesquehonite definitely has the lowest decomposition temperature of only 70–100 °C. This low thermal stability, insufficient for thermoplastic polymers or thermosetting resins, seems to make this mineral an interesting candidate for PU. In industrial practice, PU foams (PUFs) and elastomers are produced in temperatures that generally do not exceed 60 °C. Thermoplastic PUs are processed in higher temperatures (from 170 to 230 °C for products of 98 A–74 D Shore hardness) and therefore require different types of carbonate fillers (*26*). Here, hydromagnesite or huntite, with significantly higher decomposition temperatures of 240 and 400 °C, respectively, may be suitable.

Huntite naturally occurs in cavities in rocks composed of magnesite, dolomite, and deweylite. Its presence is a result of precipitation from magnesium-rich aqueous solutions or interaction of such solutions with other carbonates. In comparison with other carbonates, its density is low (*27*). As proved by scanning electron microscopy photographs, particles of huntite have fine, highly platy structures that may provide support to the emerging ash layer (*11*).

Hydromagnesite is a hydrothermal mineral occurring in serpentine and altered magnesium-rich igneous rocks. It is also as an alteration product of brucite. Its particles, while still platy, are more blocky than particles of huntite (*11, 28*).

Mixtures of natural hydromagnesite and natural huntite are commercially available as UltraCarb (weight ratio approximately 60:40 hydromagnesite:huntite). The decomposition of this material takes place in a wide temperature range. According to the manufacturer, water is released at a temperature of about 220 °C, carbon dioxide is released at 330 °C, and char is formed at about 560 °C (*29*). Overall, UltraCarb exhibits flame retardancy potential intermediate between ATHs and MDHs (*30*).

MMTs

Nanocomposites and nanofillers currently are of high interest to both researchers and industry. Natural clays are especially being widely investigated as FR and mineral fillers. The interest in these materials is caused not only by the FR aspect but also a positive effect on mechanical properties of composites.

MMT, the main representative of clays, belongs to the 2:1 layered family (a central octahedral sheet of alumina is placed between two tetrahedral sheets of silica) and is formed mainly by the action of lichens on cave deposits and waters (*31*).

Naturally occurring clays have a hydrophilic character, and polymeric materials are mainly hydrophobic. In order to avoid agglomeration of filler and any signs of phase segregation, modification should be carried out. Usually, this means intercalating organic compounds—alkyl ammonium or phosphonium salts and cationic surfactants (*32, 33*). When organically modified MMT is added to the polymer matrix, it forms a "labyrinth" that hinders penetration of oxygen. Therefore, it works as a thermal insulator and limits mass transport of the volatile products of the combustion process. It has also been found that layered silicates and aluminum phosphinate provide a synergistic effect in terms of flame retardancy of PUFs (*34–36*). The proposed mechanism of the combustion process of PU with MMT is presented in Figure 4. In general, the incorporation of MMT dilutes the polymer and produces a stable organic–inorganic interface. Under fire conditions, an MMT-rich layer protects the inner parts of composite through increasing the diffusion path of the volatiles and heat. This results in a lowering of peak heat release rate and delay of thermal degradation, but it does not lower the total heat release. Thus, the aforementioned materials are on fire over a longer period of time and burn less intensely (*28–30*).

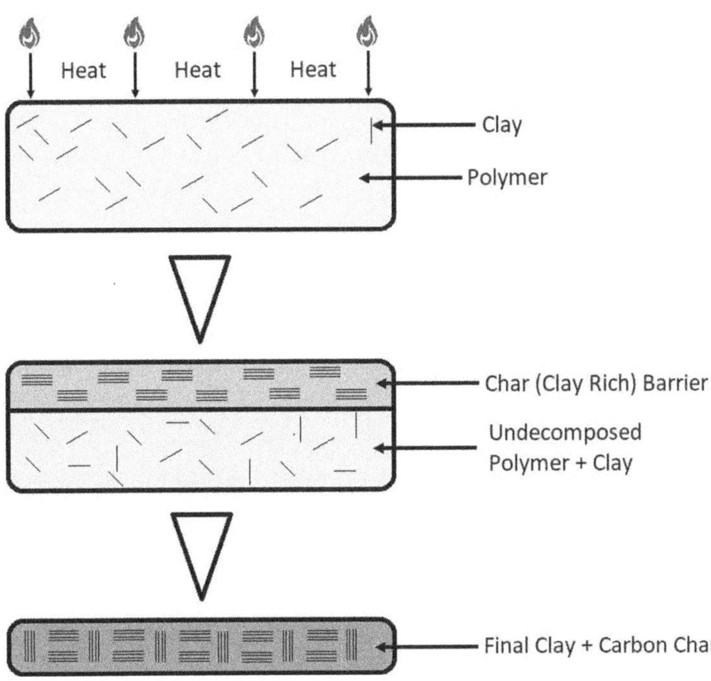

Figure 4. The schematic diagram for the combustion process of PU with MMT.

Mechanism of Action of Mineral FRs in PU Materials

Both low- and high-density PUFs have a large surface area and high air permeability. Accordingly, they burn with sufficient ignition and oxygen sources. FRs are often added to reduce this flammability. The choice of an FR depends on the intended use and the physicochemical properties of the final product. There are several properties of PU materials that are known to influence the combustion process (*20, 37, 38*):

- Quantity of flammable products of pyrolysis and their chemical composition;
- Ignition conditions determining the susceptibility of the material to ignition;
- The speed of movement of flames on the surface of the material;
- Speed and amount of heat released during combustion;
- Amount of smoke generated during combustion;
- Emission and composition of toxic gases;
- Combustion dynamics—that is, the rate of fire transfer;
- Self-extinguishing ability—that is, the speed of chemical reactions aimed at extinguishing the flame.

Flame-free PUs are flammable materials. This applies particularly to foam products, as their well-developed porous structure facilitates combustion. PUs decompose at temperatures of more than 200 °C. As the temperature rises, the proportion of HCN and carbon oxides in products of destruction increases, and nitrogen oxides and nitriles are released. Isocyanate vapors float above the surface of the burned polymer while liquid polyols undergo further decomposition. The flammability of PUFs depends on their type, the degree of cross-linking, and the physical condition of the product. Rigid PUFs have a higher flash point than flexible foams, because a higher content of carbonyl groups increases the cohesive energy and thus makes PUFs less flammable than polyether foams (*33, 38, 39*).

In the pyrolysis process (in the range of 80–150 °C), the decomposition of the polymer begins with the breaking of hydrogen bonds between oxygen atoms and HNH groups (in the urethane group). Hydrogen bonds with other chemical groups are broken in the temperature range of 130–200 °C. In polymers made of toluene diisocyanate, at more than 200 °C, the urethane bonds crack and toxic HCN is released. Of course, other compounds can be formed as well (acetonitrile, acrylonitrile, propionitrile, benzonitrile, aniline, pyridine, pyrrole, and others). During the pyrolysis of PU made of methylene diphenyl diisocyanate, the main volatile products are phenylisocyanate, p-toluene isocyanate, o-benzonitrile, isoquinoline, and hydrocarbons (benzene, toluene, xylene, biphenyl, naphthalene, carbazole, dimethylbiphenyl, etc.). Benzonitrile, the predominant nonvolatile decomposition product, undergoes further degradation to the final product, HCN. The type and content of pyrolysis products depends on the chemical structure of the PU. The most common gaseous products produced during the combustion of rigid foams are the following: CO_2, CO, HCN, NO_2, HCl, and organic compounds (*33, 38, 39*).

During the process of burning PU, significant amounts of toxic gases are generated. Therefore, it is extremely important to select the appropriate FR compounds. As mentioned before, the group of common FR-PUs include nontoxic minerals. Mineral fillers (salts and hydroxides) are an old technology that has been used since the 1920s. Some references suggest that they might have been used as early as the 17th century, although these materials still make an effective and cost-effective technology. They are also considered very environmentally friendly. Figure 5 shows the main advantages of using minerals as FRs in PUs (*28, 33*).

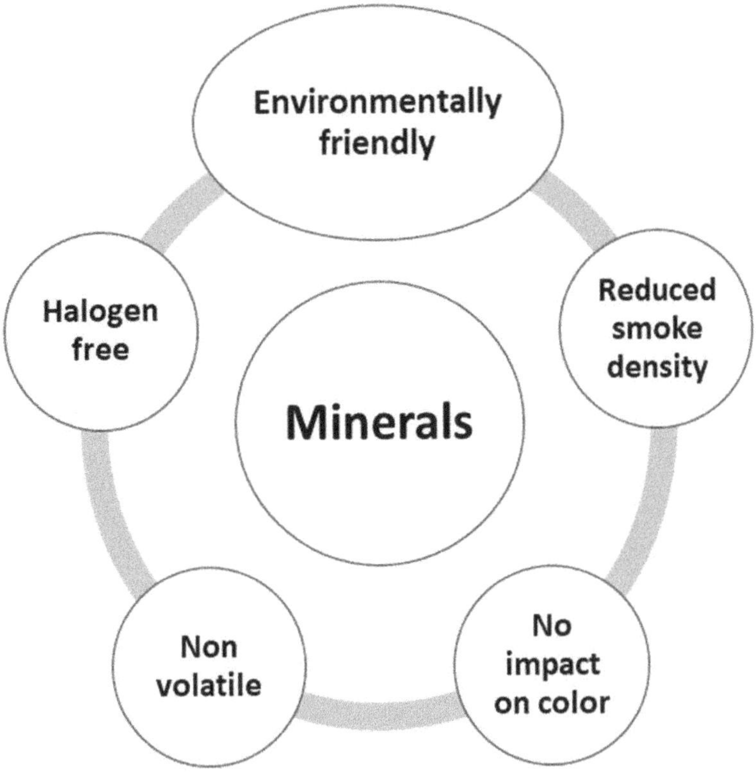

Figure 5. The primary benefits of using minerals as FRs in PUs.

Under fire conditions, FRs in PUs tend to reduce smoke and overall emission of toxic gases. These fillers are also quite inexpensive and can be easily blended with other additives used in PUs. Minerals such as salts and hydroxides can act as FRs not only for PU but also for other polymers: polyolefins, thermoplastic elastomers, poly(vinyl chloride), rubbers, thermosetting plastics, polyamides, and so forth (*38, 40*).

Unfortunately, mineral fillers also have some disadvantages. First, they have a limited so-called "fire resistance window." When a sufficient amount of heat is provided, complete degradation of the salts or hydroxides occurs, and all of the water/carbon dioxide is released. The resulting metal oxide is unable to completely protect the polymer. First, the mineral filler can significantly retard ignition and the initial flame growth but cannot completely stop them if a sufficiently constant source of fire/heat is used. Second, mineral fillers must be used in amounts exceeding 25 wt %. This may have a negative impact on the performance parameters of the final product. Their decomposition temperature is generally in the range of 200–250 °C. In such a case, mineral FRs work best with the addition of other FR materials/compounds (*14, 30, 38, 40*).

The mechanism of action of mineral FRs in PUs is complex and consists of several stages. During heating, inorganic hydroxides and salts decompose to release important products—water vapor and carbon dioxide. At elevated temperatures, hydroxides or carbonates should release water (steam) or carbon dioxide. However, this process should start before the polymer begins to decompose. During combustion, metal oxides are also formed. All the occurring reactions are exothermic (*20, 41*). Examples of the decomposition reactions of hydroxides and inorganic salts acting as FRs in PU materials are shown in Figure 6.

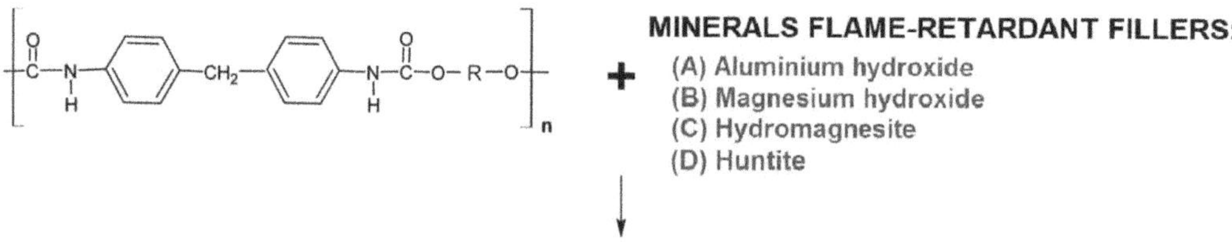

(A)

$$2Al(OH)_3 + \text{Heat } (180\text{–}200\ °C) \rightarrow Al_2O_3 + 3H_2O(g)$$

$$\text{Or: } Al_2O_3{\cdot}3H_2O + \text{Heat } (180\text{–}200\ °C) \rightarrow Al_2O_3 + 3H_2O(g)$$

(B)

$$Mg(OH)_2 + \text{Heat } (300\text{–}320\ °C) \rightarrow MgO + H_2O(g)$$

(C)

$$3MgCO_3Mg(OH)_2{\cdot}3H_2O + \text{Heat}(220\text{–}240°C) \rightarrow Mg_4(CO_3)_3(OH)_2 + 3H_2O(g)$$

$$Mg_4(CO_3)_3(OH)_2 + \text{Heat}(320\text{–}350\ °C) \rightarrow 4MgO + 3CO_2(g) + H_2O(g)$$

$$Mg_5(CO_3)_4(OH)_2{\cdot}3H_2O + \text{Heat}(320\text{–}350\ °C) \rightarrow 5MgO + 4CO_2(g) + 5H_2O(g)$$

(D)

$$Mg_3Ca(CO_3)_4 + \text{Heat } (400\text{–}450\ °C) \rightarrow 3MgO + CaO + 4CO_2(g)$$

Figure 6. Exemplary reactions of the decomposition of hydroxides and inorganic salts acting as an FR in PU materials (20, 41).

Table 2. Mineral Fillers—Relative Contribution FR Effects and Corresponding Decomposition Data[a]

Filler	*Relative Contribution FR Effects*		*Volatile Content, % (w/w)*	
	% Filler	*% Endotherm*	*H_2O*	*CO_2*
ATH	9	55	34.5	0
Calcium hydroxide	29	55	24.0	0
MDH	19	56	31.0	0
Beohmite	18	46	15.0	0
Huntite	20	58	0	50.0
Hydromagnesite	10	56	20.0	36.0
Nesquehonite	1	58	39.0	22.0

[a] Source: Data from reference (30).

The data presented in Table 2 show the amounts of gaseous products released and the endotherm of the decomposition of given mineral fillers. In each case, an endothermic decomposition of filler takes place. During the decomposition, among all the hydroxides, ATH releases the highest amount of water. This explains its frequent use in the production of flexible PUFs. During decomposition of inorganic salts, carbon dioxide is also produced (*30*). Reactions presented in Figure 6 are accompanied by the following physical processes:

1. The burned polymer is cooled and the rate of pyrolysis is reduced.
2. There is a formation of decomposition products. Water vapor and carbon dioxide reduce the supply of air necessary to maintain the combustion process (primary decomposition). During degradation, metal oxides are also formed. Because of their relatively high heat capacity, they reduce thermal energy of the polymer and thus rate of its pyrolysis (secondary decomposition).
3. Diffusion of metal oxides (e.g., Al_2O_3) together with the carbonization products form a protective, insulating layer on the surface of the burned polymer material.
4. Water vapor released during the decomposition of hydroxides and inorganic salts dilutes the volatile products of thermal destruction of the polymer. The concentration of flammable gases is reduced below the lower ignition limit (*11, 38, 41–44*).

The aforementioned processes are also shown in Figure 7.

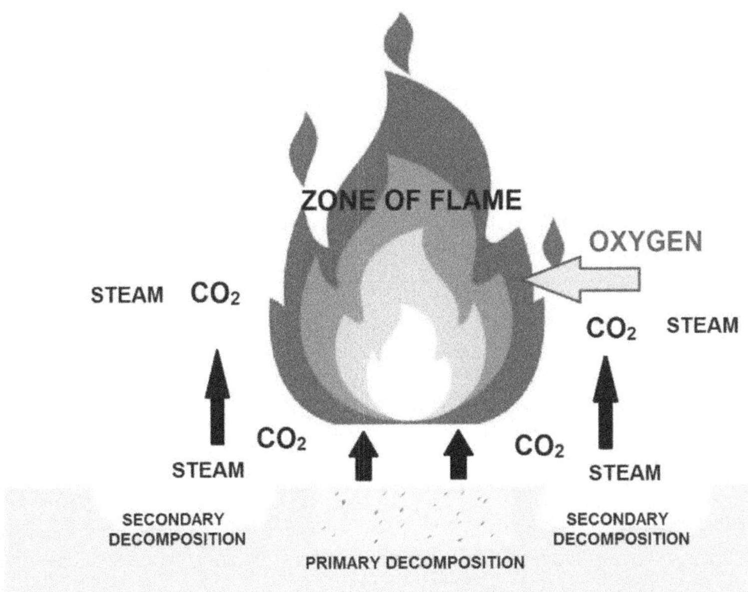

Figure 7. Diagram of the processes occurring during combustion of PU materials with mineral FRs.

Industrial Use of Mineral FRs in PUs

PUs are a subject of the growing interest of many research centers around the world. Still, their specific uses, along with growing flammability restrictions, increase the demand for FR-PU. Accordingly, FRs used in PU formulations should do the following:

1. Be compatible with all substrates;
2. Exhibit sufficient thermal stability and resistance to the alkaline environment (important during the formation of the foam);
3. Not diffuse from the finished product and cause deterioration of mechanical and thermal insulation properties (45).

$Al(OH)_3$ is currently the most commonly used inorganic, FR mineral. It is often mixed with PUs, specifically PU-forming flexible foams. Examples of trade names of the most popular mineral FRs used in PUs are presented in Tables 3 and 4.

Table 3. Examples of Trade Names of the Most Popular Mineral FRs Used in PUFs, Part One[a]

Manufacturer	*Trade Name*	*Chemical Name*	*Application*
Carl Spaeter	Hidromag Q1001	MDH (for all manufacturers and trade names)	MDH of high purity, precipitated from natural dolomite. Optimized grade for halogen free FR application in plastics and elastomers.
	Hidromag Q2001		
	Hidromag Q3001		
Lemman Laboratories International	MDH I/II/III		Acts as an inert filler, FR and smoke suppressant for plastics, synthetic rubber, and PUF. Provides additional advantages in terms of smoke density and flexibility during processing.
NYSAChem	FlameHunter MDH		FR, preventing and reducing the emission of harmful smoke. Stable at temperatures up to 330 °C.

[a] Source: Data from references (45) and (46).

An analysis of Tables 3 and 4 shows that ATH is indisputably the most commonly used mineral FR. Industrial application of this mineral may be problematic in terms of assuring the right homogeneity of the polymer melt. Figure 8 shows the scheme of the technological process enabling preparation of both rigid and flexible PUFs with the addition of mineral fillers.

In foam systems, it is necessary to use a special adapter for dosing additives. This ensures that the FR is dosed directly to the head where the polyol and isocyanate are mixed. This may be problematic, because of the necessity to design a new mixing head. In industry, the most frequently chosen processing form is direct blending of polyols with additives, including solid ATH. Proper mixing of component A (polyols with additives) in the generator tank (see Figure 8) is needed to limit the possibility of sedimentation. Continuous mixing of the polyol component makes it possible to obtain a homogeneous system. Then, components A and B are dosed (through separate channels) to the head equipped with a mixer. Then the mixed components are injected into the appropriate mold. Additionally, liquid rheological agents are often added to ensure good dispersion of the solid FR in the mixture. These additives simply delay the sedimentation process (39, 47).

Table 4. Examples of Trade Names of the Most Popular Mineral FRs Used in PUFs, Part Two[a]

Manufacturer	*Trade Name*	*Chemical Name*	*Application*
Huber Engineered Materials	MARTINAL ON-920/V	ATH (for all manufacturers and trade names)	Used as an FR and as a filler.
Nabaltec	Apyral 11		Used as an FR filler for cable sheaths and foams (insulations).
	Apyral 2		
	Apyral 60 D		Used in cross-linked elastomers, PU, urea. Products: carpet backing, flooring, foams, molded articles.
Shandong Chuanjun Chemical	ATH 1		64.5% ATH. Used as an FR, bulking agent. Used for foams and PU elastomers.
	ATH 2		64.0% ATH. Used as an FR, bulking agent, organic synthesis activator. Used for foams and PU elastomers.
	ATH 3		63.5% ATH. Used as an FR, bulking agent, organic synthesis activator. Used for foams and PU elastomers.
Sibelco Specialty Minerals	Portaflame SG 50		Recommended for flooring, roofing, wire and cable, wall covering, foam insulation, and wire bedding.
	Portaflame SG 40		
	Portaflame SG 20		
NYSAChem	FlameHunter ATH		FR, preventing and reducing the emission of harmful smoke. It is compatible with other FRs.

[a] Source: Data from references (*45*) and (*46*).

Intensive development of PU materials requires constant improvement of materials enhancing fire resistance, and sometimes the use of mineral fillers alone will give insufficient results. In order to meet the exorbitant standards, FR packages are also used. They consist not only of the mineral component (most often ATH) but melamine-, graphite-, and phosphorus-based FRs as well. Such FR mixtures are known to provide adequate protection to PUFs; therefore, the range of their industrial applications is still growing.

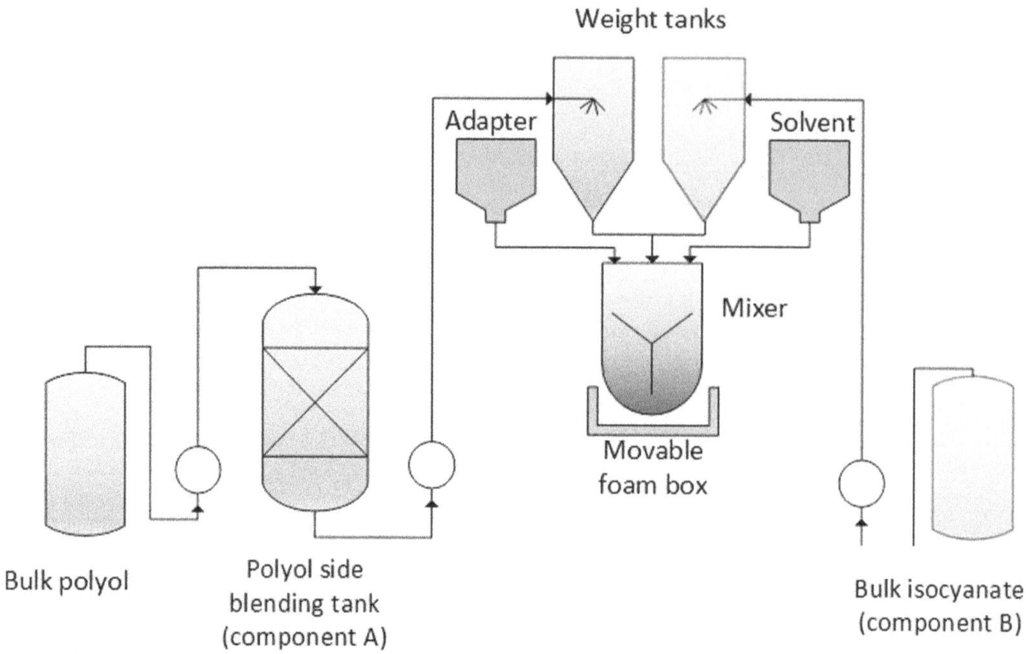

Figure 8. Diagram of an aggregate for obtaining PUFs—mixture of polyols with additives (component A), isocyanate (component B).

Conclusion

Modern polymer-based construction materials must exhibit some specific properties. There is an increasing demand for PUs with good mechanical properties, reduced flammability, and limited smoke release during the combustion process. As PUs are flammable materials, the industry must find a perfect FR additive. There are many types of mineral fillers that are already used as FRs. The most commonly used mineral fillers include ATH, calcium hydroxide, MDH, boehmite, huntite, and hydromagnesite. Some of them seem to have the potential to replace halogen-based compounds, which are currently regarded as nonecological. Also, mineral fillers are very often used for economic reasons. However, it should be remembered that use of only mineral materials may not provide the assumed flame retardancy of PU. It is believed that the combination of minerals with organic FRs could lead to the development of new, fire-resistant PU materials.

This chapter presented characteristics of numerous minerals that may be applicable as FRs. It was shown that physicochemical properties of mineral substances were strongly connected with their FR properties. The mechanism of action of mineral FRs in PUs is rather complex; therefore, processes occurring during endothermic decomposition of materials with mineral fillers were discussed. Much focus was put on industrial use of mineral FRs in PUs, and the most popular compounds used in PUFs were listed.

Acknowledgments

This work was supported by the Polish Ministry of Education and Science.

References

1. *Center for the Polyurethanes Industry*. https://polyurethane.americanchemistry.com/ (accessed March 29, 2021).
2. Hull, T. R.; Stec, A. A. Polymers and Fire. In *Fire Retardancy of Polymers: New Strategies and Mechanisms*; Hull T. R., Kandola, B. K., Eds.; Royal Society of Chemistry: Cambridge, 2009.
3. Qian, Y.; Qiao, P.; Li, L.; Han, H.; Zhao, R.; Zhang, H. Preparation of Pillared Layered Antimony Hydroxide and Its Flame Retardancy in Thermoplastic Polyurethane. *J. Therm. Anal. Calorim.* **2020,** *142* (1), 425–435. https://doi.org/10.1007/s10973-019-09028-4.
4. *Guidance.* http://chm.pops.int/Implementation/IndustrialPOPs/BDEs/Guidance/tabid/5374/Default.aspx (accessed March 29, 2021).
5. Vahabi, H.; Sonnier, R.; Ferry, L. Effects of Ageing on the Fire Behaviour of Flame-Retarded Polymers: A Review. *Polym. Int.* **2015,** *64* (3), 313–328. https://doi.org/10.1002/pi.4841.
6. Chen, L.; Wang, Y.-Z. A Review on Flame Retardant Technology in China. Part I: Development of Flame Retardants. *Polym. Adv. Technol.* **2010,** *21* (1), 1–26. https://doi.org/10.1002/pat.1550.
7. Borysiak, S. The Thermo-Oxidative Stability and Flammability of Wood/Polypropylene Composites. *J. Therm. Anal. Calorim.* **2015,** *119* (3), 1955–1962. https://doi.org/10.1007/s10973-014-4341-y.
8. Sain, M.; Park, S. H.; Suhara, F.; Law, S. Flame Retardant and Mechanical Properties of Natural Fibre-PP Composites Containing Magnesium Hydroxide. *Polym. Degrad. Stab.* **2004,** *83* (2), 363–367. https://doi.org/10.1016/S0141-3910(03)00280-5.
9. Benelli, T.; Mazzocchetti, L.; D'Angelo, E.; Lanzi, M.; Saraga, F.; Sambri, L.; Franchini, M. C.; Giorgini, L. New Nitrogen-Rich Heterocycles for Organo-Modified Bentonites as Flame Retardant Fillers in Epoxy Resin Nanocomposites. *Polym. Eng. Sci.* **2017**, *57* (6), 621–630. https://doi.org/10.1002/pen.24565.
10. Lin, H.; Yan, H.; Liu, B.; Wei, L.; Xu, B. The Influence of KH-550 on Properties of Ammonium Polyphosphate and Polypropylene Flame Retardant Composites. *Polym. Degrad. Stab.* **2011,** *96* (7), 1382–1388. https://doi.org/10.1016/j.polymdegradstab.2011.03.016.
11. Hollingbery, L. A.; Hull, T. R. The Fire Retardant Behaviour of Huntite and Hydromagnesite—A Review. *Polym. Degrad. Stab.* **2010,** *95* (12), 2213–2225. https://doi.org/10.1016/j.polymdegradstab.2010.08.019.
12. Hollingbery, L. A.; Hull, T. R. The Thermal Decomposition of Huntite and Hydromagnesite—A Review. *Thermochim. Acta* **2010,** *509* (1), 1–11. https://doi.org/10.1016/j.tca.2010.06.012.
13. Zhang, L.; Cao, J. Pyrolysis and Its Mechanism of Organomontmorillonite (OMMT) Influenced by Different Functional Groups. *J. Therm. Anal. Calorim.* **2019,** *137* (1), 1–10. https://doi.org/10.1007/s10973-018-7947-7.
14. Gajlewicz, I.; Lenartowicz, M. Nowe Kierunki Uniepalniania Tworzyw Polimerowych. *Przetwórstwo Tworzyw* **2014,** *20* (3), 216–223.
15. Chen, M.-J.; Chen, C.-R.; Tan, Y.; Huang, J.-Q.; Wang, X.-L.; Chen, L.; Wang, Y.-Z. Inherently Flame-Retardant Flexible Polyurethane Foam with Low Content of Phosphorus-Containing Cross-Linking Agent. *Ind. Eng. Chem. Res.* **2014,** 53 (3), 1160–1171. https://doi.org/10.1021/ie4036753.

16. Czupryński, B.; Paciorek-Sadowska, J.; Liszkowska, J. Modifications of the Rigid Polyurethane–Polyisocyanurate Foams. *J. Appl. Polym. Sci.* **2006**, *100* (3), 2020–2029. https://doi.org/10.1002/app.22604.
17. Thirumal, M.; Khastgir, D.; Singha, N. K.; Manjunath, B. S.; Naik, Y. P. Effect of Expandable Graphite on the Properties of Intumescent Flame-Retardant Polyurethane Foam. *J. Appl. Polym. Sci.* **2008**, *110* (5), 2586–2594. https://doi.org/10.1002/app.28763.
18. Meng, X.-Y.; Ye, L.; Zhang, X.-G.; Tang, P.-M.; Tang, J.-H.; Ji, X.; Li, Z.-M. Effects of Expandable Graphite and Ammonium Polyphosphate on the Flame-Retardant and Mechanical Properties of Rigid Polyurethane Foams. *J. Appl. Polym. Sci.* **2009**, *114* (2), 853–863. https://doi.org/10.1002/app.30485.
19. Chen, H.-B.; Shen, P.; Chen, M.-J.; Zhao, H.-B.; Schiraldi, D. A. Highly Efficient Flame Retardant Polyurethane Foam with Alginate/Clay Aerogel Coating. *ACS Appl. Mater. Interfaces* **2016**, *8* (47), 32557–32564. https://doi.org/10.1021/acsami.6b11659.
20. Hull, T. R.; Witkowski, A.; Hollingbery, L. Fire Retardant Action of Mineral Fillers. *Polym. Degrad. Stab.* **2011**, *96* (8), 1462–1469. https://doi.org/10.1016/j.polymdegradstab.2011.05.006.
21. Lee, S. O.; Jung, K. H.; Oh, C. J.; Lee, Y. H.; Tran, T.; Kim, M. J. Precipitation of Fine Aluminium Hydroxide from Bayer Liquors. *Hydrometallurgy* **2009**, *98* (1), 156–161. https://doi.org/10.1016/j.hydromet.2009.04.014.
22. Hind, A. R.; Bhargava, S. K.; Grocott, S. C. The Surface Chemistry of Bayer Process Solids: A Review. *Colloids Surfaces A Physicochem. Eng. Asp.* **1999**, *146* (1), 359–374. https://doi.org/10.1016/S0927-7757(98)00798-5.
23. Park, J.-W.; Lee, W.-K.; Lee, C.-H. Preparation of Nano-Magnesium Oxide from Seawater Bittern Using Decarboxylation/Precipitation Method. *Mol. Cryst. Liq. Cryst.* **2016**, *636* (1), 142–148. https://doi.org/10.1080/15421406.2016.1201398.
24. Pilarska, A.; Markiewicz, E.; Ciesielczyk, F.; Jesionowski, T. The Influence of Spray Drying on the Dispersive and Physicochemical Properties of Magnesium Oxide. *Dry. Technol.* **2011**, *29* (10), 1210–1218. https://doi.org/10.1080/07373937.2011.579698.
25. Zhu, C.; Wang, H.; Li, G.; An, S.; Ding, X.; Teng, H. H.; Zhao, L. CO_2 Absorption and Magnesium Carbonate Precipitation in MgCl2–NH_3–NH4Cl Solutions: Implications for Carbon Capture and Storage. *Minerals* **2017**, *7* (9), 172. https://doi.org/10.3390/min7090172.
26. *Elastollan Thermoplastic Polyurethane (TPU)*. https://www.distrupol.com/elastollan-tpu.html (accessed March 29, 2021).
27. Faust, G. T. Huntite, $Mg_3Ca(CO_3)_4$, A New Mineral. *Am. Mineral. J. Earth Planet. Mater.* **1953**, *38* (1–2), 4–24.
28. Rothon, R. *Particulate-Filled Polymer Composites;* iSmithers Rapra Publishing: Shawbury, England, 2003.
29. *Halogen-Free Fire Retardant Mineral Filler | LKAB Minerals*. https://www.lkabminerals.com/en/products/ultracarb/ (accessed March 29, 2021).
30. Hewitt, F.; Hull, T. R. Mineral Filler Fire Retardants BT. In *Fillers for Polymer Applications;* Rothon, R., Ed.; Springer International Publishing: Cham, Switzerland, 2017; pp 329–354. https://doi.org/10.1007/978-3-319-28117-9_2.

31. *Montmorillonite: Mineral information, data and localities*. https://www.mindat.org/min-2821.html (accessed March 1, 2021).
32. Xiong, J.; Zheng, Z.; Jiang, H.; Ye, S.; Wang, X. Reinforcement of Polyurethane Composites with an Organically Modified Montmorillonite. *Compos. Part A Appl. Sci. Manuf.* **2007**, *38* (1), 132–137. https://doi.org/10.1016/j.compositesa.2006.01.014.
33. Chattopadhyay, D. K.; Webster, D. C. Thermal Stability and Flame Retardancy of Polyurethanes. *Prog. Polym. Sci.* **2009**, *34* (10), 1068–1133. https://doi.org/10.1016/j.progpolymsci.2009.06.002.
34. Leszczyńska, A.; Njuguna, J.; Pielichowski, K.; Banerjee, J. R. Polymer/Montmorillonite Nanocomposites with Improved Thermal Properties: Part II. Thermal Stability of Montmorillonite Nanocomposites Based on Different Polymeric Matrixes. *Thermochim. Acta* **2007**, *454* (1), 1–22. https://doi.org/10.1016/j.tca.2006.11.003.
35. Leszczyńska, A.; Njuguna, J.; Pielichowski, K.; Banerjee, J. R. Polymer/Montmorillonite Nanocomposites with Improved Thermal Properties: Part I. Factors Influencing Thermal Stability and Mechanisms of Thermal Stability Improvement. *Thermochim. Acta* **2007**, *453* (2), 75–96. https://doi.org/10.1016/j.tca.2006.11.002.
36. Modesti, M.; Lorenzetti, A.; Besco, S.; Hrelja, D.; Semenzato, S.; Bertani, R.; Michelin, R. A. Synergism between Flame Retardant and Modified Layered Silicate on Thermal Stability and Fire Behaviour of Polyurethane Nanocomposite Foams. *Polym. Degrad. Stab.* **2008**, *93* (12), 2166–2171. https://doi.org/10.1016/j.polymdegradstab.2008.08.005.
37. Pielichowski, J.; Puszyński, A. *Chemia Polimerów*; Wydawnictwo Oświatowe FOSZE: Rzeszów, Poland, 1998.
38. G. Janowska, W. Przygocki, A. W. *Palność Polimerów i Materiałów Polimerowych*; WNT: Warszawa, 2007.
39. Herrington, R.; Hock, K.; Autenrieth, R. *Flexible Polyurethane Foams;* Dow Chemical: Midland, Michigan, 1997.
40. Baral, D.; De, P. P.; Nando, G. B. Thermal Characterization of Mica-Filled Thermoplastic Polyurethane Composites. *Polym. Degrad. Stab.* **1999**, *65* (1), 47–51. https://doi.org/10.1016/S0141-3910(98)00215-8.
41. Morgan, A. B.; Gilman, J. W. An Overview of Flame Retardancy of Polymeric Materials: Application, Technology, and Future Directions. *Fire Mater.* **2013**, 37 (4), 259–279. https://doi.org/10.1002/fam.2128.
42. Prociak, A.; Rokicki, G. *Materiały Poliuretanowe;* Ryszkowska, J., Ed.; PWN: Warszawa, Poland, 2013.
43. Almeida Pinto, U.; Visconte, L. L. Y.; Gallo, J.; Nunes, R. C. R. Flame Retardancy in Thermoplastic Polyurethane Elastomers (TPU) with Mica and Aluminum Trihydrate (ATH). *Polym. Degrad. Stab.* **2000**, *69* (3), 257–260. https://doi.org/10.1016/S0141-3910(00)00047-1.
44. Morgan, A. B. The Future of Flame Retardant Polymers—Unmet Needs and Likely New Approaches. *Polym. Rev.* **2019**, 59 (1), 25–54. https://doi.org/10.1080/15583724.2018.1454948.
45. *Additives for Polymers—The Universal Selection Source—SpecialChem*. https://polymer-additives.specialchem.com/selectors (accessed March 29, 2021).
46. *NYSAChem*. https://nysachem.com.pl/produkty1/index (accessed March 29, 2021).

47. Wilkie, C. A.; Morgan, A. B. *Fire Retardancy of Polymeric Materials*, Second ed.; CRC Press: Boca Raton, 2009.

Chapter 7

Flame-Retardant Coatings on Polyurethane Foams Deposited by Layer-by-Layer Assembly Approach

Abbas Mohammadi*[,1] and Saman Abrishamkar[1]

[1]Department of Chemistry, University of Isfahan, Isfahan 81746-73441, Iran

*Email: a.mohammadi@sci.ui.ac.ir

Polyurethane foams (PUFs) are the largest group of polyurethane products and can be classified into rigid and flexible foams. Rigid PUFs are mainly used for heat insulation in the refrigeration, construction, and automotive industries. Flexible PUFs are also widely used in the fabrication of car seats and furniture. In general, PUFs are highly flammable because of their porous structure, large surface area, and good air permeability. Flame retardants containing phosphorus, halogen, nitrogen, and silicone can partially reduce the flammability of PUFs. Also, the presence of thermally stable groups such as carbodiimide, isocyanurate, and nitrogen-containing heterocycles in the structure of PUFs can increase their fire resistance. In the last two decades a new technique called layer-by-layer (LbL) assembly has been introduced for the preparation of flame-retardant nanocoatings on the surfaces of polymeric foams. In this method, polymers, nanoparticles, and various compounds are deposited on the surface of polymeric foams in thin layers through secondary interactions such as electrostatic forces, hydrogen bonds, and van der Waals interactions. This chapter presents an overview of the combustion mechanism and flame retardancy of PUFs and introduces LbL assembly techniques, applications of this method in improving the fire resistance of PUFs, and the flame retardation mechanism of LbL coatings on the PUFs. Finally, some suggestions are discussed to improve the flame retardancy of LbL coatings.

Introduction

Polyurethane foams (PUFs) are the most widely used polyurethane (PU) products, and they are classified into rigid PUFs (RPUFs), semirigid PUFs, and flexible PUFs (FPUFs) based on their density and mechanical properties (*1*, *2*). PUFs are widely used in industrial and domestic applications, such as automotive parts, insulation materials, packaging fields, seat cushioning, and electrical and structural materials (*3*, *4*). In general, PUFs are prepared by two simultaneous reactions. The first is the polymerization reaction between isocyanate and polyol, and the latter occurs between isocyanate and water molecules, causing CO_2 production and volume expansion

(*5*). These two reactions can take place in the presence of catalysts, blowing agents, and surfactants without any external heating (*6*).

Because of the cellular structure of neat PUFs, these materials are highly flammable with a very low limiting oxygen index, which restricts their applications (*7–9*). The flammability of PUFs is strongly influenced by their raw materials, hard segment content, and physical properties, including density and porosity (*9*). The combustion of PUFs in the presence of a sufficient amount of oxygen leads to the production of combustible gases, noncombustible gases, and char residues. The heat of combustion increases the temperature of released combustible and noncombustible gases, and thereby heat transfer is raised. As a result, the heat is transferred easily from the combustion zone to adjacent areas, and further combustion, decomposition, and flame propagation are observed (*10*). Although FPUFs are chemically not very different from RPUFs, they tend to burn quickly because of their open-cell structure (*11*).

Because of the presence of large amounts of nitrogen atoms in the PUs' structure, they are prone to produce HCN gas upon pyrolysis or combustion (*12, 13*). Many studies have been conducted on the types and toxicity of the thermal decomposition and combustion products of PUFs. Accordingly, many researchers have reported that pyrolysis and combustion of FPUFs and RPUFs lead to production of PU fragments, N_2, CO_2, H_2O, $C_6H_5CH_3$, and toxic gases such as HCN and CO (*10, 14*).

Flame Retardancy in PUFs

PUFs burn immediately and release large amounts of heat and smoke when exposed to an ignition source, which can result in suffocation and irreparable damage. Therefore, it is very important to reduce heat and smoke generation of PUFs by improving their flame retardancy (*15*).

In general, the following strategies have been utilized to improve the fire safety of PUFs:

- Addition of radical scavengers such as chloro- and bromocompounds;
- Addition of fillers such as alumina trihydrate, layered double hydroxides (LDHs), and graphene oxide (GO) to reduce flame spread rate and heat of combustion;
- Addition of char-forming agents such as phosphates, which can reduce the production of combustible gases;
- Addition of flame-retardant (FR) compounds containing nitrogen (such as melamine) and silicone to increase thermal stability;
- Incorporation of thermally stable linkages such as isocyanurate bonds;
- Fabricating of coatings on the surface of flammable PUFs using FR compounds (*10, 16*).

Most FR systems that are effective for other polymers do not work well for PUFs. In the case of FR coatings, a new approach called layer-by-layer (LbL) assembly has been under consideration to promote the fire resistance of PUFs. This method often uses biocompatible materials containing positive and negative charges and has no effect on the substrate's mechanical properties (*17–21*).

LbL Assembly

Over the past two decades, a water-based technique called LbL assembly has attracted a great deal of attention as a surface modification tool to create multilayer thin films on substrates (*18, 22, 23*). Thanks to its versatility in both the component selection and deposition parameters, LbL thin coatings can be applied to a wide variety of substrates (*24*). The LbL assembly has applications in the field of flame retardancy and fire safety for various polymer matrices, including polymeric films, fabrics, and PUFs (*17, 25*). In this method, polymers, nanoparticles, and molecules are held together through different forces including electrostatic interactions, which are the most common, hydrogen bindings, van der Waals forces, and covalent bonds (*23, 26*).

In the electrostatic-based LbL process, the substrate is alternately immersed into oppositely charged nanoparticle or polyelectrolyte solutions/suspensions to deposit thin layers on the surface (*27–29*). Each positive–negative pairing layer is called a bilayer (BL). In addition to bilayers, it is possible to deposit repeating sequences of trilayers (TLs), quadlayers, and so on, to obtain the required final properties. The submerging procedure is repeated until the desired number of layers is achieved (*28, 30*) (Figure 1).

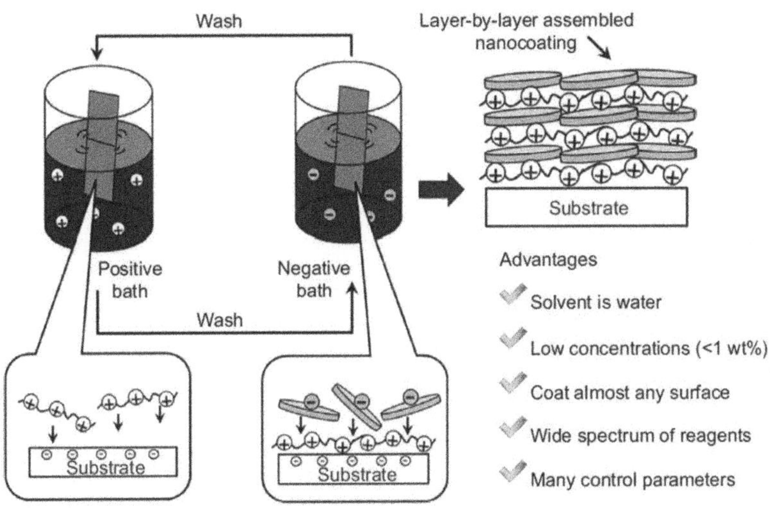

Figure 1. Outline of LbL assembly through electrostatic interaction. Reproduced with permission from reference (11). Copyright 2017 Elsevier.

The LbL process is independent from the size, shape, and topology of the substrate. Therefore, any solvent-accessible surface can be coated using this method. The final properties of the nanocoating depend on the surface properties, the employed materials, and the process parameters, such as solution concentration, ionic strength, pH, temperature, buffer content, adsorption time, and drying condition. The surface roughness, presence of impurities, and nature of functional groups play an essential role in the first steps of deposition and also the final stability and uniformity of deposited coatings (*22, 30, 31*).

Application of LbL Coatings in Flame Retardancy of PUFs

Many advantages—including ease of incorporation, processing under ambient conditions, utilizing nontoxic materials, and using water as a low-cost environmentally friendly solvent—have expanded the applications of the LbL assembly method for various kinds of substrates. In this regard, PUFs with three-dimensional (3D) structure and plenty of solvent-accessible surfaces can be well covered by this approach to improve their flame retardancy. As shown in Figure 2, during the deposition, the solution or suspension can penetrate into the PUFs, and thereby polyelectrolytes or nanoparticles are uniformly distributed on the available surfaces (*31*).

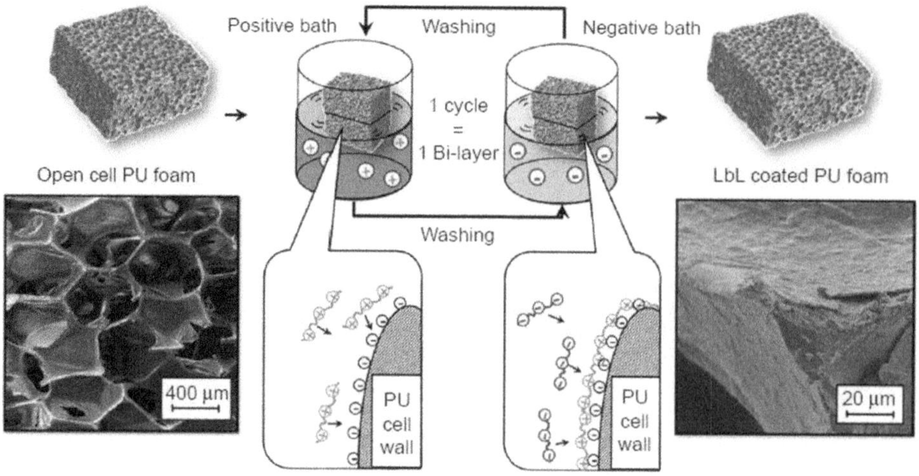

Figure 2. Schematic illustration of the LbL process through electrostatic interactions applied on PUF. Reproduced with permission from reference (11). Copyright 2017 Elsevier.

FR Mechanism of LbL Coatings in PUFs

The performance of LbL coatings in flame retardancy is directly related to the physical behavior of PUFs during cone calorimeter tests (CCTs) or flammability tests. As reported in previous studies, the combustion behavior and related fire hazards of PUFs are associated with their structure and volume change (*12*). Generally, a few seconds after combustion of PUFs, a layer of molten foam forms on the external surface, leading to melt dripping and the flame spreading along with the collapse of 3D structure of the foams. When the PUFs are surface modified according to the LbL assembly approach, an effective FR effect results. In fact, LbL coatings stabilize and protect the structure of the foams through significant reduction in melt dripping and flame spreading. In addition, the nonflammable gases released during the degradation of the LbL coatings (e.g., water, ammonia, and sulfur oxide) lead to fuel dilution (*12*, *28*).

Up to now, special polyelectrolytes and nanostructures such as LDHs, GO, carbon nanofibers (CNFs), carbon nanotubes (CNTs), and clay have been used to improve the flame retardancy of PUFs by the LbL assembly technique.

Nanoparticle-Based LbL Coatings

For practical and performance reasons, FR nanoparticles are widely used in combination with polyelectrolytes in fire-resistant LbL coatings. To date, nanoparticles such as carbon nanomaterials (e.g., CNFs, CNTs, and GO), LDHs, clays, and MXenes have been used in LbL nanocoatings. In the following sections, some of the studies on polyelectrolyte/nanoparticle LbL coatings are discussed.

Carbonaceous Nanomaterial-Based LbL Coatings

CNFs and CNTs are good candidates for use in LbL nanocoatings to improve flame retardancy of materials. This is because they form a char layer during heating, which acts as a protective shield against the fire (32). The first attempt to improve the fire protection of PUF using the LbL method was made by Kim et al. (*29*). In this study, a multilayer coating composed of CNFs stabilized with branched polyethylene imine (PEI) as a cationic layer and polyacrylic acid (PAA) as an anionic layer was fabricated. This coating completely covered all the interior and exterior surfaces of the PUF.

Although the microscopic distribution of CNF was nonuniform, the macroscopic CNF network considerably decreased the flammability of the foam. This research laid the foundation for using nanoparticles in LbL coatings deposited on foams and fabrics. Another study by Kim and Davis showed that the presence of multiwalled CNTs in a coating consisting of PAA and PEI could significantly improve the flame retardancy of PUFs. It was observed that surface covering of a PUF with four TL coatings of PAA, PEI, functionalized multiwalled CNTs, and PEI reduced the peak heat release rate (PHRR) value by 35% and prevented the formation of the melt pool during foam burning (*31*).

GO is a two-dimensional carbon nanomaterial with a unique layered structure, consisting of various reactive oxygen functional groups including carboxyl, hydroxyl, and epoxy on its edges and basal plane (*23, 33, 34*). Because of its large specific surface area and strong physical barrier effect, GO is referred to as an FR material (*35, 36*). The presence of oxygenated groups in GO makes it negatively charged in aqueous solutions. This facilitates the deposition of the GO nanostructures on the surface of different substrates by means of the LbL assembly process (*23*). Maddalena et al. reported an effective FR chitosan (CH)/GO–based LbL coating on the open-cell PUFs. The PUFs were initially immersed in PAA solution for 10 min to activate the surface and create a negatively charged layer on the surface. Afterward, the PUFs were alternatively submerged into the CH and GO solutions and rinsed with ultrapure water after each deposition step. The resulting three- and six-BL coatings were able to improve the fire resistance significantly. For example, deposition of only three-BL CH/GO coatings reduced PHRR and total smoke release (TSR) by 54 and 59%, respectively (*22*). Zhang et al. fabricated an FR mnultilayer film of CH, GO, and alginate on FPUF through a TL approach (*23*). Scanning electron microscopy (SEM) images showed that the resulting nanocoatings were uniformly distributed over the substrate surfaces. Thermogravimetric analysis indicated high char yield value of 7.16% in the best manner possible. The CCT results showed that the 10-TL coated foam exhibited a significant reduction in PHRR value (59.9%), total smoke production (30.5%), and peak carbon monoxide production (54.0%). As shown in Figure 3, this drastic reduction in flammability is made possible by the formation of a protective barrier layer on the burning surface, which not only prevents the transfer of oxygen and heat, but also diminishes flammable gases in the vapor phase (*23*).

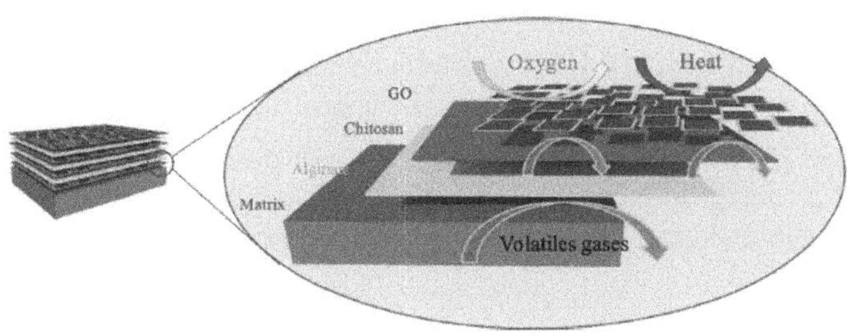

Figure 3. The scheme of the possible FR mechanism of a CH/GO/alginate coating. Reproduced with permission from reference (23). Copyright 2016 Springer Nature.

These studies suggest that GO nanosheet can serve as an effective FR agent in LbL coatings to reduce the flammability of PUFs.

LDH-Based LbL Coatings

LDHs are a class of anionic clays with unique layered structure made of positively charged layers with intercalated anions and water molecules (Figure 4) (*37, 38*). The general formula of LDHs is $[M^{2+}{}_{1-x} M^{3+}{}_x(OH)_2]^{x+}A^{n-}{}_{x/n} \cdot yH_2O$, where M^{2+} is a divalent metal cation such as Mg^{2+}, Zn^{2+}, Fe^{2+}, Ni^{2+}, or Co^{2+}; M^{3+} represents a trivalent metal cation such as Al^{3+}, Fe^{3+}, or Co^{3+}; and A^{n-} is an intercalated anion such as $NO^{-}{}_3$, $CO^{2-}{}_3$, or $SO^{2-}{}_4$ (*39–43*).

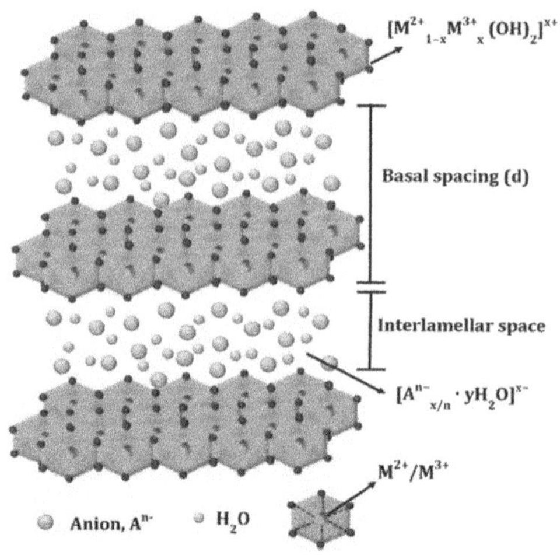

Figure 4. Structure of LDH. Reproduced with permission from reference (38). Copyright 2018 Elsevier.

Because of their high anion exchange capacity, good thermal stability, low cost, low toxicity, and desired biocompatibility, LDH nanostructures have found a wide range of applications in many fields, especially as environmentally friendly FR additives for polymer nanocomposites (*44–46*). The flame retardancy of LDHs is attributed to several mechanisms, including endothermically thermal decomposition of LDH, which leads to the heat absorption, the release of water and CO_2 that dilute the flammable gases, and a barrier effect provided by the formation of metal oxide residues (*47, 48*). In a study by Liu et al., an LDH-based LbL coating was fabricated on the surface of an FPUF to improve its flame retardancy, smoke suppression, and thermal stability (*49*). The deposition process was performed by intermittent immersion of the PUF into CH, LDH, and alginate solutions. The results of CCT analysis showed that relative to the control foam, all the coated foams exhibited a lower PHRR, total heat release, TSR, peak carbon monoxide, and peak carbon dioxide. Such a significant effect in fire resistance, smoke suppression, and reduction in toxic gas release was attributed to the reasons mentioned previously for LDH nanostructures (*49*).

Clay-Based LbL Coatings

Clays are a class of layered minerals with low cost and high thermal stability, and they are regarded as one of the most interesting halogen-free FR materials. Because of the layered structure and high aspect ratio, nanoclays lead to impressive barrier properties, which result in efficient condensed phase flame retardancy (*50*). Clays such as mica (M), vermiculite, and montmorillonite (MMT) have been used in LbL nanocoatings for PUFs.

Fahami et al. prepared an effective FR coating composed of high-aspect-ratio M stabilized by CH (CHM), and CH and PAA were used to stabilize high-aspect-ratio M stabilized by PAA (PAAM) to improve fire performance of open-cell PUF (*51*). The M was chemically neutral in water, so it was difficult to prepare a stable suspension. When M was sonicated with CH or PAA, a white suspension was formed. The results showed that the eight-BL coating of CHM/PAAM prevented the melt dripping, preserved the foam structure, and caused the foam to self-extinguish during a 10-s torch test. According to CCTs, the PHRR and TSR values were reduced by 54 and 76%, respectively. As shown in Figure 5, the char residues of the coated foams preserved their original shape. The eight-BL coated foam left more unburned portions compared with the two-BL coated foam. According to SEM images of two- and eight-BL CHM/PAAM-coated foams, M nanoplatelets were well distributed on the surface of the foams. It can be observed that by increasing the number of bilayers, the foam structure protected more effectively after combustion (*51*).

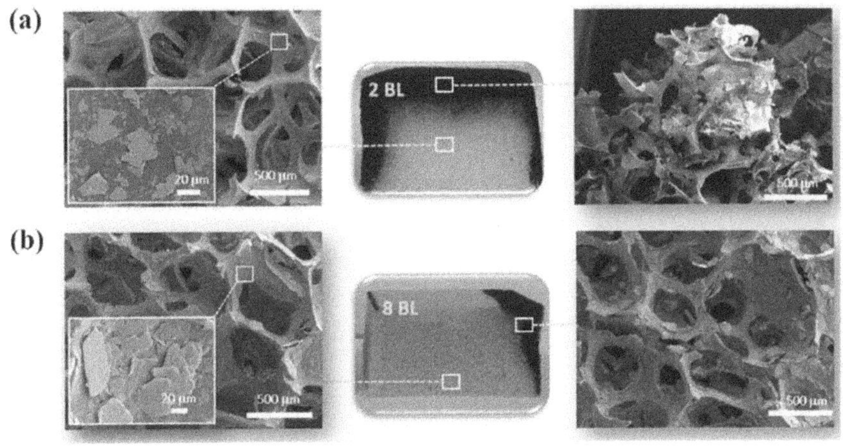

Figure 5. Cross-sectional digital images and SEM images of (a) two-BL and (b) eight-BL CHM/PAAM-coated foams after burning. Reproduced with permission from reference (51). Copyright 2020 American Chemical Society.

In another study by Lazar et al., a biobased LbL coating made of cationic CH and anionic vermiculite clay was deposited on PUFs to decrease their flammability (*52*). The results revealed that a nanocoating with eight-BL of CH/vermiculite clay exhibited thermal shielding behavior and reduced the PHRR and TSR by 53 and 63%, respectively (*52*).

MMT is a low-cost natural layered clay that is widely used in many applications such as energy storage, drug delivery, and absorption. Because of their high surface area and natural negative charge, MMT nanosheets—exfoliated forms of MMT—can be used to create thin films on the surface of different substrates through the LbL assembly method. MMT nanosheets can reduce flammability of PUFs by forming a protective intumescent char layer on the surface (*18, 26*).

Laufer et al. deposited FR coatings composed of cationic CH and anionic MMT clay nanoplatelets on the FPUFs (*53*). A 10-BL CH/MMT film with only about 30-nm thickness completely stopped the melting of the PUF exposed to a butane torch. The CCT results disclosed that the mentioned foam exhibited 52% reduction in PHRR value compared with the uncoated foam (*53*).

Cain et al. prepared a three-component nanocoating of sodium MMT, poly(allylamine hydrochloride) (PAH), and poly(sodium phosphate) (PSP) on an FPUF via the LbL approach (Figure 6) (*54*). The results demonstrated that a four-TL coating prevented melt dripping and reduced the PHRR value by 54% relative to the control foam. This improvement was because of the synergistic interaction between the PAH/PSP with intumescing effect and MMT with desired thermal shielding. In addition, the results of the SEM revealed that the nanocoating's swollen morphology was capable of maintaining the original shape and structure of the foam (54).

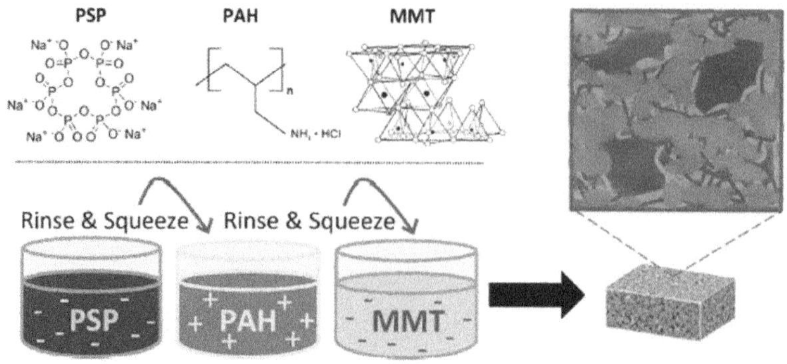

Figure 6. Schematic representation of LbL assembly of PSP, PAH, and MMT on PUF. Reproduced with permission from reference (54). Copyright 2013 Elsevier.

MXene-Based LbL Coatings

MAX phases are a large family of layered two-dimensional transition metal carbides, carbonitrides, and nitrides with a general formula of $M_{n+1}AX_n$. The M is an early transition metal (Ti, Cr, V, Sc, Mo, Nb, etc.), A stands for a group A (mainly groups 13 or 14) element, X is carbon or nitrogen, and the n value varies from 1 to 3. MXenes are generally prepared by a selective etching process. In this process, A layers from their corresponding 3D MAX phases are etched, using strong etching solutions such as hydrofluoric acid. Centrifugation and sonication are then performed to obtain single- or few-layer MXenes (Figure 7) (55, 56). These novel materials can potentially enhance fire resistance of polymers by acting as a protective layer against fire (57, 58).

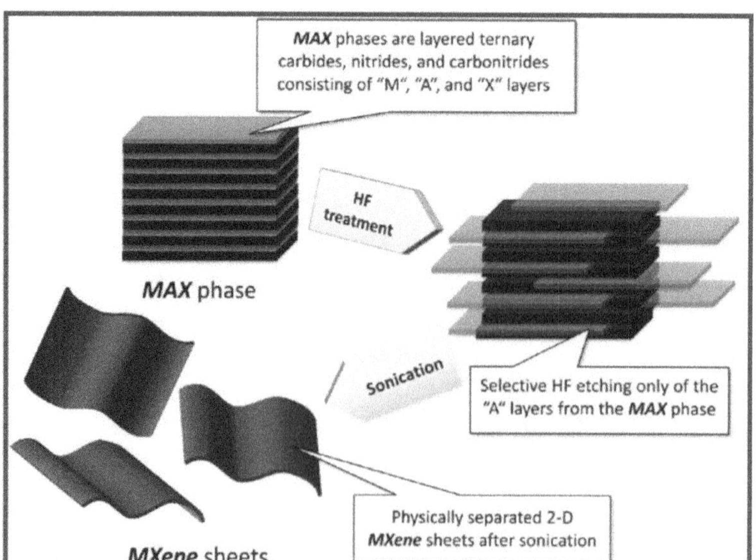

Figure 7. Schematic illustration of MXene preparation from MAX phase. In the figure, HF indicates hydrofluoric acid. Reproduced with permission from reference (56). Copyright 2012 American Chemical Society.

In this regard, Lin et al. reported the preparation of a multilayer nanocoating composed of positively charged CH and negatively charged Ti_3C_2 ultrathin nanosheets on an FPUF by using the LbL approach (*59*). MXene (Ti_3C_2) ultrathin nanosheets were prepared by exfoliation of etched Ti_3AlC_2. The resultant CH/Ti_3C_2 coatings showed remarkable reductions in PHRR, total heat release, TSR, peak smoke rate production, and the peak production of CO and CO_2 values. Because of the rapid carbonization of CH with Ti_3C_2, an outstanding char formation performance was achieved for coated foams (*59*). Another study by Lin et al. reported that a three-TL MXene/CH/phytic acid coating reduced the PHRR and TSR values of PUF by 51.1 and 84.8%, respectively. This superior improvement was the result of the barrier and charring effects of the prepared LbL coating (*60*).

Specialty Polyelectrolyte-Based LbL Coatings

Another class of LbL coatings to improve the flame retardancy of PUFs is all-polymer coatings. The presence of anionic or cationic groups in polyelectrolytes has made it possible to create coatings with FR ability on the substrate's surface through LbL assembly. To date, polymers such as CH, alginate, poly(vinyl sulfonic acid) (PVS), lignosulfonate (LS), poly(acrylic acid), PEI, PAH, poly(allylamine diphosphonate), PSP, and poly(phosphoric acid) (PPA) have been used to prepare LbL nanocoatings on PUFs.

In this regard, the first study was conducted by Laufer et al. They prepared an FR multilayered nanocoating consisted of cationic CH and anionic PVS sodium salt on an FPUF. The results showed that a 10-BL of CH-PVS film successfully suppressed melt dripping and stopped flame spread (Figure 8). This behavior was because of the fuel dilution, originating from the release of noncombustible gases such as H_2O, SO_2, and NH_3 during the coating degradation (*28*).

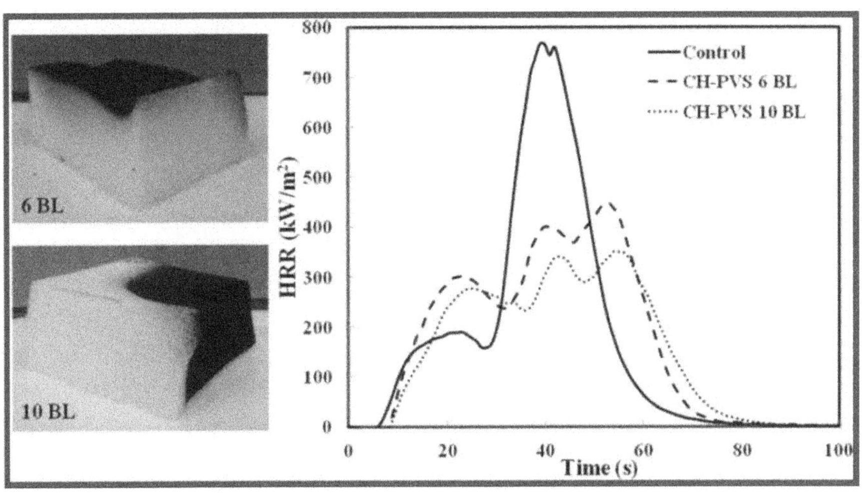

Figure 8. Images of CH-PVS coated foams after the burn test and heat release rate curves of control and coated foams. Reproduced with permission from reference (28). Copyright 2013 American Chemical Society.

In a study by Wang et al., an FR multilayered nanocoating made of CH and alginate was deposited on PUFs via the LbL assembly technique. Both alginate and CH were carbohydrate polymers that could act as FR through carbonaceous char formation. Although this coating had little effect on the thermal stability of the foam, it dramatically improved the char yield value during combustion (*61*).

Carosio et al. fabricated an effective all-polymer CH/poly(allylamine diphosphonate) coating to reduce the flammability of open-cell PUF. The results showed the presence of the LbL coating prevented the foam collapsing and stopped the melt dripping (*27*). In a subsequent study by Carosio and Alongi, CH and PPA were used to produce a fire-resistant LbL coating on FPUF. Because of the interaction between CH and PPA, a thermally stable barrier resulted, which protected the foam against mass, oxygen, and heat transfer (*17*).

LS with a high content of carbon and aromatic rings can be used as a carbonizing agent for FR systems (*62*). The LS macromolecule has a hydrophobic backbone and hydrophilic branches (carboxyl, sulfonic, and phenolic hydroxyl groups). Because of the ionizing ability of its functional groups, LS is negatively charged and acts as a polyelectrolyte in aqueous solutions. In this regard, Pan et al. prepared a fully biobased coating consisting of cationic CH and anionic LS on an FPUF using the LbL assembly method to impart thermal shielding behavior. The results showed that an eight-BL CH/LS film significantly enhanced the fire resistance of the foam and reduced the PHRR by 42%. In addition, the tensile strength of the eight-BL coated foam was increased from 0.17 to 0.19 MPa compared with the uncoated foam (*63*).

Improving the Performance of LbL Coatings

In order to improve the performance and develop the applications of LbL coatings in PUFs, some targets should be pursued, which are briefly described as follows.

Because PUFs undergo various stresses over the years, it is necessary to increase the adhesion strength, mechanical properties, and durability of LbL coatings. Another point to consider is the evaluation of how the LbL coating will age under different conditions of humidity, ultraviolet radiation, and temperature (*11*).

One of the main drawbacks of the LbL process is the large number of required deposition steps and the long deposition times. According to previous studies, the best results are obtained after an average of 8–12 deposition steps, each step lasting 1–5 min. However, studies on overcoming this problem have proved that it is possible to reduce the number of steps (*64–66*).

Toxicity and environmental problems of the FRs used in the LbL coatings are also of concern. Therefore, it is essential to evaluate the toxicological aspects of LbL coatings properly prior to planning for industrialization.

The substrate also plays an important role in the performance quality of LbL coatings. For example, an open-cell PUF with high surface area is a good substrate for LbL films. In the case of closed-cell PUFs, only their external surface is available, and they have a relatively small surface-to-volume ratio. As a consequence, it is difficult to make an effective LbL coating for them. Studies have shown that more deposition steps of LbL coatings are required to protect closed-cell PUFs against fire (*67*).

Another important factor in improving the performance of LbL coatings is the development of practical methods for deposition on the surface of PUFs. This can be considered as the last step in the industrialization of the LbL technique. In practice, with the current technology, it is not possible to use the LbL method for PUFs at an industrial scale. Therefore, more scientific studies and technical efforts are needed in order to industrialize the LbL assembly technology for PUFs (*11*).

Conclusion

The serious fire hazards of PUFs have led to many efforts to improve their flame retardancy. Over the past two decades, the LbL assembly technique has received considerable attention to fabricate FR nanocoatings on the surface of PUFs. The reported studies have shown that this approach can effectively improve the FR properties of PUFs because of the possibility of using a wide variety of high-performance nanostructures and polyelectrolytes and ease of the deposition process. However, before the LbL assembly technology can become widely used, several issues must be addressed, such as the use of natural, nontoxic, inexpensive, and efficient materials; the development of practical methods for depositing; and the increase in durability of LbL coatings. In summary, LbL assembly is a highly promising technique to enhance flame retardancy of PUFs, and it may be widely used in the future.

References

1. Kattiyaboot, T.; Thongpin, C. Effect of natural oil based polyols on the properties of flexible polyurethane foams blown by distilled water. *Energy Procedia* **2016**, *89*, 177–85.
2. Wang, C.; Zheng, Y.; Xie, Y.; Qiao, K.; Sun, Y.; Yue, L. Synthesis of bio-castor oil polyurethane flexible foams and the influence of biotic component on their performance. *Journal of Polymer Research* **2015**, *22* (8), 1–9.
3. Gómez-Fernández, S.; Ugarte, L.; Peña-Rodriguez, C.; Zubitur, M.; Corcuera, M.Á.; Eceiza, A. Flexible polyurethane foam nanocomposites with modified layered double hydroxides. *Applied Clay Science* **2016**, *123*, 109–20.
4. Kwon, O.-J.; Oh, S.-T.; Lee, S.-D.; Lee, N.-R.; Shin, C.-H.; Park, J.-S. Hydrophilic and flexible polyurethane foams using sodium alginate as polyol: effects of PEG molecular weight and cross-linking agent content on water absorbency. *Fibers and Polymers* **2007**, *8* (4), 347–55.
5. Mohammadi, A.; Lakouraj, M. M.; Barikani, M. Preparation and characterization of p-tert-butyl thiacalix [4] arene imbedded flexible polyurethane foam: an efficient novel cationic dye adsorbent. *Reactive and Functional Polymers* **2014**, *83*, 14–23.
6. Ashida, K.; Iwasaki, K. Thermosetting Foams. In *Handbook of Plastic Foams*; Landrock, A. H., , Ed.; William Andrew Publishing: Park Ridge, NJ, 1995; pp 11−220.
7. Troitzsch, J. *International Plastics Flammability Handbook: Principles-Regulations-Testing and Approval*; Hanser, 1983.
8. Käbisch, B.; Fehrenbacher, U.; Kroke, E. Hexamethoxycyclotriphosphazene as a flame retardant for polyurethane foams. *Fire and materials* **2014**, *38* (4), 462–73.
9. Wilkie, C. A.; Morgan, A. B. *Fire Retardancy of Polymeric Materials*; CRC Press, 2009.
10. Singh, H.; Jain, A. Ignition, combustion, toxicity, and fire retardancy of polyurethane foams: a comprehensive review. *Journal of Applied Polymer Science* **2009**, *111* (2), 1115–43.
11. Alongi, J.; Carosio, F. Flame Retardancy of Flexible Polyurethane Foams: Traditional Approaches versus Layer-by-Layer Assemblies. In *Novel Fire Retardant Polymers and Composite Materials*; Wang, D.-Y., Ed.; Elsevier, 2017; pp 171−200.
12. Kraemer, R. H.; Zammarano, M.; Linteris, G. T.; Gedde, U. W.; Gilman, J. W. Heat release and structural collapse of flexible polyurethane foam. *Polymer Degradation and Stability* **2010**, 95 (6), 1115–22.

13. Zammarano, M.; Krämer, R. H.; Harris, R., Jr.; Ohlemiller, T. J.; Shields, J. R.; Rahatekar, S. S.; Lacerda, S.; Gilman, J. W. Flammability reduction of flexible polyurethane foams via carbon nanofiber network formation. *Polymers for Advanced Technologies* **2008**, *19* (6), 588–95.
14. Stec, A. A.; Hull, T. R. Assessment of the fire toxicity of building insulation materials. *Energy and Buildings* **2011**, *43* (2-3), 498–506.
15. Chen, H.-B.; Shen, P.; Chen, M.-J.; Zhao, H.-B.; Schiraldi, D. A. Highly efficient flame retardant polyurethane foam with alginate/clay aerogel coating. *ACS applied materials & interfaces* **2016**, *8* (47), 32557–64.
16. Ashida, K. *Polyurethane and Related Foams: Chemistry and Technology*; CRC Press, 2006.
17. Carosio, F.; Alongi, J. Ultra-fast layer-by-layer approach for depositing flame retardant coatings on flexible PU foams within seconds. *ACS Applied Materials & Interfaces* **2016**, *8* (10), 6315–9.
18. Kim, Y. S.; Harris, R.; Davis, R. Innovative approach to rapid growth of highly clay-filled coatings on porous polyurethane foam. *ACS Macro Letters* **2012**, *1* (7), 820–4.
19. Yang, Y. H.; Li, Y. C.; Shields, J.; Davis, R. D. Layer double hydroxide and sodium montmorillonite multilayer coatings for the flammability reduction of flexible polyurethane foams. *Journal of Applied Polymer Science* **2015**, *132* (14)
20. Pan, H.; Wang, W.; Pan, Y.; Song, L.; Hu, Y.; Liew, K. M. Formation of layer-by-layer assembled titanate nanotubes filled coating on flexible polyurethane foam with improved flame retardant and smoke suppression properties. *ACS Applied Materials & Interfaces* **2015**, *7* (1), 101–11.
21. Liu, Q.; Gao, S.; Zhao, Y.; Tao, W.; Yu, X.; Zhi, M. Review of layer-by-layer self-assembly technology for fire protection of flexible polyurethane foam. *Journal of Materials Science* **2021**, *56* (16), 9605–43.
22. Maddalena, L.; Carosio, F.; Gomez, J.; Saracco, G.; Fina, A. Layer-by-layer assembly of efficient flame retardant coatings based on high aspect ratio graphene oxide and chitosan capable of preventing ignition of PU foam. *Polymer Degradation and Stability* **2018**, *152*, 1–9.
23. Zhang, X.; Shen, Q.; Zhang, X.; Pan, H.; Lu, Y. Graphene oxide-filled multilayer coating to improve flame-retardant and smoke suppression properties of flexible polyurethane foam. *Journal of Materials Science* **2016**, *51* (23), 10361–74.
24. Carosio, F.; Fina, A. Three organic/inorganic nanolayers on flexible foam allow retaining superior flame retardancy performance upon mechanical compression cycles. *Frontiers in Materials* **2019**, *6*, 20.
25. Li, Y.-C.; Kim, Y. S.; Shields, J.; Davis, R. Controlling polyurethane foam flammability and mechanical behaviour by tailoring the composition of clay-based multilayer nanocoatings. *Journal of Materials Chemistry A* **2013**, *1* (41), 12987–97.
26. Chen, P.; Zhao, Y.; Wang, W.; Zhang, T.; Song, S. Correlation of Montmorillonite Sheet Thickness and Flame Retardant Behavior of a Chitosan–Montmorillonite Nanosheet Membrane Assembled on Flexible Polyurethane Foam. *Polymers* **2019**, *11* (2), 213.
27. Carosio, F.; Negrell-Guirao, C.; Alongi, J.; David, G.; Camino, G. All-polymer layer by layer coating as efficient solution to polyurethane foam flame retardancy. *European Polymer Journal* **2015**, *70*, 94–103.

28. Laufer, G.; Kirkland, C.; Morgan, A. B.; Grunlan, J. C. Exceptionally flame retardant sulfur-based multilayer nanocoating for polyurethane prepared from aqueous polyelectrolyte solutions. *ACS Macro Letters* **2013**, 2 (5), 361–5.
29. Kim, Y. S.; Davis, R.; Cain, A. A.; Grunlan, J. C. Development of layer-by-layer assembled carbon nanofiber-filled coatings to reduce polyurethane foam flammability. *Polymer* **2011**, 52 (13), 2847–55.
30. Holder, K. M. *Flame Retardant Nanocoatings for the Protection of Polyurethane Foam*. Ph.D. Dissertation, Texas A&M, College Station, TX, 2016. https://oaktrust.library.tamu.edu/handle/1969.1/174265 (accessed Aug 21, 2021).
31. Kim, Y. S.; Davis, R. Multi-walled carbon nanotube layer-by-layer coatings with a trilayer structure to reduce foam flammability. *Thin Solid Films* **2014**, *550*, 184–9.
32. Cain, A.; Ellis, E.; Grunlan, J. Flame Retardant Carbon Nanofiber and Carbon Nanotube Coatings. *Microscopy and Microanalysis* **2011**, *17* (S2), 1470–1.
33. Mohammadi, A.; Barikani, M.; Doctorsafaei, A. H.; Isfahani, A. P.; Shams, E.; Ghalei, B. Aqueous dispersion of polyurethane nanocomposites based on calix [4] arenes modified graphene oxide nanosheets: preparation, characterization, and anti-corrosion properties. *Chemical Engineering Journal* **2018**, *349*, 466–80.
34. Mohammadi, A.; Doctorsafaei, A. H.; Zia, K. M. Alginate/calix [4] arenes modified graphene oxide nanocomposite beads: Preparation, characterization, and dye adsorption studies. *International Journal of Biological Macromolecules* **2018**, *120*, 1353–61.
35. Sang, B.; Li, Z.-w.; Li, X.-h.; Yu, L.-g.; Zhang, Z.-j. Graphene-Based Flame Retardants: A Review. *Journal of Materials Science* **2016**, *51* (18), 8271–95.
36. Bao, C.; Guo, Y.; Yuan, B.; Hu, Y.; Song, L. Functionalized graphene oxide for fire safety applications of polymers: a combination of condensed phase flame retardant strategies. *Journal of Materials Chemistry* **2012**, *22* (43), 23057–63.
37. Li, L.; Jiang, K.; Qian, Y.; Han, H.; Qiao, P.; Zhang, H. Effect of organically intercalation modified layered double hydroxides-graphene oxide hybrids on flame retardancy of thermoplastic polyurethane nanocomposites. *Journal of Thermal Analysis and Calorimetry* **2020**, 1–11.
38. Mishra, G.; Dash, B.; Pandey, S. Layered double hydroxides: A brief review from fundamentals to application as evolving biomaterials. *Applied Clay Science* **2018**, *153*, 172–186.
39. Poonoosamy, J.; Brandt, F.; Stekiel, M.; Kegler, P.; Klinkenberg, M.; Winkler, B.; Vinograd, V.; Bosbach, D.; Deissman, G. Zr-containing layered double hydroxides: Synthesis, characterization, and evaluation of thermodynamic properties. *Applied Clay Science* **2018**, *151*, 54–65.
40. Kuila, T.; Acharya, H.; Srivastava, S.; Bhowmick, A. Effect of vinyl acetate content on the mechanical and thermal properties of ethylene vinyl acetate/MgAl layered double hydroxide nanocomposites. *Journal of Applied Polymer Science* **2008**, *108* (2), 1329–35.
41. Conterosito, E.; Gianotti, V.; Palin, L.; Boccaleri, E.; Viterbo, D.; Milanesio, M. Facile preparation methods of hydrotalcite layered materials and their structural characterization by combined techniques. *Inorganica Chimica Acta* **2018**, *470*, 36–50.

42. Takehira, K. Recent development of layered double hydroxide-derived catalysts— Rehydration, reconstitution, and supporting, aiming at commercial application—. *Applied Clay Science* **2017**, *136*, 112–41.
43. Babu, H.; Coluccini, C.; Wang, D.-Y. Functional Layered Double Hydroxides and Their Use in Fire-Retardant Polymeric Materials. In *Novel Fire Retardant Polymers and Composite Materials*; Elsevier, 2017; pp 201–38.
44. Shanmuganathan, K.; Ellison, C. Layered Double Hydroxides: An Emerging Class of Flame Retardants. In *Polymer Green Flame Retardants*; Elsevier, 2014; pp 675–707.
45. Liu, Y.; Gao, Y.; Wang, Q.; Lin, W. The synergistic effect of layered double hydroxides with other flame retardant additives for polymer nanocomposites: a critical review. *Dalton Transactions* **2018**, *47* (42), 14827–40.
46. Zhang, F.; Li, M.; Yang, L.; Ye, S.; Huang, L. Ni-Mg-Mn-Fe-O catalyst derived from layered double hydroxide for hydrogen production by auto-thermal reforming of ethanol. *Catalysis Communications* **2014**, *43*, 6–10.
47. Morgan A. B., Wilkie C. A., Nelson, G. L., Eds.; *Fire and Polymers VI: New Advances in Flame Retardant Chemistry and Science*: American Chemical Society, 2012; Vol. 1118.
48. Mohammadi, A.; Wang, D.-Y.; Hosseini, A. S.; De La Vega, J. Effect of intercalation of layered double hydroxides with sulfonate-containing calix [4] arenes on the flame retardancy of castor oil-based flexible polyurethane foams. *Polymer Testing* **2019**, *79*, 106055.
49. Liu, L.; Wang, W.; Hu, Y. Layered double hydroxide-decorated flexible polyurethane foam: significantly improved toxic effluent elimination. *RSC Advances* **2015**, *5* (118), 97458–66.
50. Kandemir, A.; Kaynak, C. Contribution of Nanoclays to the Performance of Traditional Flame Retardants in ABS. *Polymer Composites* **2012**, *33*, 420–429.
51. Fahami, A.; Lee, J.; Lazar, S.; Grunlan, J. C. Mica-based multilayer nanocoating as a highly effective flame retardant and smoke suppressant. *ACS Applied Materials & Interfaces* **2020**, *12* (17), 19938–43.
52. Lazar, S.; Carosio, F.; Davesne, A.-L.; Jimenez, M.; Bourbigot, S.; Grunlan, J. Extreme heat shielding of clay/chitosan nanobrick wall on flexible foam. *ACS Applied Materials & Interfaces* **2018**, *10* (37), 31686–96.
53. Laufer, G.; Kirkland, C.; Cain, A. A.; Grunlan, J. C. Clay–chitosan nanobrick walls: completely renewable gas barrier and flame-retardant nanocoatings. *ACS Applied Materials & Interfaces* **2012**, *4* (3), 1643–9.
54. Cain, A. A.; Nolen, C. R.; Li, Y.-C.; Davis, R.; Grunlan, J. C. Phosphorous-filled nanobrick wall multilayer thin film eliminates polyurethane melt dripping and reduces heat release associated with fire. *Polymer Degradation and Stability* **2013**, *98* (12), 2645–52.
55. Lei, J.-C.; Zhang, X.; Zhou, Z. Recent advances in MXene: Preparation, properties, and applications. *Frontiers of Physics* **2015**, *10* (3), 276–86.
56. Naguib, M.; Mashtalir, O.; Carle, J.; Presser, V.; Lu, J.; Hultman, L.; Gogotsi, Y.; Barsoum, M. W. Two-Dimensional Transition Metal Carbides. *ACS Nano* **2012**, *6* (2), 1322–31.
57. Shinde, P. V.; Singh, M. K. Synthesis, Characterization, and Properties of Graphene Analogs of 2D Material. In *Fundamentals and Sensing Applications of 2D Materials*; Hywel, M., Rout, C. S., Late, D. J., Eds.; Woodhead Publishing, 2019; Chapter 4, pp 91–143.

58. Naguib, M.; Kurtoglu, M.; Presser, V.; Lu, J.; Niu, J.; Heon, M.; Hultman, L.; Gogotsi, Y.; Barsoum, M. W. Two-dimensional nanocrystals produced by exfoliation of Ti3AlC2. *Advanced Materials* **2011**, *23* (37), 4248–53.
59. Lin, B.; Yuen, A. C. Y.; Li, A.; Zhang, Y.; Chen, T. B. Y.; Yu, B.; Lee, E. W. M.; Peng, S.; Yang, W.; Lu, D.-D.; Chan, Q. N.; Yeoh, G. H.; Wang, C. H. MXene/chitosan nanocoating for flexible polyurethane foam towards remarkable fire hazards reductions. *Journal of Hazardous Materials* **2020**, *381*, 120952.
60. Lin, B.; Yuen, A. C. Y.; Chen, T. B. Y.; Yu, B.; Yang, W.; Zhang, J.; Yao, Y.; Wu, S.; Wang, C. H.; Yeoh, G. H. Experimental and numerical perspective on the fire performance of MXene/Chitosan/Phytic acid coated flexible polyurethane foam. *Scientific Reports* **2021**, *11* (1), 1–13.
61. Wang, X.; Pan, Y.-T.; Wan, J.-T.; Wang, D.-Y. An eco-friendly way to fire retardant flexible polyurethane foam: layer-by-layer assembly of fully bio-based substances. *RSC Advances* **2014**, *4* (86), 46164–9.
62. Lu, W.; Li, Q.; Zhang, Y.; Yu, H.; Hirose, S.; Hatakeyama, H.; Matsumoto, Y.; Jin, Z. Lignosulfonate/APP IFR and its flame retardancy in lignosulfonate-based rigid polyurethane foams. *Journal of Wood Science* **2018**, *64* (3), 287–93.
63. Pan, Y.; Zhan, J.; Pan, H.; Wang, W.; Tang, G.; Song, L.; Hu, Y. Effect of fully biobased coatings constructed via layer-by-layer assembly of chitosan and lignosulfonate on the thermal, flame retardant, and mechanical properties of flexible polyurethane foam. *ACS Sustainable Chemistry & Engineering* **2016**, *4* (3), 1431–8.
64. Cain, A.; Plummer, M.; Murray, S.; Bolling, L.; Regev, O.; Grunlan, J. C. Iron-containing, high aspect ratio clay as nanoarmor that imparts substantial thermal/flame protection to polyurethane with a single electrostatically-deposited bilayer. *Journal of Materials Chemistry A* **2014**, *2* (41), 17609–17.
65. Patra, D.; Vangal, P.; Cain, A. A.; Cho, C.; Regev, O.; Grunlan, J. C. Inorganic nanoparticle thin film that suppresses flammability of polyurethane with only a single electrostatically-assembled bilayer. *ACS Applied Materials & Interfaces* **2014**, *6* (19), 16903–8.
66. Davis, R.; Li, Y.-C.; Gervasio, M.; Luu, J.; Kim, Y. S. One-pot, bioinspired coatings to reduce the flammability of flexible polyurethane foams. *ACS Applied Materials & Interfaces* **2015**, *7* (11), 6082–92.
67. Carosio, F.; Cuttica, F.; Di Blasio, A.; Alongi, J.; Malucelli, G. Layer by layer assembly of flame retardant thin films on closed cell PET foams: Efficiency of ammonium polyphosphate versus DNA. *Polymer Degradation and Stability* **2015**, *113*, 189–96.

Chapter 8

Metal Oxide-Based Compounds as Flame Retardants for Polyurethanes

Anil M. Palve,[*,1] Omkar V. Vani,[1] and Ram K. Gupta[2]

[1]Department of Chemistry, Mahatma Phule Arts, Science, and Commerce College, Panvel, Navi Mumbai, MH 410206, India

[2]Department of Chemistry and Kansas Polymer Research Center, Pittsburg State University, Pittsburg, Kansas 66762, United States

[*]Email: palve_anil@yahoo.com

A broad overview of metal oxide-based flame-retardant polyurethanes (FR-PUs) is offered in this chapter. The beginning of the chapter discusses the needs for the FR materials and provides examples. The chapter also briefly describes different types of FR-PUs. The purpose of the chapter is to show the result of incorporating various additives such as transition metal oxides, metal–organic frameworks, phosphorus, and silica-based materials. This chapter also explains their synergistic effect on polymeric materials and their properties such as flame retardancy, thermal ability, and mechanical ability. This chapter mainly highlights the application of metal oxide-based PU materials in the FR process. The future outlook of metal oxide-based PUs as FRs is mentioned at the end of this chapter.

Introduction

In the last few decades, flame retardant (FR) compounds have shown marvelous growth because of increasing obligatory guidelines and voluntary participation in fire protection. Noteworthy awareness has been paid to controlling the inherent flammability of combustible polymer compounds. Polyvinyl chloride, poly-tetrafluoroethylene, aromatic polymers, and aliphatic polymers contain additives or reactive FR systems. These are used to delay or suppress a combustion reaction. FR materials are important, as they save lives because of this delay and nonflammability. This gives them time to depart from transport vehicles safely. FRs make material tougher when exposed to fire, and they measurably decrease and slow down fire spread. Otto Bayer, in Germany in the 1930s, developed a polymer called polyurethane (PU) (*1*). In 1849 Charles Adolphe Wurtz and his coworkers from the French Chemical Institute revealed the reaction between alcohol and isocyanate, which produces urethane (carbamate) groups (2) and is given in eq 1. Polyols, diisocyanates, chain extenders, and catalysts are the key components of PU composite materials:

$$R' - OH + R - N = C = O \rightarrow R' - O - CO - NH - R \; ..... (1)$$

Some have tried to substitute the earlier polymer material polyamide fibers (nylon), which were protected by a U.S. patent. This formed new organic polymers, which had the polyvalency property. The ability to replace other unusual materials created growth for plentiful applications (3).

Polyols and diisocyanates are the two basic materials used for the preparation of commercial PUs. The high reactivity of isocyanates helps them to react with alcohol, more or less rapidly, depending on their nature (4). PU is one of the most produced polymers in the world because of the rapid development of construction and automotive sectors, shoes, and industrial furnishing. Maximum times used for polyols and isocyanates in industry are presented in Figure 1 and Figure 2, respectively. These structures can be combined to obtain materials with desired properties.

Polyol	Structural Formula
Poly(ethylene glycol)	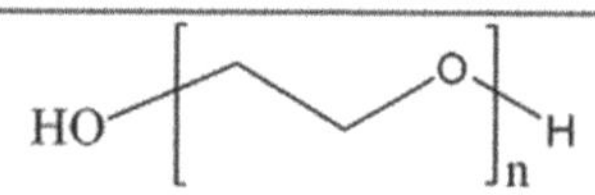
Poly(propylene glycol)	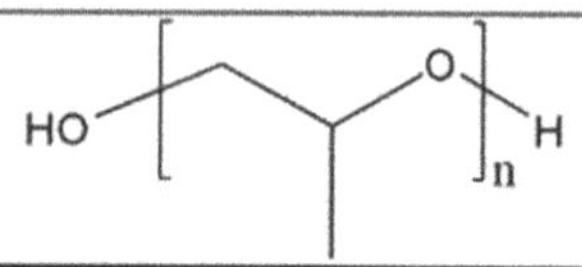
Poly(tetramethylene glycol)	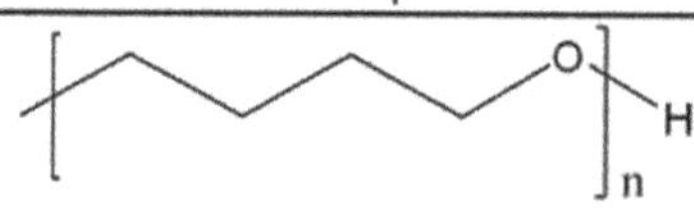
Poly[di(ethylene glycol) adipate]	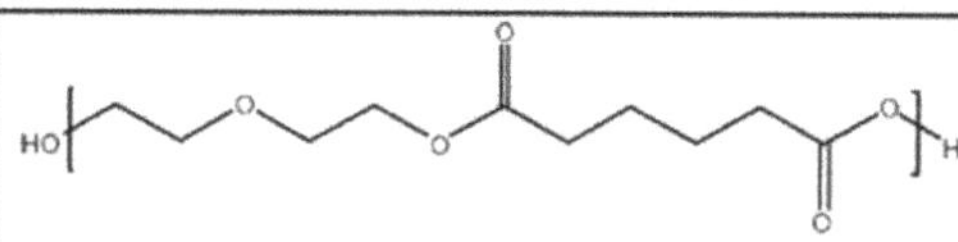

Figure 1. Various examples of the polyols used in industry.

Various types of PUs are given in Figure 3. The main concern of PU materials is their less sufficient heat resistance and high flammability. The chemistry of the thermal disintegration of PUs is different, depending on starting materials used. PUs produce smoke while burning and create environmental pollution. In addition, CO_2, CO, HCN, and other toxic contents are released in the environment with the combustion of these materials.

In the modern era, fire hazard has become a negotiable topic. This is because it is the era of modern science, which has until this point made ample progress in fire safety. Society members also use modern devices such as furnaces or ovens, which are well controlled in terms of fire. In big towers, society members are using up-to-date machines such as smoke alarms and creating awareness by using exit signs, sprinkler systems, and many other fire-preventing tools and pieces of equipment. The fire issue is being resolved with the help of modern science, using safety measures such as modern FRs. Their progress continues with the help of engineers all around the world. Before entering into this chapter, we must first see some of the important terms related to fire science and fire hazards. It is also important to understand the necessity of FR materials in various applications. A fire hazard is a broad concept; it includes the decrease or stopping of fire and also involves fire safety behavior (5).

Nature	Isocyanate name	Structural Formula
Aliphatic	Hexamethylene diisocyanate	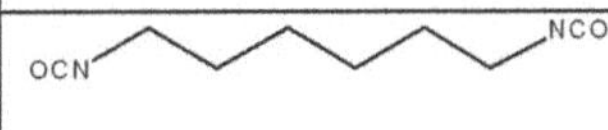
	4,4'-Methylene dicyclohexyl diisocyanate	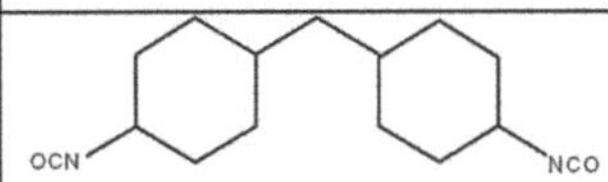
	Isophoronediisocyanate	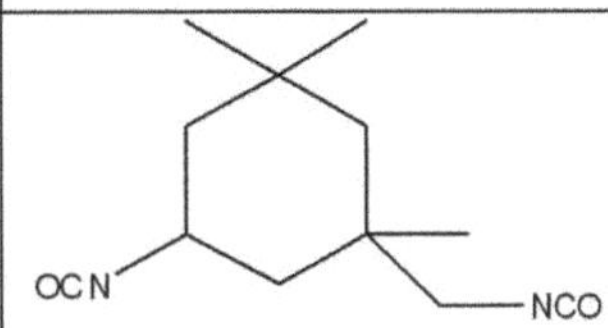
Aromatic	Methylene diisocyanate	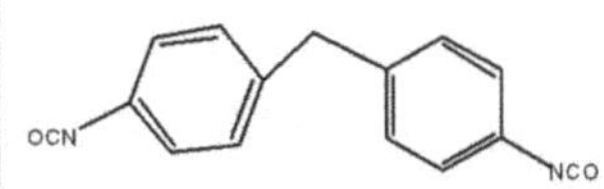
	Toluene diisocyanate	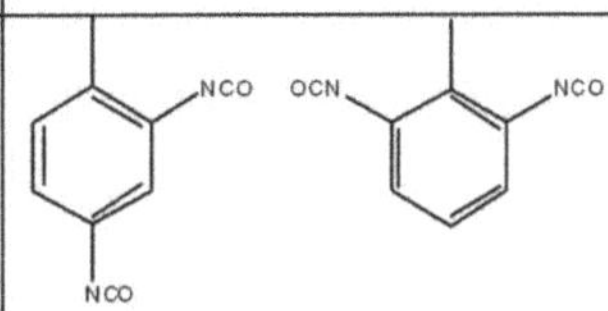

Figure 2. Predominantly used aliphatic and aromatic diisocyanates in PU preparation.

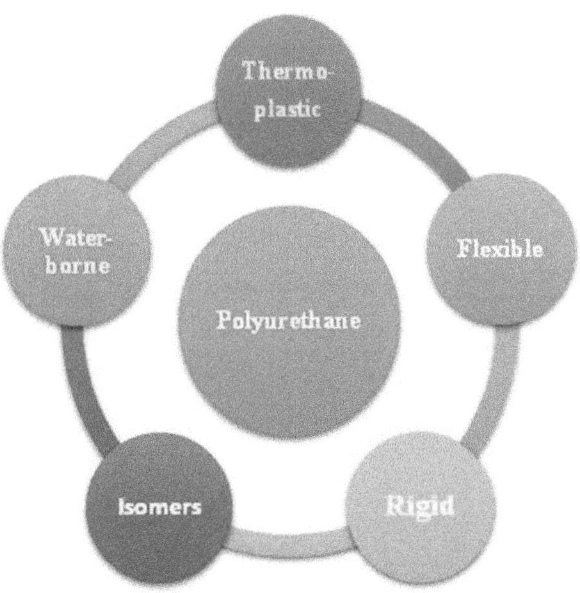

Figure 3. Schematic representation of types of PUs.

The growth or start of fire can be controlled in PUs by FR compounds. FRs are chemical materials inserted to inhibit or retard the ignition or burning of polymer materials. Thermal stability is an important key for the FR and should be introduced in PU composites to stop combustion under fire. These FR materials have been developed by many researchers and used in industry to decrease or stop the flame process (6). They have been added or used in many products to reduce the fire risk in home furnishings, foams, mattresses, carpets, flooring, and many more home appliances. They have added to electronic appliances such as desktop computers, laptops, mobile phones, TVs, electrical ropes, and wires. They are also used in various items and appliances such as chairs, hair coats, and additional parts of vehicles, cars, ships, and trains (7).

Classification of FRs

Categories of FR compounds depend on the mechanism involved, constituents used, and properties of the polymer. Halogenated FR hydrocarbons have chlorine- and bromine-containing compounds. Inorganic FRs include boron-based oxides, antimony oxides, aluminum hydroxide, molybdenum-based oxides, zinc and magnesium oxides, and so forth. Phosphorous-containing compounds include phosphates, halogenated and inorganic phosphorous-containing salts, organic phosphate ester, and so forth. Three main mechanisms are involved: gas-phase flame retardancy, endothermic FRs, and char-forming FRs (Figure 4) (8). Initially, halogen-based FRs were largely used in many applications. These materials show the vapor-phase mechanism of free radical scavenging, and toxic gases are given out during this process. Industries are interested in the production of halogen-free intumescent FRs because of environmental concerns. An overview of PU-based nanocomposites for FRs are explored in a book chapter by Palve and Gupta (9). The main types of FRs are illustrated in Figure 4.

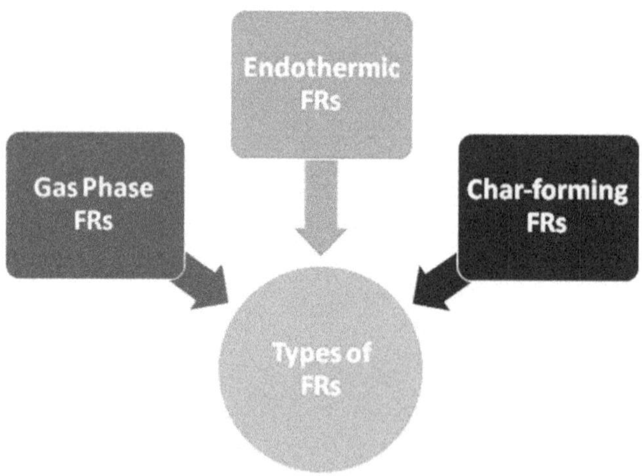

Figure 4. Schematic representation of types of FRs.

Halogenated FRs

Organo-halogen chemicals mainly include organo-chlorine and organo-bromine compounds. This means the chemical contains chlorine or bromine bonded to the carbon atom. Brominated FRs are a well-known group of FRs and are useful in the electronic and textile industries. They also are used in construction and coating materials. Polybrominated phenyl was the first FR. It was synthesized by the Velsicol Chemical Corp. The firing process was delayed because of the gas-phase reaction of bromine atoms before the PUs reached their detonation temperature, which stopped the chemical results from taking place within the fire. Initially, halogen-based FRs were very popular because of their applications in various fields. After 1970, many of the halogenated FRs were considered contaminants and have since been banned in several countries.

Endothermic FRs

Endothermic FRs undergo an endothermic thermal decomposition at which combustion takes place. The decomposition temperature of the FRs and PU is important in the endothermic FR process. If the decomposition temperature of both is the same, then FRs help to take out heat from the substrate by forming a metal-oxide layer. This involves the endothermic decomposition of the FR additives. The lower substrate temperature slows the decomposition rate. A ceramic-like residue is formed. which helps to protect the PUs. Mainly, aluminum and magnesium hydroxides are used in endothermic FRs. Both materials release nonflammable gases, like water vapor and carbon dioxide, which dilute the combustible gases and cool the system.

Char-Forming FRs

Additives react with FRs to form a carbon layer on the surface of the substance. This is known as a char-forming FR. These layers help in decreasing the rate of decomposition and also help to lower the evolution of gases. This mechanism mainly involves phosphorous and nonhalogenated FRs. Char can act as thermal insulation during this process. A nitrogen–phosphorous substance like melamine is used in this type of FR. The formation of cross-linked structures such as melem and melon in melamine helps in char stability during the condensed phase. The formation of the nitrogen–phosphorous compound during this process increases char stability. Some reports have utilized environmentally friendly FRs such as phosphonatetriazine compounds (*10*). Instead of the traditional method of synthesizing char-forming FRs, a Schiff base made up of a polyamide-containing phosphonate oligomer has been reported for char-forming FRs. The combination of pentaerythritol octahydrogen tetraphosphate with tris(2-hydroxyethyl) isocyanurate produced an interesting char-forming agent, which could increase the quality of char (*11*).

Inorganic FRs

In an inorganic FR system, numerous inorganic fillers are used to increase the characteristics of FRs. To obtain the required properties, these materials should be used in large concentrations in the case of FRs. The efficiency and properties of FRs can be increased by the addition of other various inorganic FRs through synergism. Antimony oxides are not often used as FRs, but they can act as synergists, along with halogenated FRs like chlorine and bromine. This proves that antimony oxides work like catalysts to break chlorine or bromine more quickly, resulting in the active bromine atoms releasing more quickly into the gas phase than they did before. Antimony halogen compounds possess FR properties and remove radicals, causing a fire risk. Antimony oxide, antimony pentoxide, and sodium antimonite could be used as FRs. The use of the antimony FRs yields antimony chloride with an oxychloride as a highly reactive intermediate. Antimony is nontoxic, but antimony compounds may contain small amounts of arsenic, so precautions should be taken while handling these compounds. Aluminum and magnesium hydroxides are inorganic FRs that can be used independently. Mainly inert gases like water vapor can help in creating the protective layer on PU material during the burning process. The fire that may spread can be reduced by absorbing these evolved gases. Boron compounds are also used in PU as FR material.

Nitrogen-Based FRs

Nitrogen-based FRs are an interesting class of additives with many applications. These FRs have good thermal stability. They decompose at a high-temperature range. Researchers are currently working on newer nitrogen-based FRs. They are combining nitrogen and organic and inorganic acids (mainly cyanuric acid, phosphoric acid, boric acid, etc.). The main intention of using these acids is to encourage the generation of shielding char. The choice of acid is also important for relevant applications and properties. The most commonly used types of nitrogen FRs are melamine-based products. The cross-linked structure of the melamine is obtained during the FR process. This transformation supports the development of char, which slows down oxygen delivery. Work on the formation of ethylene glycol with a melamine–formaldehyde resin shows excellent mechanical properties, FR properties, and thermal stability. The FR effect can be increased by a mixture of ammonium polyphosphate (APP) (known for smoke suppression) and dimethyl methylphosphonate (*12*). The incorporation of nitrogen contents in FRs has many advantages, such as environmental friendliness, low toxicity, low evolution of smoke, recyclability, and good efficiency with low cost. Nitrogen-based FRs are eco-friendly because they do not create additional gases or toxic elements in the environment. As they contain only nitrogen and phosphorous, waste disposal is not an important concern in nitrogen-based FRs (*13*).

Phosphorous-Based FRs

Phosphorus never exists in free elemental form because of its reactivity. It is a multivalent nonmetal element of the nitrogen group. Elemental phosphorus occurs in two major forms: white and red phosphorus. Organo-phosphorous FRs contain a bond between phosphorous and carbon. Intumescent FRs are made of phosphorous combined with nitrogen fillers. Many commercial products are now synthesized using phosphorus-based FRs instead of halogenated FRs.

In general, the mode of the mechanism of FRs is in the condensed or gas phase, or both phases. The phosphorous-based FR mechanism involves various radicals—like $PO^{\bullet}$, $PO_2^{\bullet}$, and $HPO^{\bullet}$—that can act as scavengers for $H^{\bullet}$ and $OH^{\bullet}$ in the gas phase. Ammonium phosphate on canvas was the first phosphorus-based FR. Phosphorous-containing FRs consist of ammonium polyphosphate, ethylene diamine phosphate, melamine polyphosphate, piperazine pyrophosphate, and so forth. APP is used as a phosphorous FR. Phosphorous-containing FRs include organic and inorganic phosphates. Phosphorous FRs are used in PUs, polyethylene terephthalate, resins, and so forth. To get the flame retardancy and eco-friendliness needed, many phosphorus-based FRs have been prepared with nonhalogen-containing compounds and show excellent FR properties, low toxicity, and low amounts of smoke during the burning process. Liquid phosphorus-containing imidazole, whose phosphorus-containing group and imidazole ring are attached by an aminopropyl chain, has been investigated (*14*). Aliphatic phosphorous-based FRs contain a large percentage of phosphorous because of their lower molecular weight and sensitivity to hydrolysis. This disadvantage being overcome by using nitrogen–phosphorous-based FRs with dimethyl *N,N′*-1,3-phenylenebis(*P*-methylphosphonamidate) was reported (*15*). An intumescent FR feature for PU foams has been created using phosphorus and triazole groups (*16*). The synthetic route involves simple azide–alkyl click chemistry. In this product, phosphate acts as a char-forming group. Such a type of material can be used to replace halogen-based FRs. Black phosphorous has become a research interest because of its layered structure. Black phosphorous can be used as a good additive in PUs for an FR agent, because it shows good thermal stability, mechanical stability, and characteristic dimension effects (*17*).

Eco-Friendly/Organic-Based FRs

Organic-based FR-PUs are manly prepared by hydrocarbon compounds. Conventional FRs, like halogenated and metal ones, have limitations related to environmental issues. Carbon-based FRs are the new eco-friendly class that has better mechanical, thermal, and electrical properties. The use of hydrocarbons enables an increase in surface to mass ratio, and they can be used as innovative FRs. Many groups are currently working on FRs that have fewer fire hazards, produce less gas, and are environmentally benign. For making benign FRs, the use of the additive and reactive FR compounds in PU foams has been attempted (*18*). Bhoyate et al. have used aliphatic and aromatic reactive bromine polyols for organic FR-Pus (*18*). Such types of bromine-based polyols can be used as successful FRs for rigid PU foams. Cellulose nanofiber is another material used in the presence of sodium bicarbonate. Reports show an excellent application of cellulose nanofiberaero gels synthesized by incorporating sodium bicarbonate as an FR (*19*). Other materials reported to have better FR properties include: magnesium hydroxide nanoscale material in cellulose gelled in alkaline urea solution (*20*); cellulose-based composite aero gels in the presence of aluminum hydroxide (*21*); cellulose-based aerogels with graphene oxide; clay mineral sepiolite nanorods (*22*); and multilayer zirconium phosphate-based graphene oxide incorporated into cellulose nanofibers (*23*). Therefore, mechanical and thermal stability is increased because of the presence of carbon nanoadditives.

Transition Metal Oxides as FRs

Metal oxides with FRs have shown effectiveness in reducing PUs' flammability. Nanoscale metal oxides such as TiO_2 and Al_2O_3—along with poly(methyl methacrylate) and metal oxides such as Al_2O_3 and Fe_2O_3, in combination with red phosphorus—are used in recycled polyethylene terephthalate, which is used as an FR (*24*). Molybdenum trioxide and cuprous oxide are branded for smoke control in metal oxides. However, if these oxides combine with graphene, they are added into a PU matrix, and enhancement in physical and chemical properties—such as thermal stability, flame retardancy, and fire control—is observed. These enhancements in properties are because of the combined effects of stacking space in graphene sheets and catalytic charring of molybdenum and cuprous oxide. This has also been supported using thermogravimetric analysis of 2 wt % hybrids, heat release rate, and smoke density studies (*25*). A schematic illustration for graphene-based molybdenum trioxide and cuprous oxide PU composites is given in Figure 5.

Some interesting studies have revealed the effective use of metal oxides—such as manganese dioxide, zinc oxide, nickel(III) oxide, and so forth—incorporating into blends of polypropylene/APP/dipentaerythritol for FRs (*26*). The dehydration and oxidation reactions in metal oxides lead to an increase in FR properties.

Metal–Organic Framework-Based FRs

Two-dimensional metal–organic frameworks (MOFs) have created attention in many fields such as catalysis, gas sensing, electrode materials for supercapacitors, drug deliveries, and so forth. MOFs are prepared by metal holding units with organic linkers. Easy synthesis and tunable functionalities are the keys to increasing their popularity for different applications. Similarly, MOFs show potential as FRs for PU materials. These properties may be achieved because of exhibition of adjustable structures, metal and ligands being able to vary as needed, organic portions improving compatibility, and so forth. Materials of Institut Lavoisier (MIL) materials, zeoliticimidazolate frameworks, and University of Oslo materials are the most popular MOFs. When mixed with PUs, they show flame retardancy (27). In the last decades, the aforementioned MOFs have been reported in literature with different metals for flame retardancy.

Cobalt–organic framework material is one of the materials used as an FR. The mixture of 4.5 wt % APP and 1.5 wt % cobalt–organic framework improves the FR capacity of PUs. Compared with thermoplastic PU (TPU), Co–MOFs show better results because of the limiting oxygen index (LOI), which is 28.2 (28). In addition, a greener approach for the synthesis of the MOF and its use in smoke and fire control is a demand of the present era. A greener Ni–MOF is synthesized in water–ethanol at room temperature. This has shown to be a potentially promising substitute for APP by nickel phosphate, which creates smoke and fire control (29). Synthesis, smoke and flame retardancy, and interfacial charring effects of this material are shown in Figure 6.

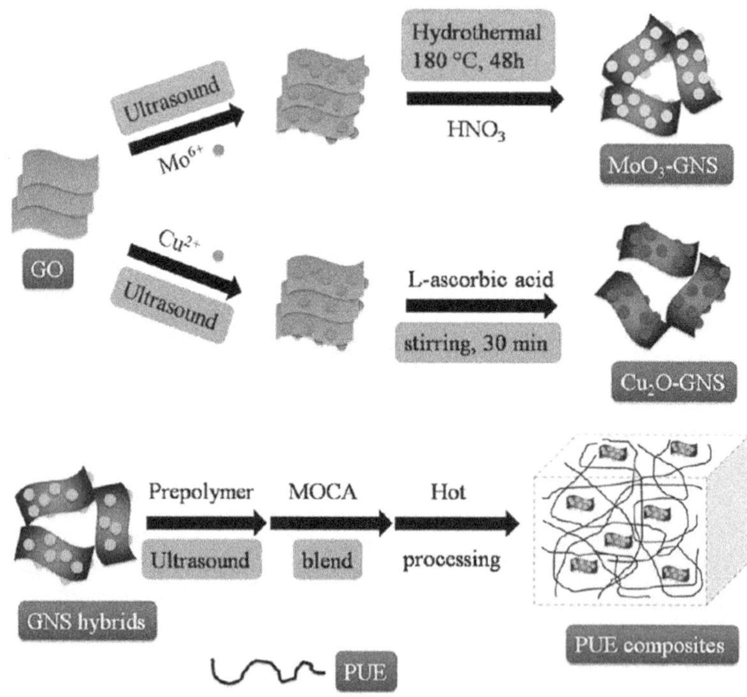

Figure 5. A schematic representation for the modification of graphene-based molybdenum trioxide and cuprous oxide PU elastomer (PUE) composites. GO stands for graphene oxide; GNS stands for graphene; MOCA stands for 3,3'-dichloro-4,4'-diaminodiphenylmethane. Adapted with permission from reference (25). Copyright 2016 American Chemical Society.

Other MOFs—including Fe- and Co-based ones incorporated in polystyrene (30), polyphosphazene-wrapped Fe 2-methylimidazole (31), Co-based ones (28), Ni-based ones (32), MIL-53 (Fe)@C/graphite carbon nitride hybrid (MFeCN) (33), and so forth—have been reported as FR applications.

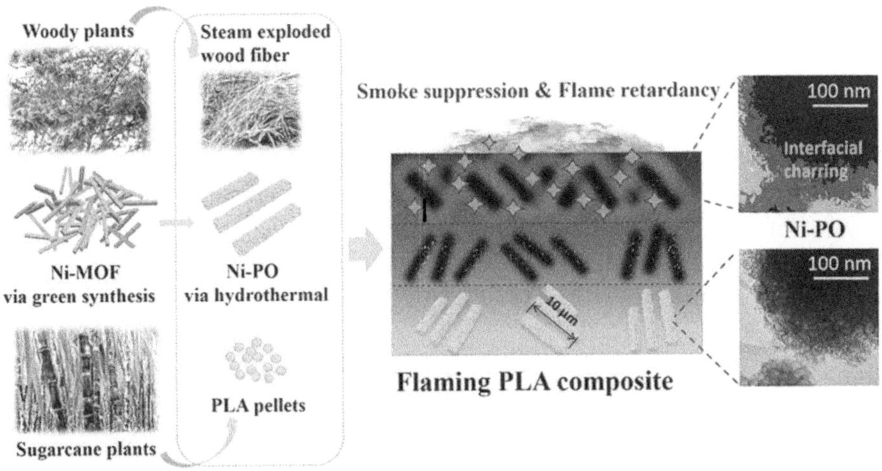

Figure 6. Smoke suppression and flame control processes using Ni–MOF. PLA stands for poly(lactic acid). Adapted with permission from reference (29). Copyright 2019 American Chemical Society.

Aluminum-Based FRs

Aluminiumtrihydrate, phosphate, hydroxide, and so forth—in combination with other materials—can be used in aluminum-based FR-PUs. Aluminum-based hydroxides are cheap, are nontoxic, and are used as flame suppression agents. Aluminum hydroxide can also be used as a synergistic FR agent with expandable graphite. Aluminum hydroxide decomposes to create aluminum oxide and water as side products (eq 2). The calculated theoretical loss of water in this reaction is 34.6 wt %. This helps in the slowdown of fire hazards:

$$2Al(OH)_3 \rightarrow Al_2O_3 + 3H_2O \quad .....(2)$$

Literature reports show that the use of the aluminum-based FRs in combination with carbon allotrope or other materials increases FR properties. The use of aluminum hydroxide in combination with expandable graphite and polyisocyanurate–PU foams shows excellent FR properties (*34*). This is an eco-friendly FR and can be used for commercial purposes. Enhanced FR properties were reported with aluminum phosphate microcapsules and smoke suppression properties of flexible PU foams (*35*). Extraordinary increases in FRs in combination with dimethyl methylphosphonate, aluminum hydroxide, ammonium polyphosphate, and graphite have been reported, and thermal insulation properties remain constant (*36*).

Magnesium Hydroxide/Oxide-Based FRs

Magnesium hydroxide is made up of halogen (chlorine and bromine) and acid-free FR additives used in PUs. Magnesium hydroxide decomposes to create magnesium oxide and water as side products (eq 3):

$$Mg(OH)_3 \rightarrow MgO + H_2O \quad .....(3)$$

It has 31% water-binding capacity, and it adsorbs large amounts of energy during thermal decomposition. The important criteria in developing FRs are that they should have small amounts of smoke, should have small amounts of toxicity, should be halogen-free, should have high thermal stability, and so forth. All of the aforementioned properties are inherent in magnesium hydroxide, so it is favored in industry as an inorganic FR. The following factors determine the effectiveness of magnesium hydroxide as an FR: (1) endothermic disintegration starts at around 300 °C for magnesium hydroxide when taking out heat from the substrate, (2) the discharge of water vapor upon the disintegration of magnesium hydroxide weakens the fire in the gas phase, and (3) there is a reduction of thermal energy to degrade PUs.

Ai and coworkers reported the organic/inorganic hexakis(4-boronic acid-penoxy)-cyclophosphazene-6B)/magnesium hydroxide FR. This composite material helps in the cost of FRs. During the process, the magnesium hydroxide disintegrates endothermically and gives oxide. Magnesium oxide is known for its formation of a physical protective layer on the material. However, if the decomposition continues, the external carbon layer may become cracked. The formation of cracks and holes is avoided with a mixture of cyclophosphazene-6B and magnesium hydroxide (Figure 7) (37).

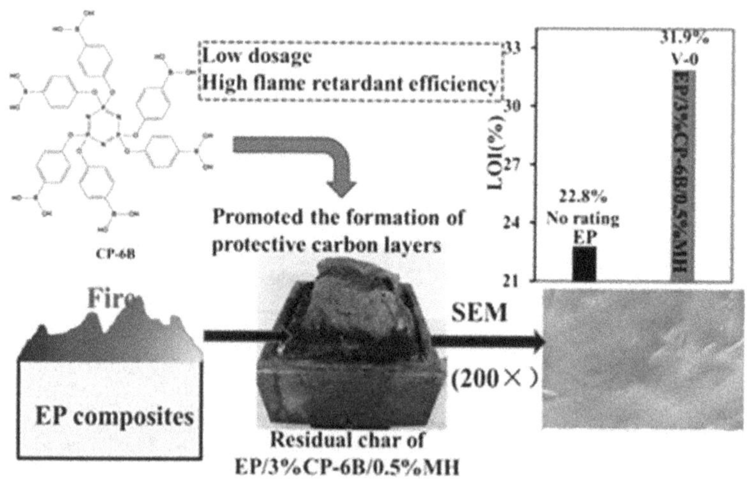

Figure 7. Synergistic effect of organic/inorganic FRs. EP stands for epoxy resin; SEM stands for scanning electron microscopy. Adapted with permission from reference (37). Copyright 2019 American Chemical Society.

Magnesium hydroxide with ceramic precursors has also been reported. It overcomes the noncohesive characteristics of magnesium oxide (38). The cross-linking effect of polycarbosilane increases the thermal stability in the presence of magnesium hydroxide. Also, the charring effect of polycarbosilane is minimized, which helps in avoiding heat and mass shift.

Zinc Oxide-Based FRs

Zinc oxide is eco-friendly and has ecological and engineering applications. Zinc oxide is a widely studied and useful material in many applications such as buffer layers in copper indium gallium solar cells, antibacterial materials, sunscreen lotions, baby powders, light-emitting diodes, paints, and so forth. In addition to the aforementioned applications, zinc oxide is widely used in FR strategies in the cotton industry. Earlier, halogenated FRs were used for fabrics. Because of the toxicity of and formation of smoke by halogenated FRs, eco-friendly zinc oxide has been employed on cotton fabrics. This consists of an FR polymer or compound forming a composite material (Figure 8) (39).

Figure 8. Zinc oxide particle used as an FR on cotton fabrics. Adapted with permission from reference (39). Copyright 2018 American Chemical Society.

Like zinc oxide, phosphorus FRs are efficient FRs. The effectiveness of phosphorus FRs can improve with the addition of nanoscale zinc oxide for rigid PU foam. This nanoscale zinc oxide can add to the efficiency of phosphorus FRs and also offer insulation. This idea was implemented in literature with nanoscale zinc oxide, porous zeolite and layered montmorillonite, and phosphorus compounds (*40*). Zinc oxide incorporated into polypropylene was synthesized and used for FR applications (*41*).

Silica-Based FRs

In FR material, silicone-based products are famous, as they form shielding exterior coatings while burning. The silicone in PU has been reported to advance its LOI and standard of safety flammability performance (UL-94). The use of halogen-based FR has been restricted in industry. Therefore, silica-based FRs are captivating compared with halogen-based FRs. Silica-based FRs can be synthesized from naturally occurring precursors, like rice husk ash, with high surface areas. Brannum and coworkers explored the use of cheap and ecologically benign ingredients such as *Cladophora* spp. algae coated with silica diatoms. This was a silica sol–gel method for the synthesis of silica nanocoated FR-PUs. This material has improved FR properties, like the fire resistance of common PUs, and can be self-extinguishable. Interestingly, the mechanical properties of the final foam are sustained during synthesis and show moderate flammability. These properties are supported by LOI and UL-94 tests (*42*). Time-lapse photos of the burning of silica-treated PU foam, with approximately 15 s of time lapse, are shown in Figure 9.

Figure 9. Time-lapse (~15 s) photos of the burning of silica-treated PU foam. Adapted with permission from reference (42). Copyright 2019 American Chemical Society.

The maintenance of mechanical properties is important in FR-PUs for various applications. Melamine is known for its flame retardancy in PUs. FR-PU foam with melamine, silica, and ionic liquid has been reported for FRs with good mechanical properties (*43*). The LOI has been increased by the addition of 15% SiO_2 nanomaterials with phenethyl-bridged 9,10-dihydro-9-oxa-10-phosphaphenanthrene-10-oxide (*44*). This material is further tested for char-forming agents in FRs.

Synergistic Effects of Metal Oxides

It has been well established that the combination of two or more metal oxides may create a synergistic effect and offer a surprising improvement in fire resistance for PUs. Unaccompanied Sb_2O_2 does not show flame retardancy, but its compounds with Br and Cl give rise to its effects. Different synergetic systems with various metal oxides, phosphorous, nitrogen, and hydroxides of metal have been commercialized. A few interesting examples have been reviewed in this section of the chapter. Combinations of metal oxides, such as molybdenum trioxide and cuprous oxide with graphene, lead to a better synergetic effect for fire retardancy (*25*). The synergistic effect in laser marking has been accomplished by the presence of Bi_2O_3 with TPU composites. This effect was because of carbonization via thermoplastic thermal decomposition and reduction of bismuth oxide to black bismuth metal (Figure 10) (*45*).

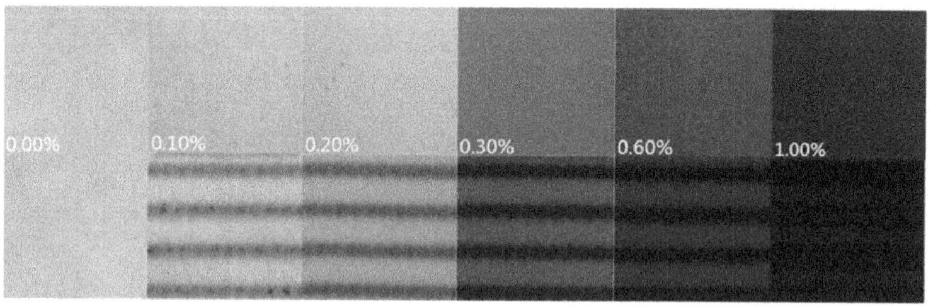

Figure 10. Laser marking stereomicroscope micrographs of Bi_2O_3 with TPU composites before (top row) and after (bottom row). Adapted with permission from reference (45). Copyright 2015 American Chemical Society.

Over the past decade, many researchers have contributed to the development of the synergistic effect of metal chalcogenides (especially oxides) in FR systems. Magnesium oxide, zinc oxide, and iron(III) oxide have been reported for FR properties (*46*). In ternary metal oxide, magnesium oxide shows an effective synergistic agent, which increases the char residue of FRs by 25%. The improved flame retardancy was observed in iron and lanthanum containing $LaFeO_3$ (*47*). The synergy effect concerning smoke suppression and low toxicity has also been reported with carbon-encapsulated Cu@C, Fe_2O_3@C, and Cu_2O@Fe_2O_3@C (*48*). The synergetic effect in FR properties has been increased in the mentioned hybrid material because of the presence of various metals with carbon.

Conclusion and Future Outlooks

This chapter gave a general introduction of FR-PUs and their types. It also highlighted the importance of metal oxide-based PUs and their flammability. The literature studies about the mechanisms for organic, inorganic, and carbon-based FRs were reported. This chapter illustrated FR-PUs based in transition metal oxides, magnesium hydroxide, MOFs, aluminum, magnesium, zinc oxide, silica, and synergistic metal oxides. An important area of research of FR-PU–like effects of different loadings of the fillers—the action mechanism of adding the synergists on FRs—was covered in this chapter. The importance of LOI and thermogravimetric analysis in the formation of char during the burning process of PUs was illustrated wherever necessary. The impacts of graphene sheets and MOFs, along with metal oxides, shows a synergistic effect, which was illustrated as smoke suppression and other properties. Metal oxide-based PU is an emerging area in FR additive materials that involves solid-state chemistry, and it provokes either char generation or efficient vapor-phase free radical inhibition for burning PUs.

Because of the massive consumption of PUs, their safety-related concerns—such as smoke generation, toxicity, and toxic gases during a fire—are important. The future of FR-PUs is extremely attractive and is based on that material and how they might have less impact on the environment. An inventive synthetic root should be attempted for the synthesis of greener additives for the commercial manufacture of FRs. The future generation of FRs will have to surpass barriers, such as cost and high yield, without impacting other properties.

References

1. Bayer, O. Das Di-Isocyanat-Polyadditionsverfahren (Polyurethane). *Angew. Chemie* **1947**, *59*, 257–272.
2. Raspoet, G.; Nguyen, M. T.; McGarraghy, M.; Hegarty, A. F. The Alcoholysis Reaction of Isocyanates Giving Urethanes: Evidence for a Multimolecular Mechanism. *J. Org. Chem.* **1998**, *63*, 6878–6885.
3. Yilgör, I.; Yilgör, E.; Wilkes, G. L. Critical Parameters in Designing Segmented Polyurethanes and Their Effect on Morphology and Properties: A Comprehensive Review. *Polymer (Guildf).* **2015**, *58*, A1–A36.
4. Akindoyo, J. O.; Beg, M. D. H.; Ghazali, S.; Islam, M. R.; Jeyaratnam, N.; Yuvaraj, A. R. Polyurethane Types, Synthesis and Applications-a Review. *RSC Adv.* **2016**, *6*, 114453–114482.
5. M. de Souza, F.; Choi, J.; Bhoyate, S.; Kahol, P. K.; Gupta, R. K. Expendable Graphite as an Efficient Flame-Retardant for Novel Partial Bio-Based Rigid Polyurethane Foams. *C — J. Carbon Res.* **2020**, *6*, 27.
6. Vahabi, H.; Rastin, H.; Movahedifar, E.; Antoun, K.; Brosse, N.; Saeb, M. R. Flame Retardancy of Bio-Based Polyurethanes: Opportunities and Challenges. *Polymers* **2020**, *12* (6), 1234MDPI AG.
7. Altun, Y.; Doğan, M.; Bayramlı, E. The Effect of Red Phosphorus on the Fire Properties of Intumescent Pine Wood Flour - LDPE Composites. *Fire Mater.* **2016**, *40*, 697–703.
8. Lewin, M. Synergism and Catalysis in Flame Retardancy of Polymers. *Polym. Adv. Technol.* **2001**, *12*, 215–222.
9. Palve, A. M.; Gupta, R. K. Polyurethane-Based Nanocomposites and Their Applications. In *Polyurethane Chemistry: Renewable Polyols and Isocyanates*; Gupta, R. K., Kahol, P. K., Eds.; ACS Symposium Series; American Chemical Society: Washington, D.C., 2021; Vol. 1380, pp 225–255.
10. You, G.; Cheng, Z.; Tang, Y.; He, H. Functional Group Effect on Char Formation, Flame Retardancy and Mechanical Properties of Phosphonate-Triazine-Based Compound as Flame Retardant in Epoxy Resin. *Ind. Eng. Chem. Res* . **2015**, *54*, 7309–7319.

11. Dong, X.; Yang, J.; Hua, X.; Nie, S.; Kong, F. Synthesis of a Novel Char-Forming Agent (PEIC): Improvement in Flame Retardancy, Thermal Stability, and Smoke Suppression for Intumescent Flame-Retardant Polypropylene Composites. *J. Appl. Polym. Sci.* **2020**, *137*, 1–11.
12. Zhu, H.; Xu, S. Preparation of Flame-Retardant Rigid Polyurethane Foams by Combining Modified Melamine–Formaldehyde Resin and Phosphorus Flame Retardants. *ACS Omega* **2020**, *5*, 9658–9667.
13. Horacek, H.; Grabner, R. Advantages of Flame Retardants Based on Nitrogen Compounds. *Polym. Degrad. Stab.* **1996**, *54*, 205–215.
14. Huo, S.; Yang, S.; Wang, J.; Cheng, J.; Zhang, Q.; Hu, Y.; Ding, G.; Zhang, Q.; Song, P. A Liquid Phosphorus-Containing Imidazole Derivative as Flame-Retardant Curing Agent for Epoxy Resin with Enhanced Thermal Latency, Mechanical, and Flame-Retardant Performances. *J. Hazard. Mater.* **2020**, *386*, 121984.
15. Vothi, H.; Nguyen, C.; Pham, L. H.; Hoang, D.; Kim, J. Novel Nitrogen-Phosphorus Flame Retardant Based on Phosphonamidate: Thermal Stability and Flame Retardancy. *ACS Omega* **2019**, *4*, 17791–17797.
16. Sykam, K.; Meka, K. K. R.; Donempudi, S. Intumescent Phosphorus and Triazole-Based Flame-Retardant Polyurethane Foams from Castor Oil. *ACS Omega* **2019**, *4*, 1086–1094.
17. He, L.; Zhou, X.; Cai, W.; Xiao, Y.; Chu, F.; Mu, X.; Fu, X.; Hu, Y.; Song, L. Electrochemical Exfoliation and Functionalization of Black Phosphorene to Enhance Mechanical Properties and Flame Retardancy of Waterborne Polyurethane. *Compos. Part B Eng.* **2020**, *202*, 108446.
18. Bhoyate, S.; Ionescu, M.; Radojcic, D.; Kahol, P. K.; Chen, J.; Mishra, S. R.; Gupta, R. K. Highly Flame-Retardant Bio-Based Polyurethanes Using Novel Reactive Polyols. *J. Appl. Polym. Sci.* **2018**, *135*, 46027.
19. Farooq, M.; Sipponen, M. H.; Seppälä, A.; Österberg, M. Eco-Friendly Flame-Retardant Cellulose Nanofibril Aerogels by Incorporating Sodium Bicarbonate. *ACS Appl. Mater. Interfaces* **2018**, *10*, 27407–27415.
20. Han, Y.; Zhang, X.; Wu, X.; Lu, C. Flame Retardant, Heat Insulating Cellulose Aerogels from Waste Cotton Fabrics by in Situ Formation of Magnesium Hydroxide Nanoparticles in Cellulose Gel Nanostructures. *ACS Sustain. Chem. Eng.* **2015**, 3, 1853–1859.
21. He, C.; Huang, J.; Li, S.; Meng, K.; Zhang, L.; Chen, Z.; Lai, Y. Mechanically Resistant and Sustainable Cellulose-Based Composite Aerogels with Excellent Flame Retardant, Sound-Absorption, and Superantiwetting Ability for Advanced Engineering Materials. *ACS Sustain. Chem. Eng.* **2018**, *6*, 927–936.
22. Wicklein, B.; Kocjan, A.; Salazar-Alvarez, G.; Carosio, F.; Camino, G.; Antonietti, M.; Bergström, L. Thermally Insulating and Fire-Retardant Lightweight Anisotropic Foams Based on Nanocellulose and Graphene Oxide. *Nat. Nanotechnol.* **2015**, *10*, 277–283.
23. Wang, D.; Peng, H.; Wu, Y.; Zhang, L.; Li, M.; Liu, M.; Zhu, Y.; Tian, A.; Fu, S. Bioinspired Lamellar Barriers for Significantly Improving the Flame-Retardant Properties of Nanocellulose Composites. *ACS Sustain. Chem. Eng.* **2020**, *8*, 4331–4336.
24. Morgan, A. B. A Review of Transition Metal-Based Flame Retardants: Transition Metal Oxide/ Salts, and Complexes. In *Fire and Polymers V*; ACS Symposium Series; American Chemical Society: Washington, D.C., 2009; Vol. 1013.

25. Xu, W.; Liu, L.; Zhang, B.; Hu, Y.; Xu, B. Effect of Molybdenum Trioxide-Loaded Graphene and Cuprous Oxide-Loaded Graphene on Flame Retardancy and Smoke Suppression of Polyurethane Elastomer. *Ind. Eng. Chem. Res.* **2016**, *55*, 4930–4941.
26. Wu, N.; Yang, R. Effects of Metal Oxides on Intumescent Flame-Retardant Polypropylene. *Polym. Adv. Technol.* **2011**, *22*, 495–501.
27. Pan, Y.; Zhang, Z.; Yang, R. The Rise of MOFs and Their Derivatives for Flame Retardant Polymeric Materials: A Critical Review. *Compos. Part B* **2020**, *199*, 108265.
28. Wang, H.; Qiao, H.; Guo, J.; Sun, J.; Li, H.; Zhang, S.; Gu, X. Preparation of Cobalt-Based Metal Organic Framework and Its Application as Synergistic Flame Retardant in Thermoplastic Polyurethane (TPU). *Compos. Part B* **2019**, *182* (June), 107498107498.
29. Zhang, L.; Chen, S.; Pan, Y. T.; Zhang, S.; Nie, S.; Wei, P.; Zhang, X.; Wang, R.; Wang, D. Y. Nickel Metal-Organic Framework Derived Hierarchically Mesoporous Nickel Phosphate toward Smoke Suppression and Mechanical Enhancement of Intumescent Flame Retardant Wood Fiber/Poly(Lactic Acid) Composites. *ACS Sustain. Chem. Eng.* **2019**, 7, 9272–9280.
30. Hou, Y.; Hu, W.; Gui, Z.; Hu, Y. Preparation of Metal-Organic Frameworks and Their Application as Flame Retardants for Polystyrene. *Ind. Eng. Chem. Res.* **2017**, *56*, 2036–2045.
31. Sang, L.; Cheng, Y.; Yang, R.; Li, J.; Kong, Q.; Zhang, J. Polyphosphazene-Wrapped Fe–MOF for Improving Flame Retardancy and Smoke Suppression of Epoxy Resins. *J. Therm. Anal. Calorim.* **2021**, *144*, 51–59.
32. Wang, X.; Wang, S.; Wang, W.; Li, H.; Liu, X.; Gu, X.; Bourbigot, S.; Wang, Z.; Sun, J.; Zhang, S. The Flammability and Mechanical Properties of Poly (Lactic Acid) Composites Containing Ni-MOF Nanosheets with Polyhydroxy Groups. *Compos. Part B Eng.* **2020**, *183*, 107568.
33. Chen, Z.; Chen, T.; Yu, Y.; Zhang, Q.; Chen, Z.; Jiang, J. Metal-Organic Framework MIL-53 (Fe)@C/Graphite Carbon Nitride Hybrids with Enhanced Thermal Stability, Flame Retardancy, and Smoke Suppression for Unsaturated Polyester Resin. *Polym. Adv. Technol.* **2019**, *30*, 2458–2467.
34. Jin, J.; Dong, Q. X.; Shu, Z. J.; Wang, W. J.; He, K. Flame Retardant Properties of Polyurethane/Expandable Praphite Composites. *Procedia Eng.* **2014**, *71*, 304–309.
35. Liu, H.; Zhang, B.; Han, J. Flame Retardancy and Smoke Suppression Properties of Flexible Polyurethane Foams Containing an Aluminum Phosphate Microcapsule. *RSC Adv.* **2017**, 7, 35320–35329.
36. Liu, Y.; He, J.; Yang, R. Effects of Dimethyl Methylphosphonate, Aluminum Hydroxide, Ammonium Polyphosphate, and Expandable Graphite on the Flame Retardancy and Thermal Properties of Polyisocyanurate-Polyurethane Foams. *Ind. Eng. Chem. Res.* **2015**, 54, 5876–5884.
37. Ai, L.; Chen, S.; Zeng, J.; Yang, L.; Liu, P. Synergistic Flame Retardant Effect of an Intumescent Flame Retardant Containing Boron and Magnesium Hydroxide. *ACS Omega* **2019**, 4, 3314–3321.
38. Wang, C.; Wang, Y.; Han, Z. Enhanced Flame Retardancy of Polyethylene/Magnesium Hydroxide with Polycarbosilane. *Sci. Rep.* **2018**, *8*, 1–10.
39. Wang, Y. W.; Shen, R.; Wang, Q.; Vasquez, Y. ZnO Microstructures as Flame-Retardant Coatings on Cotton Fabrics. *ACS Omega* **2018**, 3, 6330–6338.

40. Xu, W.; Wang, G.; Zheng, X. Research on Highly Flame-Retardant Rigid PU Foams by Combination of Nanostructured Additives and Phosphorus Flame Retardants. *Polym. Degrad. Stab.* **2015**, *111*, 142–150.
41. Guo, J.; Liu, G.; Guo, Y.; Tian, L.; Bao, X.; Zhang, X.; Yang, B.; Cui, J. Enhanced Flame Retardancy and Smoke Suppression of Polypropylene by Incorporating Zinc Oxide Nanowires. *J. Polym. Res.* **2019**, *26*, 19.
42. Brannum, D. J.; Price, E. J.; Villamil, D.; Kozawa, S.; Brannum, M.; Berry, C.; Semco, R.; Wnek, G. E. Flame-Retardant Polyurethane Foams: One-Pot, Bioinspired Silica Nanoparticle Coating. *ACS Appl. Polym. Mater.* **2019**, *1*, 2015–2022.
43. Członka, S.; Strąkowska, A.; Strzelec, K.; Kairytė, A.; Kremensas, A. Melamine, Silica, and Ionic Liquid as a Novel Flame Retardant for Rigid Polyurethane Foams with Enhanced Flame Retardancy and Mechanical Properties. *Polym. Test.* **2020**, *87*, 106511.
44. Wang, K.; Liu, H.; Wang, C.; Huang, W.; Tian, Q.; Fu, Q.; Yan, W. Flame-Retardant Performance of Epoxy Resin Composites with SiO2 Nanoparticles and Phenethyl-Bridged DOPO Derivative. *ACS Omega* **2021**, *6*, 666–674.
45. Zhong, W.; Cao, Z.; Qiu, P.; Wu, D.; Liu, C.; Li, H.; Zhu, H. Laser-Marking Mechanism of Thermoplastic Polyurethane/Bi2O3 Composites. *ACS Appl. Mater. Interfaces* **2015**, *7*, 24142–24149.
46. Lin, M.; Li, B.; Li, Q.; Li, S.; Zhang, S. Synergistic Effect of Metal Oxides on the Flame Retardancy and Thermal Degradation of Novel Intumescent Flame-Retardant Thermoplastic Polyurethanes. *J. Appl. Polym. Sci.* **2011**, *121*, 1951–1960.
47. Wang, L.; Yang, W.; Wang, B.; Wu, Y.; Hu, Y.; Song, L.; Yuen, R. K. K. The Impact of Metal Oxides on the Combustion Behavior of Ethylene-Vinyl Acetate Coploymers Containing an Intumenscent Flame Retardant. *Ind. Eng. Chem. Res.* **2012**, *51*, 7884–7890.
48. Xu, Z.; Duan, L.; Hou, Y.; Chu, F.; Jiang, S.; Hu, W.; Song, L. The Influence of Carbon-Encapsulated Transition Metal Oxide Microparticles on Reducing Toxic Gases Release and Smoke Suppression of Rigid Polyurethane Foam Composites. *Compos. Part A Appl. Sci. Manuf.* **2020**, *131*, 105815.

Chapter 9

Improved Flame Retardancy in Polyurethanes Using Layered Double Hydroxides

Henri Vahabi,[*,1] Elnaz Movahedifar,[1] Maryam Jouyandeh,[1] Mohammad Reza Saeb,[1] and Sabu Thomas[2]

[1]Université de Lorraine, CentraleSupélec, LMOPS, F-57000 Metz, France

[2]International and Inter University Centre for Nanoscience and Nanotechnology, School of Energy Materials, Mahatma Gandhi University, Kottayam 686560, India

[*]Email: henri.vahabi@univ-lorraine.fr

Clays are inexpensive fillers and additives extensively used in polymer systems because of their low risk for health and the environment. They can be found in nature in large quantities with diverse types, shapes, and chemical structures. Layered double hydroxides (LDHs) are among the most promising clays to improve thermal stability and flame retardancy of polymers ranging from thermoplastics to thermosets. LDH platelets are usually combined with conventional flame retardants, mainly phosphorus ones, resulting in synergistic effects. The dispersion state of LDH in polymers is a serious challenge, such that both the flame-retardant and mechanical properties of the resulting nanocomposites should be optimized for practical use. Therefore, chemical modification of LDH has been considered to improve the intercalation of organic molecules in the presence of LDH via enlargement of clay galleries. Polyurethanes (PUs) are finding new dimensions year by year, and their composites with clays are becoming the most important parts of fire protection engineering ranging from automotive uses and coatings to construction and biomedical applications. However, the high flammability of PU limits its use in advanced systems. In this chapter, the mode of action of LDH in flame retardancy, as well as its ability to reduce the flammability of PU, are discussed through a survey of the literature.

Introduction

As in other areas of research and applications, the concept of nanotechnology has led to continued developments in the field of flame retardancy, both academically and industrially. A nanofiller is defined as a nano-object with a length ranging between 0 and 100 nm in at least one dimension (*1–3*). Various nanofillers have been employed in polymers to improve their flame retardancy, which is systematically classified in terms of performance based on the dimensionless Flame Retardancy Index (*4*). Different types of nanoclay (*5–8*), carbon nanotubes (CNTs) (*9*), graphene (*10–12*), polyhedral oligomeric silsesquioxanes (*13–15*), and metallic oxide nanoparticles (*16–20*) have been examined as flame retardants (FRs). Finding the best nanoparticle for flammability prevention in polymers depends on several material, processing, and economic factors. Sometimes a combination of two or more FRs is required to achieve a practical level of flame retardancy.

Nanoparticles can enhance the FR properties of polymers in different ways. During combustion, they may retard the thermal degradation of polymers or form a char or residue protective layer that blocks the transfer of heat and mass as well as gas and oxygen by decelerating the mobility of polymer chains (*21*). Nanoparticles also can reduce heat conductivity and diffusivity of polymer materials or can bring about catalytic action to facilitate char formation (*22*). To achieve a satisfactory level of flame retardancy performance, conventional FRs other than nanofillers should be used. This leads to a substantial improvement in flame retardancy in terms of cone calorimetry parameters, shown by significant reduction in the value of heat release rate (HRR). On the other hand, the total heat release (THR) is not governed by nanofillers, and this is a signature of slow burning of the material after polymer ignition (*23, 24*).

Among nanofillers, layered double hydroxides (LDHs) represent a promising class of layered materials to be used in the flame retardancy of polymers (*25–28*). Generally, there are two classes of LDH clays, namely cationic clays (also known as clay minerals) and anionic clays (or LDH family minerals) (*29, 30*). The clay minerals are cationic layers negatively charged with alumina–silicate layers in which small cations occupy the space between layers that keeps the charge of the molecule neutral (*31–33*). By contrast, anionic ones with positive charges between brucitelike metal hydroxide layers are balanced, and water molecules occupy interlayer space. The guest species can change their arrangement and exchange their charges between clay layers, which determines order–disorder transformation degree. This makes it possible to have diverse LDH structures (*34–36*). Figure 1 compares anionic and cationic clay structures.

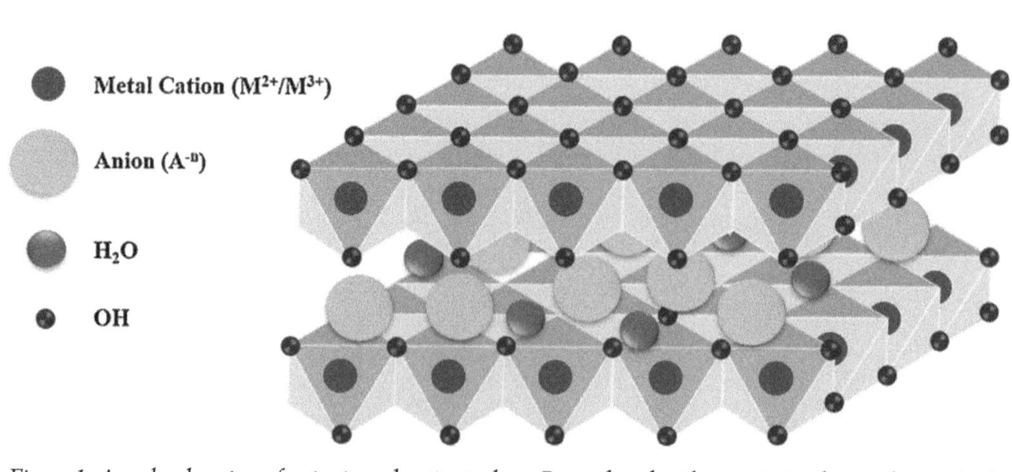

Figure 1. A molecular view of anionic and cationic clays. Reproduced with permission from reference (37). Copyright 2019 Elsevier.

The following in eq 1 is typical of LDH structure:

$$[M^{2+}{}_{1-x}M^{3+}{}_{x}(OH)_2]^{x+}\,(A^{n-})_{x/n}.yH_2O \quad (1)$$

where M^{2+} is representative of a divalent metal ion (such as Mg^{2+}, Ca^{2+}, or Zn^{2+}), M^{3+} is a trivalent metal ion (such as Al^{3+}, Cr^{3+}, Fe^{3+}, or Co^{3+}), and A^{n-} is an anion (such as Cl^-, $CO_3{}^{2-}$, or $NO_3{}^-$) (*38, 39*).

Figure 2 shows a view of compositional changes in the LDH structure in which one can achieve a variety of structures by changing metals in the brucitelike layer of $[M^{2+}{}_{1-x}M^{3+}{}_{x}(OH)_2]^{x+}$ as well as the basal spacing *d*. The use of an intercalating agent in interlamellar space *l* forms new LDH structures.

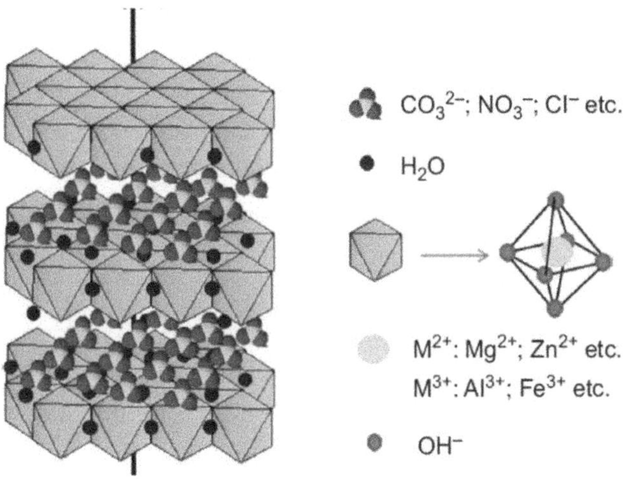

Figure 2. Compositional view of an LDH structure. Reproduced with permission from reference (40). Copyright 2017 Elsevier.

By using an intercalating agent, intercalated LDH with a general formula expressed by eq 2 is formed:

$$[M^{2+}{}_{1-x}M^{3+}{}_{x}(OH)_2]^{x+}\,[(A^{n-})_{x/n}.yH_2O]^{x-} \quad (2)$$

with A^{n-} as the intercalating anion(s).

LDH is used in different applications including biochemistry (*41*), catalysts (*42*), curing of thermoset resins (*31, 36*), ion exchange (*43*), drug delivery (*44*), and photochemistry (*45*) Figure 3.

Polymer/LDH composites have different applications (*46, 47*). The key parameter in the preparation of polymer/LDH composites is the state of dispersion of LDH inside the polymer matrix. The presence of a large number of hydroxyl groups on the surface of LDHs makes them hydrophilic, a negative parameter for dispersion in highly hydrophobic polymers. Strong electrostatic interactions exist between the hydroxide layers as well as between the intercalated anions and water molecules, which makes intercalation or exfoliation of LDH in polymers cumbersome (*48*). Generally, in situ polymerization and melt blending are used to prepare polymer/LDH composites (*49*). However, LDH can be used as a top-coating layer to resist flames.

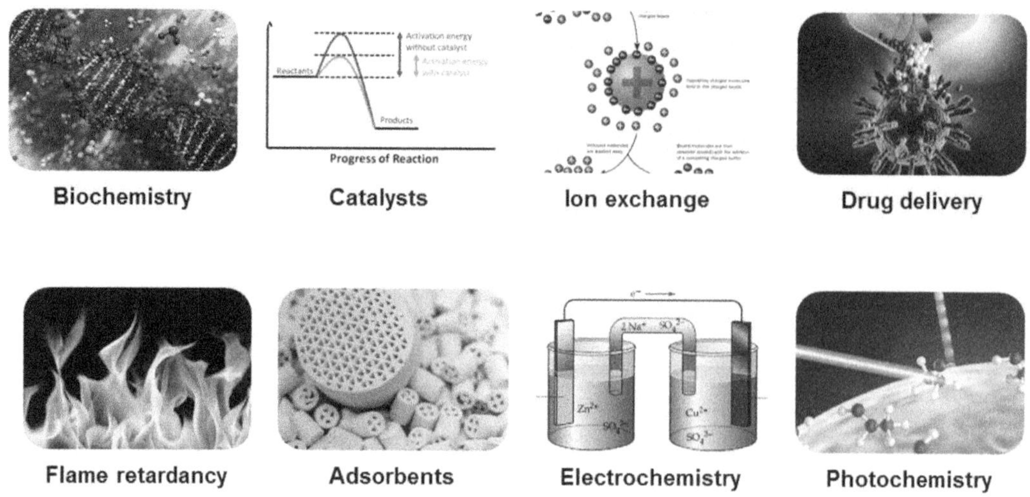

Figure 3. Applications of LDH including biochemistry, catalysts, ion exchange, drug delivery, flame retardancy, adsorbents, electrochemistry, and photochemistry.

LDHs also have been used as FR systems in several polymers including epoxy (*50, 51*), polypropylene (*52–54*), polyethylene (*55, 56*), polystyrene (*57, 58*), ethylene vinyl acetate copolymer (*59, 60*), and acrylonitrile–butadiene–styrene (*61*). The mode of action of LDH is relatively similar to that of other nanofillers and includes the endothermic reaction during its decomposition; the release of water and carbon monoxide, which dilutes the gas phase; the formation of a protecting barrier against heat; the diffusion of oxygen into the material; and the release of gases to the gas phase. The introduction of some elements such as Zn^{2+} in the form of cation inside the LDH layer can promote polymer charring (*62–64*). Compared with other members of the clay family, LDH can more efficiently improve the flame retardancy of polymers. For instance, halloysite nanotube incorporation into epoxy was less effective in terms of peak of HRR (pHRR) compared with LDH. The tensile strength of the sample containing LDH was higher than that of the one containing Halloysite nanotube (*65–68*).

As mentioned earlier, the use of LDH only is not sufficient to improve the fire behavior of polymers. Its combination with conventional FRs or its chemical modification (*40*) to introduce chemicals such as phosphorus-based molecules is necessary. Various fillers and methods have been used (*69, 70*) to improve the flame retardancy of different types of polyurethanes (PUs) (*71–73*). This chapter discusses developments in the use of various LDHs, their chemical modification with FR elements, their combination with conventional FRs, and their use in coating systems for improving the flame retardancy of PUs.

Thermal Decomposition of LDHs

Because there are a multitude of LDHs, their thermal decomposition pathway can change depending on their nature, crystallinity, cations molar ratio (M^{2+}/M^{3+}), and interlayer anion (*74*). However, two steps of decomposition can be detected for all LDHs. The first step, from a low temperature up to 200 °C, corresponds to the release of adsorbed and interlamellar water and the release of organic molecules if there are any. Afterward, the second step begins and is related to lamellar hydroxyl group (dehydroxylation) loss and also anion loss (*36, 75*). Figures 4a and 4b display the thermogravimetric analysis (TGA) curves for two types of LDHs—MgAl-CO_3-LDH (a) and ZnAl-LDH (b). Figure 4b shows two stages of decomposition at the beginning of the curve up to 200 °C, which are related to the loss of water (76). The remaining residue at the end of the TGA test are between 50 and 60 wt %.

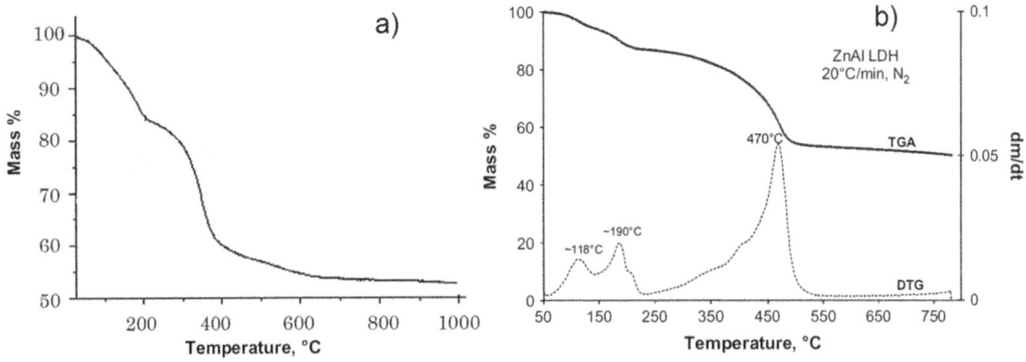

Figure 4. TGA curves of two LDHs. (a) MgAl-CO_3-LDH. (b) ZnAl-LDH (74, 76). Part b: Adapted with permission from reference (76). Copyright 2008 Elsevier.

It is clear that the chemicals in LDH can modify its degradation pathway. For example, Figure 5a shows the TGA curves for three modified LDHs containing nitrogen, phosphorus, silane, and nitrate groups (77). Thermal stability and the rate of remaining residue are significantly modified. Figure 5b displays the TGA curves of two LDHs modified with two calix[4]arenes (62). The onset temperature and the rate of remaining residue can be decreased or increased by the chemical used in the modification of LDH.

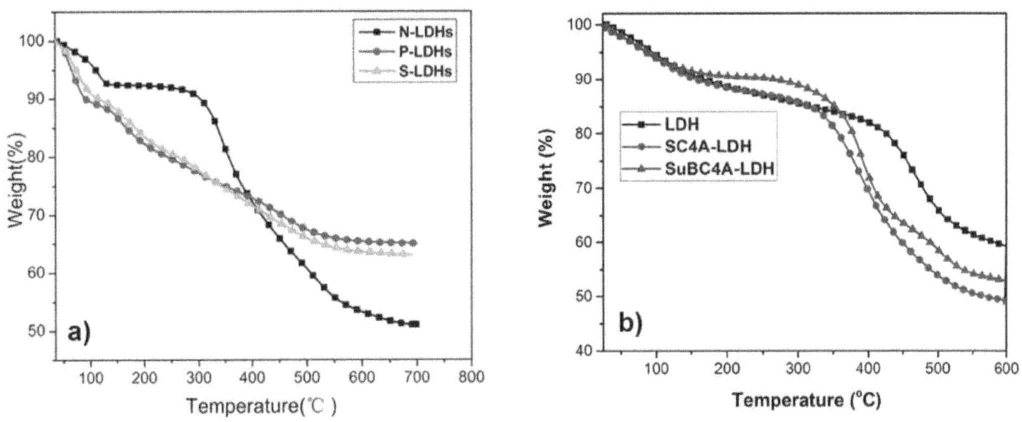

Figure 5. TGA curves of three LDHs modified with (a) phosphorus and (b) sodium p-sulfonatocalix[4]arene (SC4A) and sodium tetrakis O-(4-butoxysulfonato)-p-tert-butyl calix[4]arene (SuBC4A). (a) Reproduced with permission from reference (77). Copyright 2016 American Chemical Society. (b) Reproduced with permission from reference (78). Copyright 2019 Elsevier.

Various Types of LDHs Used in PU for FR Purposes

A wide variety of LDH structures have been used in the preparation of FR polymers. Table 1 displays various types of LDHs used for flame retardancy in PUs. Magnesium–aluminum LDH (MgAl–LDH) is the LDH used most for FR purposes. One reason for the abundance of this particular LDH in different studies can be related to its ease of synthesis and its availability. Figure 6 shows the effect of different LDHs (modified or unmodified) on the reduction of pHRR obtained in cone calorimeter tests, with respect to pure PU, in the function of its loading percentage. The efficiency of LDH in the reduction of pHRR is important even at a low loading percentage. However, the modification of LDH with some chemicals as well as its physical combination with conventional FRs, especially phosphorus ones, are necessary to achieve a high reduction of pHRR. Numerous works have been dedicated to these issues. A summary of these works will be presented in three sections: chemically modified LDHs, combinations of LDH and conventional FRs, and the use of LDH in coating systems to improve the flame behavior of PU.

Table 1. LDHs and Modified LDHs Used To Improve the Flame Retardancy of PUs

Type of LDH	*Ref.*
NO_3-MgAl–LDH	*(79)*
Heptamolybdate (Mo_7O_{24})-intercalated MgAl–LDH	*(79)*
MgAl–LDH with nitrate in its interlayer	*(77)*
MgAl–LDH-containing tripolyphosphate (P-LDH)	*(77)*
Aminopropyltriethoxysilane (APTS) grafted onto the surface of P_3O_{10}–MgAl–LDH	*(77)*
MgAl–LDH-loaded graphene hybrid	*(80)*
NO_3MgAl–LDH	*(48)*
Sodium dodecyl sulfate (SDS)–modified NO_3–MgAl–LDH	*(48)*
SC4A-modified MgAl–LDH	*(78)*
SuBC4A-modified MgAl–LDH	*(78)*
Lanthanum-doped MgAl–LDH (La–MgAl–LDH)	*(81)*
Sodium dodecyl sulfate–modified CoAl–LDH	*(82)*
Ammonium polyphosphate (APP)@CoAl–LDH	*(83)*
Stearate-intercalated MgAl–LDH	*(84)*

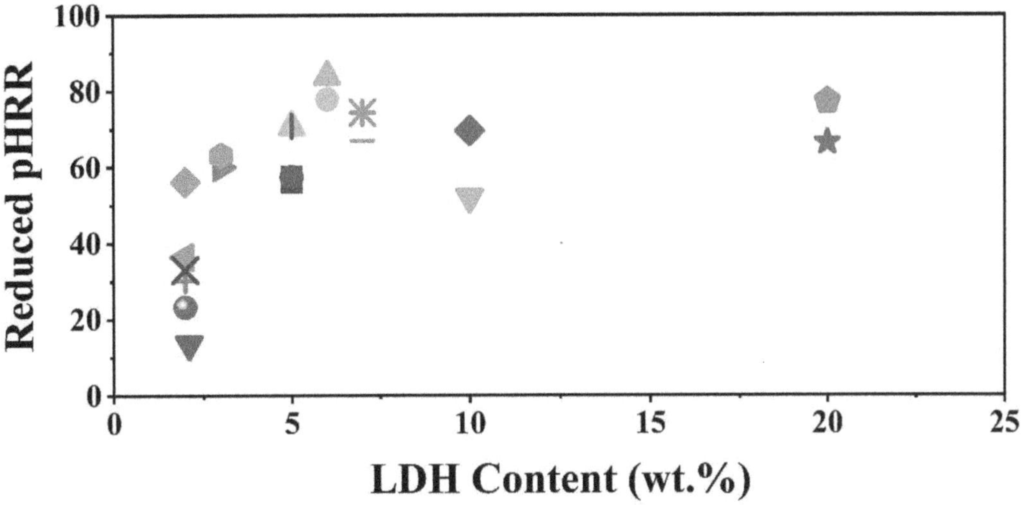

Figure 6. Reduced pHRR rate for PUs filled with LDH wt %, with respect to the pure PU. The square on the left represents N–MgAl–LDH (61), the circle in the middle represents P3O10–MgAl–LDH (61), the triangle in the center (with the line running through) represents APTS–P3O10–MgAl–LDH (61), the triangle pointed downward in the center represents NO3–MgAl–LDH (63), the diamond in the center represents Mo-MgAl-LDH (63), the triangle pointed sideways to the left represents MgAl–LDH (64), the triangle pointing sideways to the right represents reduced GO (RGO)–MgAl–LDH (64), the hexagon represents Mo–RGO–MgAl–LDH (64), the star represents SDS–NO3–MgAl–LDH (69), the pentagon represents SDS–NO3–MgAl–LDH/graphene oxide (GO) (69), the circle on the bottom represents MgAl–LDH (65), the plus sign (below the "x") represents La–MgAl–LDH (65), the "x" represents La–MgAl–LDH/GO (65), the asterisk represents APP–6@CoAl–LDH (67), the solid sideways line represents APP/CoAl–LDH (67), the solid vertical line represents P–CoAl–LDH (70), the circle on the top represents phosphoryl polyethyleneimine amide (PPEIA)–LDH (71), the top triangle pointed upward represents PPEIA–LDH (71), the bottom triangle pointed downward represents CNT/ZnAl–LDH (72), and the diamond on the left represents Sb–LDH (73).

Chemically Modified LDH Used in PUs

A survey of the literature shows that chemical modification of LDH is usually performed in order to improve its dispersion in PU or improve the flame retardancy. For the second case, the chemicals added to the interlayer of LDH are commonly phosphorus-, nitrogen-, or silicon-based molecules. Some of these studies are reviewed from a flame-retardancy performance point of view. Table 2 displays a summary of the results obtained from fire tests on PU containing modified LDH. Xu et al. (79) introduced $Mo_7O_{24}^{6-}$ into the interlayer space of NO_3–MgAl–LDH (Mo–MgAl–LDH). Mo–MgAl–LDH/PU elastomer (PUE) composite was prepared by in situ polymerization. The authors showed that the incorporation of MgAl–LDH and Mo–MgAl–LDH at 10 wt % significantly decreased pHRR in a cone calorimeter test, 52 and 69% respectively, with respect to pure PU. The barrier effect and dilution of gas phases explained the improvement in flame retardancy of PU in the presence of these LDHs. This study also demonstrated that the mechanical properties (tensile strength) of samples increased by increasing LDH loading percentage up to 3 wt % and afterward decreased, for both pure and modified LDH. This behavior was explained by the exfoliation of LDH nanolayers at a low percentage, which created a strong interfacial interaction between the hydroxyl groups on the surface of LDH and the polar group on PU macromolecular chains. The agglomeration of LDH at a higher loading percentage could also explain the decrease in mechanical properties.

Three types of modified LDH were prepared by Xu et al. (77) to improve the flame retardancy of PUE. The authors used three methods: the ion exchange method to prepare MgAl–LDH-containing tripolyphosphate ($P_3O_{10}^{5-}$) (P-LDH), the hydrolysis silylation method to graft aminopropyltriethoxysilane (APTS) onto the surface of P-LDH and to prepare APTS grafted onto the surface of P-LDH (S-LDH), and finally, the hydrothermal method to obtain nitrogen-containing MgAl–LDH with nitrate in its interlayers. These LDHs, at 1, 3 and 5 wt %, were used to prepare in situ polymerization of PUE composites. The results obtained in a cone calorimeter test showed that the best results in terms of reduction in pHRR were obtained when 5 wt % of P-LDH and S-LDH were incorporated into PUE, 58 and 70%, respectively, of reduction compared with pure PUE. The characterization of char residue remaining at the end of cone calorimetry allowed the authors to conclude that in the presence of the aforementioned LDH, a graphitic crystalline form of char, was produced instead of a disordered graphite, which was beneficial for the barrier effect during combustion. The presence of some stable structures such as −P(=O)−O−C− and −P(=O)−O−Si− had a positive effect on thermal oxidative resistance of char layer and contributed to improving the flame retardancy of PUE.

Huang et al. (85) used a simple method to modify COAl−LDH with phosphorus elements. Red phosphorus (RP) and LDH were mixed in a ball-milling machine at 700 rpm for 48 h (Figure 7a). Then the unreacted RP was removed by the extraction method. The LDH obtained (P-LDH) was used to prepare thermoplastic PU (TPU) via melt blending at 190 °C. The authors showed that the incorporation of the modified LDH, P-LDH, at 5 wt % led to an increase in the limiting oxygen index (LOI) of 28 as compared with 23 for pure TPU. Also, the UL-94 was rated V-0, whereas it was V-2 for pure TPU. The cone calorimetry results confirmed the aforementioned tendency: the pHRR was significantly decreased in the presence of P-LDH at 5 wt %, from 1623 kW/m^2 for pure TPU to 473 kW/m^2 for TPU/ P-LDH; 5 wt % was formed in the presence of P-LDH. A simple physical combination of RP and LDH was not as efficient as was modified LDH (P-LDH) to improve the flame retardancy behavior. The same research group (86) also investigated the effect of incorporation of a modified LDH by phosphoryl polyethyleneimine amide (PPEIA)–LDH on the flame retardancy of TPU. The incorporation of 6 wt % PPEIA–LDH led to an LOI of 29 and drastically reduced the pHRR to 228 kW/m^2. The condensed phase mechanism was the main reason for the improvement in flame retardancy because of the synergistic effect between LDH and PPEIA to improve the barrier effect.

Figure 7. Schematic of the preparation method of P-LDH (a) and HRR curves (b) obtained in cone calorimeter tests for TPU containing P-LDH and combination of LDH and RP. Reproduced with permission from reference (85). Copyright 2019 American Chemical Society.

Li et al. (*48*) performed the intercalation of graphene oxide (GO) and sodium dodecyl sulfate (SDS) in LDH. According to the authors, SDS has been used to improve the dispersion of LDH inside PU, and GO is able to improve the flame retardancy thanks to its monolayer network of sp^2-hybridsized carbon atoms, which can create strong interactions with the polymer matrix. The modified LDH (SDS-LDH-GO) was incorporated into TPU via melt blending at 20 wt %. They also modified the percentage of GO inside SDS-LDH-GO, from 1 to 5 wt %. The comparison of pHRR in cone calorimeter tests clearly showed the effect of modification of LDH because a decrease of 32% in pHRR was observed in the presence of 20 wt % SDS-LDH-5GO (containing 5 wt % GO) compared with TPU containing 20 wt % unmodified LDH. The modified LDH was rated V-0 in a UL-94 test, whereas it was nonrated for the TPU/unmodified LDH sample. The char residue mass was increased in the presence of modified LDH, which explained a better flame behavior. GO also was used by Xu et al. (*80*) to modify MgAl–LDH. First, the authors oxidized the graphite to obtain GO. Then, reduced GO (RGO)–LDH was obtained via the precipitation method. $Mo_7O_{24}^{6-}$-modified RGO–LDH hybrid (RGO–LDH/Mo) was prepared using the ion exchange method. RGO–LDH/Mo was added at 0.5, 1, 2, and 3 wt % to the reactive to prepare PUE composite (Figure 8). The authors also prepared pure PUE, PUE/RGO, PUE/LDH, and PUE/RGO–LDH samples and compared their fire behavior performance in cone calorimetry tests. The authors claimed that better flame retardancy was obtained in the presence of 3 wt % RGO–LDH/Mo, demonstrated by the reduction of pHRR (63%) compared with pure PUE and explained by the action of graphene and LDH layers in the condensed phase as barrier effect. It was concluded that the degradation of RGO–LDH/Mo generates MoO_3, which promotes char production by catalytic carbonization reaction. Nevertheless, the comparison of all HRR curves showed that the performance of RGO–LDH is quite similar to that of RGO–LDH/Mo.

Calix[4]arene molecules were used for modification of LDH by Mohammadi et al. (*78*). Two types of anionic calix[4]arenes derivatives including sodium *p*-sulfonatocalix[4]arene (SC4A) and sodium tetrakis *O*-(4-butoxysulfonato)-*p*-tert-butyl calix[4]arene (SuBC4A) were synthesized and then intercalated between LDH layers to obtain SC4A–LDH and SuBC4–LDH (Figure 9). The objective of the intercalation of these molecules was to increase the interlayer distance of LDHs in order to improve the dispersion state in PU foam and thus improve the flame retardancy behavior. The PU foam composites were prepared by mixing modified LDH (5 php of polyols) and polyol, and then this blend was added to 4,4′-methylene diphenyl isocyanate, and the polymerization was performed to obtain corresponding PU composites. The dispersion state was improved compared with modified LDH and SC4A–LDH. Consequently, the flame retardancy of samples containing SuBC4–LDH was better than the other ones. The pHRR decreased from 743 kW/m^2 for PU/unmodified LDH to 468 kW/m^2 for PU/SuBC4–LDH. This behavior was explained by a better dispersion of SuBC4–LDH in PU and formation of a better char residue, which enabled the reduction of the release of gases to the gas phase and was a better heat barrier.

Xie et al. (*87*) prepared CNT/ZnAl–LDH filler by embedding CNT into ZnAl–LDH nanosheets using a simple coprecipitation method. Then a PU foam with dimensions of $100 \times 100 \times 25$ mm^3 was immersed into the CNT/ZnAl–LDH suspension several times (two, four, and six times) and dried at 80 °C for 15 min after each immersion. The authors claimed that the PU foam treated six times with CNT/ZnAl–LDH showed a better flame retardancy, demonstrated by the decrease of pHRR to 434 kW/m^2, compared with 502 kW/m^2 for untreated PU. However, the comparison of

HRR curves revealed that the difference between other samples was not important in terms of flame retardancy, and the action of CNT/ZnAl–LDH was not notably relevant.

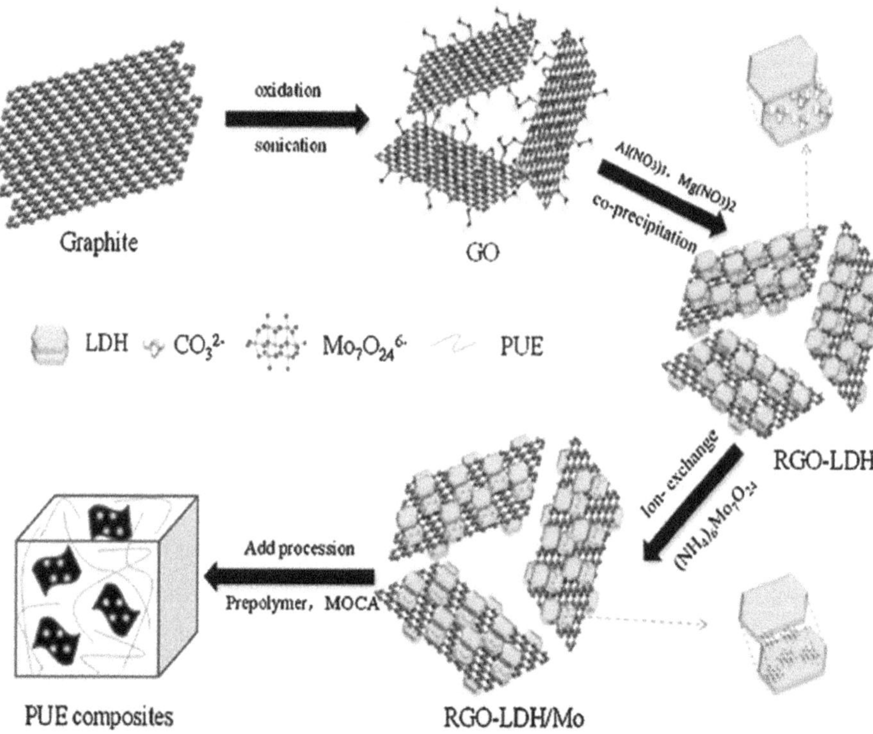

Figure 8. Schematic representation of the preparation of RGO–LDH modified with $Mo_7O_{24}^{6-}$, namely RGO–LDH/Mo, and its incorporation into PUE, 3,3'-dic hloro-4,4'-diaminodiphenyl-methane (MOCA in figure). Reproduced with permission from reference (80). Copyright 2016 Elsevier.

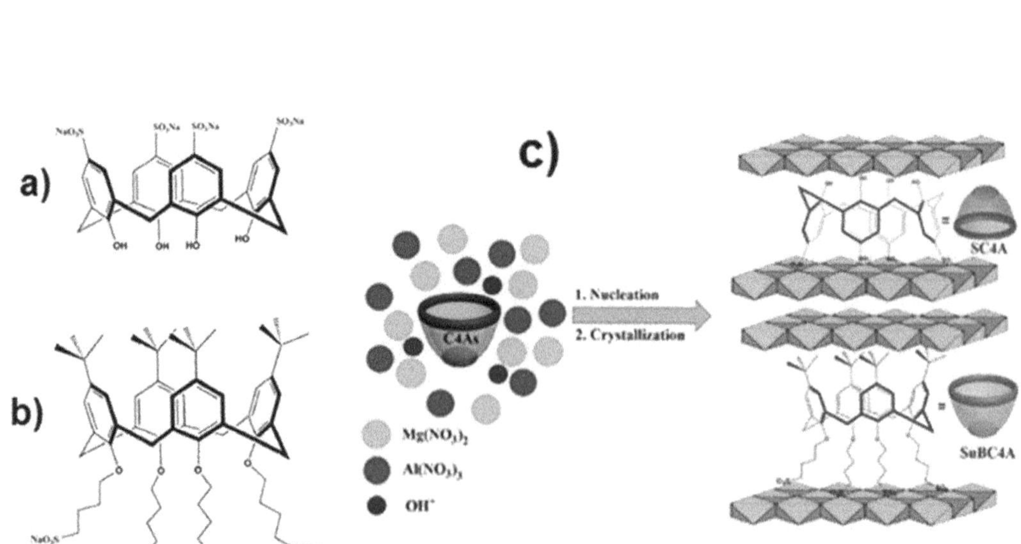

Figure 9. Chemical structure of two calix[4]arenes: (a) SC4A, (b) SuBC4A. (c) Intercalation of these molecules into LDH. Reproduced with permission from reference (78). Copyright 2019 Elsevier.

Table 2. Summary Results of Flame Tests in Which Modified LDH Was Used in PUs[a]

Sample Code	*FR (wt %)*	*Heat Flux (kW/m²)*	*Time to Ignition (TTI) (s)*	*pHRR (kW/m²)*	*THR (MJ/m²)*	*UL-94*	*LOI (%)*	*Ref.*
PUE	0	50	-	913	-	-	-	(77)
PUE/Gal–LDH with nitrate in its interlayer (*N*-MgAl–LDH)	5	50	-	401	-	-	-	(77)
PUE/P-LDH (P_3O_{10}–MgAl–LDH)	5	50	-	389	-	-	-	(77)
PUE/APTS grafted onto the surface of P_3O_{10}–MgAl–LDH	5	50	-	267	-	-	-	(77)
PUE	0	50	-	941	-	-	-	(79)
PUE/NO_3–MgAl–LDH	10	50	-	452	-	-	-	(79)
PUE/Mo_7O_{24}-intercalated MgAl–LDH	10	50	-	286	-	-	-	(79)
PUE	0	50	-	1025	-	-	-	(80)
PUE/MgAl–LDH	2	50	-	651	-	-	-	(80)
PUE/MgAl–LDH–loaded graphene	3	50	-	409	-	-	-	(80)
PUE/Mo_7O_{24}-intercalated RGO–LDH	3	50	-	378	-	-	-	(80)
TPU	0	50	-	1046.7	134.5	Not reported (NR)	17.0	(48)
TPU/NO_3–MgAl–LDH	20	50	-	-	97.9	NR	26.2	(48)
TPU/SDS-modified NO_3–MgAl–LDH	20	50	-	351.4	94.6	NR	27.3	(48)
TPU/SDS-modified NO_3–MgAl–LDH–GO	20	50	-	238.5	82.2	V-0	28.9	(48)
TPU	0	50	37	1623	104.6	V-2	23.0	(85)
TPU/RP-modified CoAl–LDH 5	5	50	31	473	63.7	V-0	28.0	(85)
TPU/P-3.4/CoAl–LDH-1.6	5	50	28	690	75.1	V-2	25.0	(85)
TPU	0	50	34	1423	104.6	V-2	23	(86)
TPU/PPEIA/LDH (PPEIA-5.2/LDH-0.8)	6	50	19	315	63.5	V-2	27	(86)
TPU/PPEIA-modified LDH (PPEIA–LDH-6)	6	50	18	228	52.4	V-0	29	(86)

Table 2. (Continued). Summary Results of Flame Tests in Which Modified LDH Was Used in PUs[a]

Sample Code	*FR (wt %)*	*Heat Flux (kW/m²)*	*Time to Ignition (TTI) (s)*	*pHRR (kW/m²)*	*THR (MJ/m²)*	*UL-94*	*LOI (%)*	*Ref.*
Flexible PU foam (FPUF)	0	35	-	502	-	-	-	(*87*)
FPUF/CNTs/ZnAl–LDH	2.1	35	-	434	-	-	-	(*87*)

[a] All obtained in cone calorimetry.

Combination of LDH and Conventional FRs in PUs

Gao et al. (*88*) investigated the combination effect of expandable graphite (EG), melamine polyphosphate (MPP), and NiAl–LDH on the flame retardancy of a rosin-based rigid PU foam (RPUF). They fixed the loading percentage of EG and MPP at 10 php of polyol by weight and varied that of LDH from 0.2 to 5 php. The LOI tests showed that the combination of EG and MPP at 10 php and LDH at 3 php led to an increased LOI value of 28, compared with 26.5 for the sample containing only EG and MPP at 10 php. Then they compared these sample using a cone calorimeter test. The authors claimed that the presence of LDH improved the flame retardancy in the presence of EG and MPP. However, the HRR curves seem to be relatively similar, which is indicative of a very modest change in behavior. In another work (*82*), the same research group worked on a combination of EG, diethyl ethylphosphonate (DEEP), and LDH or montmorillonite (MMT) on the flame retardancy of rigid polyisocyanurate-PU (PIR) foams. The authors varied the weight ratio between EG and DEEP (fixed at a total of 45 php) and varied the percentage of LDH or MMT from 1 to 5 php. Once again, based on LOI tests, the best samples were selected to be analyzed in cone calorimeter tests. Figure 10 displays the HRR curves for neat PIR and PIR containing EG and DEEP at 45 php (ratio 2:1) and LDH at 2 php or MMT at 3 php. These curves clearly show the effect of a combination of EG and DEEP on the flame retardancy of PIR, demonstrated by the decrease of pHRR. However, it is difficult to estimate the efficiency of LDH compared with MMT because the loading percentages of these additives were not equal. The comparison of the HRR curves of PIR-FR45 and PIR-FR45-LDH reveals that the presence of LDH had relatively no effect on flame retardancy because the pHRR was similar and the THR was higher for the sample containing LDH.

The combination of LDH with lignin and Exolit OP560 (EX) in the synthesis of flexible PU foams (FPUFs) was investigated by Gómez-Fernández et al. (*89*). Surprisingly, the results of LOI tests showed that there was no effect for the presence of LDH in combination with lignin. Nevertheless, in cone calorimetry tests, this combination showed some effects, proved by the decrease of pHRR when LDH at 3 php was combined with lignin and EX both at 5 php. pHRR was decreased to 645 kW/m^2 for this sample, compared with 851 kW/m^2 for the sample containing solely lignin and EX. Qian et al. (*81*) studied the combination of a lanthanum-doped MgAl–LDH (La–MgAl–LDH) and GO on the flame retardancy of TPU prepared in an internal mixer at 180 °C for 3 min. The percentage of loading was fixed at 2 wt % for all samples. LOI value increased to 23.2 for the La–MgAl–LDH/GO sample compared with 21.4 for pure TPU, whereas it was 21.8 for the sample containing solely MgAl–LDH and 23.2 for the La–MgAl–LDH sample. These results clearly showed a modest improvement in flame behavior, even in the case of a combination of La–MgAl–LDH and GO. This result was in accordance with that obtained in cone calorimeter tests in which the shape and pHRR were quite similar for all samples containing additives. Therefore, it seems that the combination had no effect on flame retardancy performance.

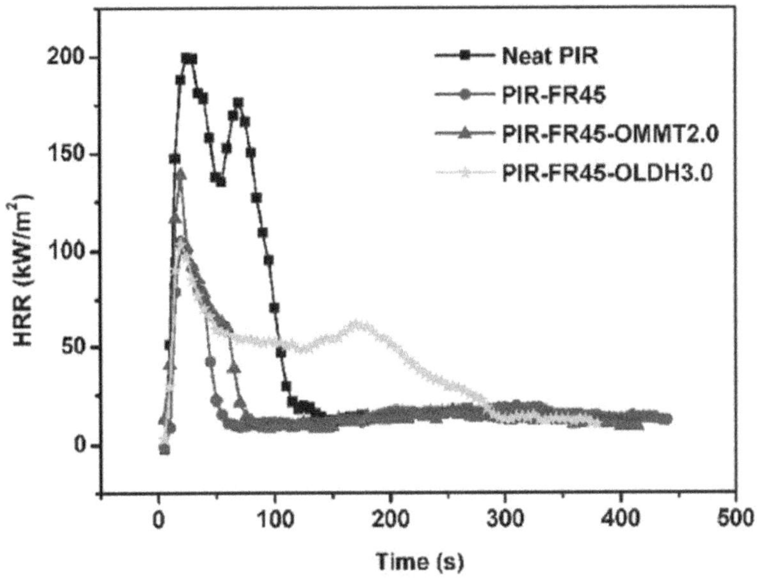

Figure 10. HRR curves obtained in cone calorimeter tests for PIR foams containing EG, DEEP at 45 php, namely FR45, and samples containing LDH at 2 php and MMT at 3 php, namely PIR-FR45-OMMT2.0 and PIR-FR45-OLDH3.0 (82). Reproduced with permission from reference (82). Copyright 2014 Elsevier.

Ammonium polyphosphate (APP) is a well-known conventional polymeric FR at micrometric size (*90*). Its synergistic effect in combination with LDH on flame retardancy in different polymers has been reported by several authors (*91–93*). However, the incorporation of APP into polymers generally deteriorates their mechanical properties because of the poor interfacial adhesion between APP and polymer. To overcome this problem, Huang et al. (*83*) prepared a modified APP on which LDH nanoparticles were deposited using electrostatic action (APP@LDH) to benefit the synergistic effect and cause no deterioration of mechanical properties. First, CoAl–LDH was dispersed in water and then added to a suspension containing APP under stirring. Afterward, the precipitation was centrifuged and dried (Figure 11). Different ratios of APP and LDH were prepared: 2.6:0.4 ($APP_{2.6}@LDH_{0.4}$), 4.3:0.7 ($APP_{4.3}@LDH_{0.7}$), and 6:1 ($APP_6@LDH_1$). The surface of APP was coated with LDH nanosheets. Finally, the modified APP as well as the APP/LDH combination (without treatment, only a physical mixing, ratio 6:1 [APP_6/LDH_1]) were incorporated into TPU at 190 °C in a Banbury mixer for 6 min and samples obtained via an injection-molding machine. A UL-94 test showed that all materials were classified V-2, except the TPU sample containing $APP_6@LDH_1$, which was V-0. The LOI value of pure TPU was 22. The incorporation of additives, regardless of the type of additive, led to increased LOI values—28 for $TPU/APP_{2.6}@LDH_{0.4}$, 28.7 for $TPU/APP_{4.3}@LDH_{0.7}$, and 29 for $TPU/APP_{6.0}@LDH_{1.0}$. The LOI of the sample containing physically mixed APP/LDH was 28. According to the results obtained in cone calorimeter tests, the best flame behavior was observed for the $TPU/APP_{6.0}@LDH_{1.0}$ sample, in which the pHRR decreased to 219 kW/m^2, compared with 282 kW/m^2 for the $TPU/APP_{6.0}/LDH_{1.0}$ sample. These results showed that APP modified with LDH performed better than does a simple mixing of them to improve the flame retardancy, which was explained by the creation of a more stable char layer induced during combustion caused by the accumulation of LDH nanosheets on the surface of the material. However, we can observe that the difference between the two samples in terms of flame behavior is not significant. The authors also reported that the tensile strength value was 32.5% higher for TPU/$APP_{6.0}$@$LDH_{1.0}$ compared with that of TPU/$APP_{6.0}$/$LDH_{1.0}$ thanks to a better interfacial adhesion because of the presence of LDH nanosheets at the surface of APP. Table 3 displays a summary of the literature dealing with the combination of LDH and conventional FRs in different polymers.

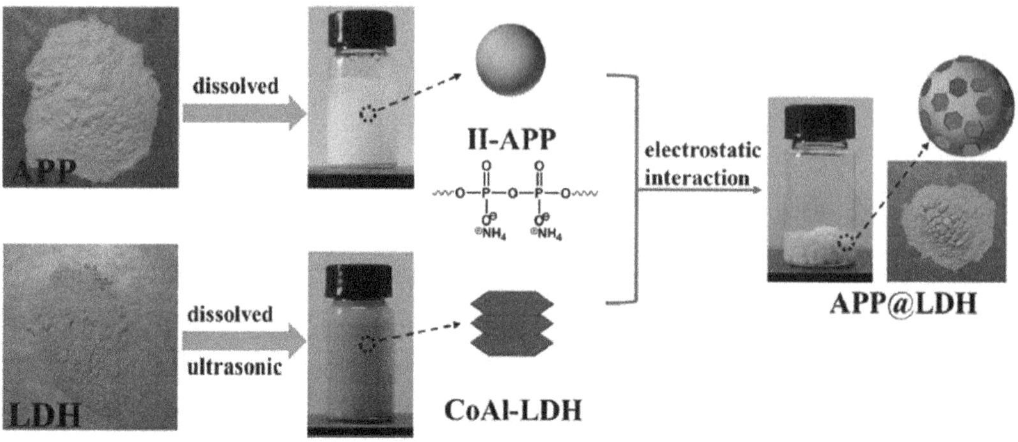

Figure 11. Preparation method of APP coated at the surface with CoAl–LDH (APP@LDH). Reproduced with permission from reference (83). Copyright 2019 Elsevier.

Table 3. Summary Results of Flame Tests in Which LDH Was Combined with Conventional FRs To Improve Fire Behavior of PUs[a]

Sample Code	*FR (wt %)*	*Heat Flux (kW/m²)*	*TTI (s)*	*pHRR (kW/m²)*	*THR (MJ/m²)*	*UL-94*	*LOI (%)*	*Ref.*
FPUF	0	35	9	838	46.54	NR	18.9	(78)
FPUF/MgAl–LDH	5 php	35	5	743	45.45	NR	19.2	(78)
FPUF/SC4A-modified MgAl–LDH	5 php	35	4	648	46.51	NR	20.4	(78)
SuBC4A-modified MgAl–LDH	5 php	35	5	468	33.11	V-1	22.5	(78)
TPU	0	50	-	1103	139.1	-	21.4	(81)
TPU/MgAl–LDH	2	50	-	846	139.6	-	21.8	(81)
TPU/La–MgAl–LDH	2	50	-	769	140.3	-	22	(81)
TPU/La–MgAl–LDH/GO	2	50	-	738	140.0	-	23.2	(81)
Rigid PIR foam	0	35	3	199	15.3	-	22.1	(82)
Rigid polyisocyanurate–PU foam/EG/DEEP/SDS-modified CoAl–LDH	48 php	35	4	103	14.0	-	32.5	(82)

Table 3. (Continued). Summary Results of Flame Tests in Which LDH Was Combined with Conventional FRs To Improve Fire Behavior of PUs[a]

Sample Code	*FR (wt %)*	*Heat Flux (kW/m²)*	*TTI (s)*	*pHRR (kW/m²)*	*THR (MJ/m²)*	*UL-94*	*LOI (%)*	*Ref.*
TPU	0	50	28	855	88.4	V-2	22.0	(83)
TPU/APP@CoAl–LDH	7	50	18	219	59.9	V-0	29.2	(83)
TPU/APP_6/CoAl–LDH_1	7	50	16	282	68.2	V-2	28	(83)
RPUF	0	35	2	250	19.3	-	19.1	(88)
RPUF/EG/MPP/CO_3^{2-}-intercalated Ni–Al LDH	23 php	35	5	153	18.3	-	28.0	(88)
FPUF	0 php	50	-	1209	56	-	18.2	(89)
FPUF/CO_3–MgAl–LDH	3 php	50	-	1214	53	-	17.1	(89)
FPUF/CO_3–MgAl–LDH/lignin	8 php	50	-	798	50	-	17.2	(89)
FPUF/CO_3–MgAl–LDH/EX	8 php	50	-	852	49	-	18.2	(89)
FPUF/CO_3–MgAl–LDH/EX/lignin	13 php	50	-	645	48	-	18.4	(89)
TPU	0	50	-	1046	134.5	-	21.4	(94)
Pillared layered antimony hydroxide	2	50	-	459	96.5	-	22.5	(94)

[a] All obtained in cone calorimetry.

Use of LDH in Coating Systems to Improve the Flame Retardancy of PUs

All studies reviewed earlier in this chapter deal with LDH as an additive incorporated into bulk PU. It is worth mentioning that LDH was also used in coating FR systems for PU. However, research is limited on this subject. Yang et al. (95) investigated the flame retardancy of PU foam using a layer-by-layer (LbL) assembly method by a multicomponent system containing branched polyethyleneimine, poly(acrylic acid), sodium MMT (Na-MMT), and LDH. Several samples were prepared using these materials and by bi-, tri-, and quadlayer LbL coating sequences. The authors found that a five bilayer poly(acrylic acid)/branched polyethyleneimine/LDH coating decreased the pHRR by 40% and the THR by 70%. They also showed that the combination of LDH and Na-MMT reduced the flammability of PU and led to a reduction in the mass of coating materials. The LbL deposition process was also used by Liu et al. (96) to improve the flame retardancy of FPUF. They impregnated PU foam successively in three solutions containing chitosan, NiAl–LDH, or MgAl–LDH and alginate. These three layers were repeated for 3, 6, or 12 times. The cone calorimetry results revealed that the pHRR was decreased by increasing the number of layers, so that regardless of the type of LDH, the best results were achieved for samples with 12 times coating. NiAl–LDH had more impact on the decrease of pHRR than did MgAl–LDH. The pHRR was decreased from 801 kW/m² for pure FPUF to 197 kW/m² for an FPUF sample coated 12 times with NiAl-LDH/chitosan/alginate for (FPUF/NiAl–LDH12) compared with 276 kW/m^2 for an FPUF sample coated 12 times with MgAl–LDH/chitosan/alginate (FPUF/MgAl–LDH12). The mechanism of action was a physical barrier effect for all samples. In the case of the FPUF/NiAl–LDH12 sample, the catalytic action of NiAl–LDH, which promoted carbonization, was the main reason for its superiority compared with MgAl–LDH. Table 4 summarizes the results obtained in cone calorimeter tests for all aforementioned samples.

Table 4. Summary Results of Flame Tests in Which LDH Was Used in FR Coating Systems To Improve Fire Behavior of PUs[a]

Sample Code	*FR (wt %)*	*Heat Flux (kW/m^2)*	*TTI (s)*	*pHRR (kW/m^2)*	*THR (MJ/m^2)*	*Ref.*
FPUF	0	35	-	451	43	(95)
FPUF/bilayer coatings containing LDH	0.5	35	-	271	37	(95)
FPUF/bilayer coatings containing LDH1/Na-MMT0.5	1	35	-	329	40	(95)
FPUF/trilayer coatings containing LDH1/Na-MMT0.5	1.5	35	-	320	36	(95)
FPUF/quadlayer coatings containing LDH0.5/Na-MMT1	1.5	35	-	303	41	(95)
FPUF	0	35	2	801	22.1	(96)
FPUF/trilayer coatings containing MgAl–LDH	4 (mg/mL)	35	3	276	16.7	(96)
FPUF/trilayer coatings containing NiAl–LDH	4 (mg/mL)	35	4	197	16.2	(96)

[a] All obtained in cone calorimetry.

Concluding Remarks and Future Perspectives

Various clays are used to improve the flame behavior of polymeric materials. These clays are inexpensive and without proven risk for health and the environment. LDH is one of the most attractive clays, with a high versatility in structure that offers many possibilities for chemical modification. However, its dispersion in polymers is a principal key to achieve the optimal properties including fire behavior. In this chapter, a review of the literature on the use of different LDHs in PU was presented. Our survey of the published research showed that LDH can be used in its initial state or in combination with conventional FRs. However, it seems that the chemical modification of LDH via the intercalation of organic molecules, including phosphorus, is beneficial for both the flame and the mechanical properties of PU. It is also shown that LDH can be used in FR coating systems on PU. The number of studies on the use of LDH in biobased PU is limited, which can be a perspective point for future works. The combination of LDH and biobased FRs in biobased or nonbiobased PU should be developed in the future. Finally, the combination of LDH and conventional FRs is restricted to a limited number of conventional FRs (for example, different phosphorus-based FRs with different degrees of oxidation and different LDH types). Therefore, research on these aspects is necessary in the future to elucidate the potential of LDH in the flame retardancy of PU. It should be also noted that dispersion of LDH in thermosetting PU systems could be a challenge for future developments.

The most promising aspect of future works in this field is that there is a possibility for synthesis of LDH based on phosphorus and carbonaceous ingredients aimed at flame retardancy. Because LDH has a wide variety of structures, highly efficient FRs based on LDH and conventional or nanoscale additives can be developed.

References

1. Rabiee, N.; Bagherzadeh, M.; Ghadiri, A. M.; Kiani, M.; Fatahi, Y.; Tavakolizadeh, M.; Pourjavadi, A.; Jouyandeh, M.; Saeb, M. R.; Mozafari, M.; Shokouhimehr, M.; Varma, R. S. Multifunctional 3D Hierarchical Bioactive Green Carbon-Based Nanocomposites. *ACS Sustainable Chemistry & Engineering* **2021**, *9* (26), 8706–8720.
2. Rabiee, N.; Bagherzadeh, M.; Ghadiri, A. M.; Fatahi, Y.; Aldhaher, A.; Makvandi, P.; Dinarvand, R.; Jouyandeh, M.; Saeb, M. R.; Mozafari, M.; Shokouhimehr, M.; Hamblin, M. R.; Varma, R. S. Turning Toxic Nanomaterials into a Safe and Bioactive Nanocarrier for Co-delivery of DOX/pCRISPR. *ACS Applied Bio Materials* **2021**, *4* (6), 5336–5351.
3. Rabiee, N.; Bagherzadeh, M.; Jouyandeh, M.; Zarrintaj, P.; Saeb, M. R.; Mozafari, M.; Shokouhimehr, M.; Varma, R. S. Natural Polymers Decorated MOF-MXene Nanocarriers for Co-delivery of Doxorubicin/pCRISPR. *ACS Applied Bio Materials* **2021**, *4* (6), 5106–5121.
4. Vahabi H., Movahedifar E., Ganjali M. R., Saeb M. R. Chapter 3 - Polymer nanocomposites from the flame retardancy viewpoint: A comprehensive classification of nanoparticle performance using the flame retardancy index. In *Handbook of Polymer Nanocomposites for Industrial Applications*; Hussain C. M., Ed.; Elsevier, 2021; pp 61−146.
5. Vahabi, H.; Jouyandeh, M.; Cochez, M.; Khalili, R.; Vagner, C.; Ferriol, M.; Movahedifar, E.; Ramezanzadeh, B.; Rostami, M.; Ranjbar, Z.; Hadavand, B. S.; Saeb, M. R. Short-lasting fire in partially and completely cured epoxy coatings containing expandable graphite and halloysite nanotube additives. *Progress in Organic Coatings* **2018**, *123*, 160–167.
6. Vahabi, H.; Saeb, M. R.; Formela, K.; Cuesta, J.-M. L. Flame retardant epoxy/halloysite nanotubes nanocomposite coatings: Exploring low-concentration threshold for flammability compared to expandable graphite as superior fire retardant. *Progress in Organic Coatings* **2018**, *119*, 8–14.
7. Jouyandeh, M.; Hadavand, B. S.; Tikhani, F.; Khalili, R.; Bagheri, B.; Zarrintaj, P.; Formela, K.; Vahabi, H.; Saeb, M. R. Thermal-Resistant Polyurethane/Nanoclay Powder Coatings: Degradation Kinetics Study. *Coatings* **2020**, *10* (9), 871.
8. Seidi, F.; Jouyandeh, M.; Akbari, V.; Paran, S. M. R.; Livi, S.; Ducos, F.; Vahabi, H.; Ganjali, M. R.; Saeb, M. R. Super-crosslinked ionic liquid-intercalated montmorillonite/epoxy nanocomposites: Cure kinetics, viscoelastic behavior and thermal degradation mechanism. *Polymer Engineering & Science* **2020**, *60* (8), 1940–1957.
9. Kashiwagi, T.; Grulke, E.; Hilding, J.; Groth, K.; Harris, R.; Butler, K.; Shields, J.; Kharchenko, S.; Douglas, J. Thermal and flammability properties of polypropylene/carbon nanotube nanocomposites. *Polymer* **2004**, *45* (12), 4227–4239.
10. Babu, K.; Rendén, G.; Afriyie Mensah, R.; Kim, N. K.; Jiang, L.; Xu, Q.; Restás, Á.; Esmaeely Neisiany, R.; Hedenqvist, M. S.; Försth, M. A review on the flammability properties of carbon-based polymeric composites: State-of-the-art and future trends. *Polymers* **2020**, *12* (7), 1518.

11. Vahabi, H.; Raveshtian, A.; Fasihi, M.; Sonnier, R.; Saeb, M. R.; Dumazert, L.; Kandola, B. K. Competitiveness and synergy between three flame retardants in poly (ethylene-co-vinyl acetate). *Polymer Degradation and Stability* **2017**, *143*, 164–175.
12. Bayat, S.; Moini Jazani, O.; Molla-Abbasi, P.; Jouyandeh, M.; Saeb, M. R. Thin films of epoxy adhesives containing recycled polymers and graphene oxide nanoflakes for metal/polymer composite interface. *Progress in Organic Coatings* **2019**, *136*, 105201.
13. Vahabi, H.; Ferry, L.; Longuet, C.; Otazaghine, B.; Negrell-Guirao, C.; David, G.; Lopez-Cuesta, J. M. Combination effect of polyhedral oligomeric silsesquioxane (POSS) and a phosphorus modified PMMA, flammability and thermal stability properties. *Materials Chemistry and Physics* **2012**, *136* (2), 762–770.
14. Vahabi, H.; Eterradossi, O.; Ferry, L.; Longuet, C.; Sonnier, R.; Lopez-Cuesta, J. M. Polycarbonate nanocomposite with improved fire behavior, physical and psychophysical transparency. *European Polymer Journal* **2013**, *49* (2), 319–327.
15. Seidi, F.; Jouyandeh, M.; Taghizadeh, A.; Taghizadeh, M.; Habibzadeh, S.; Jin, Y.; Xiao, H.; Zarrintaj, P.; Saeb, M. R. Polyhedral oligomeric silsesquioxane/epoxy coatings: a review. *Surface Innovations* **2020**, *9* (1), 3–16.
16. Lin, M.; Li, B.; Li, Q.; Li, S.; Zhang, S. Synergistic effect of metal oxides on the flame retardancy and thermal degradation of novel intumescent flame-retardant thermoplastic polyurethanes. *Journal of Applied Polymer Science* **2011**, *121* (4), 1951–1960.
17. Vahabi, H.; Laoutid, F.; Mehrpouya, M.; Saeb, M. R.; Dubois, P. Flame retardant polymer materials: An update and the future for 3D printing developments. *Materials Science and Engineering: R: Reports* **2021**, *144*, 100604.
18. Jouyandeh, M.; Jazani, O. M.; Navarchian, A. H.; Shabanian, M.; Vahabi, H.; Saeb, M. R. Bushy-surface hybrid nanoparticles for developing epoxy superadhesives. *Applied Surface Science* **2019**, *479*, 1148–1160.
19. Jouyandeh, M.; Ganjali, M. R.; Ali, J. A.; Aghazadeh, M.; Stadler, F. J.; Saeb, M. R. Curing epoxy with electrochemically synthesized NixFe3-xO4 magnetic nanoparticles. *Progress in Organic Coatings* **2019**, *136*, 105198.
20. Jouyandeh, M.; Zarrintaj, P.; Ganjali, M. R.; Ali, J. A.; Karimzadeh, I.; Aghazadeh, M.; Ghaffari, M.; Saeb, M. R. Curing epoxy with electrochemically synthesized GdxFe3-xO4 magnetic nanoparticles. *Progress in Organic Coatings* **2019**, *136*, 105245.
21. Vahabi, H.; Wu, H.; Saeb, M. R.; Koo, J. H.; Ramakrishna, S. Electrospinning for developing flame retardant polymer materials: current status and future perspectives. *Polymer* **2021**, 123466.
22. Morgan A., Wilkie C. *Considerations Regarding Specific Impacts of the Principal Fire Retardancy Mechanisms in Nanocomposites, Flame Retardant Polymer Nanocomposites*; Wiley Inter-Science: New York, 2007.
23. Vahabi H., Movahedifar E., Saeb M. *Flame Retardancy of Reactive and Functional Polymers. Reactive and Functional Polymers Volume Three*; Springer, 2020; pp 165—195.
24. Vahabi H., Sonnier R., Taguet A., Otazaghine B., Saeb M. R., Beyer G. Halloysite nanotubes (HNTs)/polymer nanocomposites: thermal degradation and flame retardancy. In *Clay Nanoparticles*; Elsevier, 2020; pp67—93.

25. Karami, Z.; Jouyandeh, M.; Ali, J. A.; Ganjali, M. R.; Aghazadeh, M.; Maadani, M.; Rallini, M.; Luzi, F.; Torre, L.; Puglia, D.; Saeb, M. R. Development of Mg-Zn-Al-CO3 ternary LDH and its curability in epoxy/amine system. *Progress in Organic Coatings.* **2019**, *136*, 105264.
26. Karami, Z.; Jouyandeh, M.; Hamad, S. M.; Ganjali, M. R.; Aghazadeh, M.; Torre, L.; Puglia, D.; Saeb, M. R. Curing epoxy with Mg-Al LDH nanoplatelets intercalated with carbonate ion. *Progress in Organic Coatings* **2019**, *136*, 105278.
27. Karami, Z.; Jouyandeh, M.; Ghiyasi, S.; Ali, J. A.; Ganjali, M. R.; Aghazadeh, M.; Maadani, M.; Rallini, M.; Luzi, F.; Torre, L.; Puglia, D.; Saeb, M. R. Exploring curing potential of epoxy nanocomposites containing nitrate anion intercalated Mg–Al–LDH with Cure Index. *Progress in Organic Coatings* **2020**, *139*, 105255.
28. Karami, Z.; Paran, S. M. R.; Vijayan, P. P.; Ganjali, M. R.; Jouyandeh, M.; Esmaeili, A.; Habibzadeh, S.; J. Stadler, F.; Saeb, M. R. A Comparative Study on Cure Kinetics of Layered Double Hydroxide (LDH)/Epoxy Nanocomposites. *Journal of Composites Science* **2020**, *4* (3), 111.
29. Costa, F. R.; Saphiannikova, M.; Wagenknecht, U.; Heinrich, G. Layered double hydroxide based polymer nanocomposites. *Wax Crystal Control - Nanocomposites - Stimuli-Responsive Polymers* **2007**, 101–168.
30. Ray, S. S.; Okamoto, M. Polymer/layered silicate nanocomposites: a review from preparation to processing. *Progress in Polymer Science* **2003**, *28* (11), 1539–641.
31. Karami, Z.; Aghazadeh, M.; Jouyandeh, M.; Zarrintaj, P.; Vahabi, H.; Ganjali, M. R.; Torre, L.; Puglia, D.; Saeb, M. R. Epoxy/Zn-Al-CO3 LDH nanocomposites: Curability assessment. *Progress in Organic Coatings* **2020**, *138*, 105355.
32. Abdi, A. A.; Jouyandeh, M.; Vahabi, H.; Shabanian, M.; Lafon-Pham, D.; Gabrion, X.; Laheurte, P.; Nahavandi, A. M.; Saeb, M. R. Correlating the Photophysical Properties with the Cure Index of Epoxy Nanocomposite Coatings. *Journal of Inorganic and Organometallic Polymers and Materials* **2021**, *31* (3), 923–933.
33. Karami, Z.; Jouyandeh, M.; Ali, J. A.; Ganjali, M. R.; Aghazadeh, M.; Maadani, M.; Rallini, M.; Luzi, F.; Torre, L.; Puglia, D.; Akbari, V.; Saeb, M. R. Cure Index for labeling curing potential of epoxy/LDH nanocomposites: A case study on nitrate anion intercalated Ni-Al-LDH. *Progress in Organic Coatings* **2019**, *136*, 105228.
34. Mishra, G.; Dash, B.; Pandey, S. Layered double hydroxides: A brief review from fundamentals to application as evolving biomaterials. *Applied Clay Science* **2018**, *153*, 172–186.
35. Karami, Z.; Ganjali, M. R.; Zarghami Dehaghani, M.; Aghazadeh, M.; Jouyandeh, M.; Esmaeili, A.; Habibzadeh, S.; Mohaddespour, A.; Formela, K.; Haponiuk, J. T. Kinetics of cross-linking reaction of epoxy resin with hydroxyapatite-functionalized layered double hydroxides. *Polymers* **2020**, *12* (5), 1157.
36. Seidi, F.; Jouyandeh, M.; Paran, S. M. R.; Esmaeili, A.; Karami, Z.; Livi, S.; Habibzadeh, S.; Vahabi, H.; Ganjali, M. R.; Saeb, M. R. Imidazole-functionalized nitrogen-rich Mg-Al-CO3 layered double hydroxide for developing highly crosslinkable epoxy with high thermal and mechanical properties. *Colloids and Surfaces A: Physicochemical and Engineering Aspects* **2021**, *611*, 125826.
37. Karami, Z.; Jouyandeh, M.; Ali, J. A.; Ganjali, M. R.; Aghazadeh, M.; Paran, S. M. R.; Naderi, G.; Puglia, D.; Saeb, M. R. Epoxy/layered double hydroxide (LDH) nanocomposites:

Synthesis, characterization, and Excellent cure feature of nitrate anion intercalated Zn-Al LDH. *Progress in Organic Coatings* **2019**, *136*, 105218.

38. Wang, Y.; Gao, H. Compositional and structural control on anion sorption capability of layered double hydroxides (LDHs). *Journal of Colloid and Interface Science* **2006**, *301* (1), 19–26.
39. Evans D. G., Slade R. C. T. Structural Aspects of Layered Double Hydroxides. In: Duan X., Evans D. G., Eds.; *Layered Double Hydroxides*; Springer Berlin Heidelberg: Berlin, Heidelberg, 2006; pp 1–87.
40. Babu H. V., Coluccini C., Wang D. Y. 8 - Functional layered double hydroxides and their use in fire-retardant polymeric materials. In *Novel Fire Retardant Polymers and Composite Materials*; Wang D.-Y. , Ed.; Woodhead Publishing, 2017; pp 201–238.
41. Arrabito, G.; Bonasera, A.; Prestopino, G.; Orsini, A.; Mattoccia, A.; Martinelli, E.; Pignataro, B.; Medaglia, P. G. Layered double hydroxides: a toolbox for chemistry and biology. *Crystals* **2019**, *9* (7), 361.
42. Yan, K.; Liu, Y.; Lu, Y.; Chai, J.; Sun, L. Catalytic application of layered double hydroxide-derived catalysts for the conversion of biomass-derived molecules. *Catalysis Science & Technology* **2017**, *7* (8), 1622–1645.
43. Costantino, U.; Vivani, R.; Bastianini, M.; Costantino, F.; Nocchetti, M. Ion exchange and intercalation properties of layered double hydroxides towards halide anions. *Dalton Transactions* **2014**, *43* (30), 11587–11596.
44. Ladewig, K.; Xu, Z. P.; Lu, G. Q. Layered double hydroxide nanoparticles in gene and drug delivery. *Expert Opinion on Drug Delivery* **2009**, *6* (9), 907–922.
45. Mohapatra, L.; Parida, K. A review on the recent progress, challenges and perspective of layered double hydroxides as promising photocatalysts. *Journal of Materials Chemistry A* **2016**, *4* (28), 10744–10766.
46. Daniel S., Thomas S. Layered double hydroxides: fundamentals to applications. In *Layered Double Hydroxide Polymer Nanocomposites*; Elsevier, 2020; pp 1–76.
47. Leroux, F.; Besse, J.-P. Polymer interleaved layered double hydroxide: a new emerging class of nanocomposites. *Chemistry of Materials* **2001**, *13* (10), 3507–3515.
48. Li, L.; Jiang, K.; Qian, Y.; Han, H.; Qiao, P.; Zhang, H. Effect of organically intercalation modified layered double hydroxides-graphene oxide hybrids on flame retardancy of thermoplastic polyurethane nanocomposites. *Journal of Thermal Analysis and Calorimetry* **2020**, *142* (2), 723–733.
49. Martínez-Gallegos, S.; Herrero, M.; Barriga, C.; Labajos, F.; Rives, V. Dispersion of layered double hydroxides in poly (ethylene terephthalate) by in situ polymerization and mechanical grinding. *Applied Clay Science* **2009**, *45* (1–2), 44–49.
50. Kalali, E. N.; Wang, X.; Wang, D.-Y. Functionalized layered double hydroxide-based epoxy nanocomposites with improved flame retardancy and mechanical properties. *Journal of Materials Chemistry A* **2015**, *3* (13), 6819–6826.
51. Becker, C. M.; Gabbardo, A. D.; Wypych, F.; Amico, S. C. Mechanical and flame-retardant properties of epoxy/Mg–Al LDH composites. *Composites Part A: Applied Science and Manufacturing* **2011**, *42* (2), 196–202.
52. Feng, Y.; Tang, P.; Xi, J.; Jiang, Y.; Li, D. Layered double hydroxides as flame retardant and thermal stabilizer for polymers. *Recent Patents on Nanotechnology* **2012**, *6* (3), 231–237.

53. Jiang, T.; Liu, C.; Liu, L.; Hong, J.; Dong, M.; Deng, X. Synergistic flame retardant properties of a layered double hydroxide in combination with zirconium phosphonate in polypropylene. *RSC Advances* **2016**, *6* (94), 91720–91727.
54. Wang, D.-Y.; Das, A.; Costa, F. R.; Leuteritz, A.; Wang, Y.-Z.; Wagenknecht, U.; Heinrich, G. Synthesis of organo cobalt– aluminum layered double hydroxide via a novel single-step self-assembling method and its use as flame retardant nanofiller in PP. *Langmuir* **2010**, *26* (17), 14162–14169.
55. Costa, F. R.; Wagenknecht, U.; Heinrich, G. LDPE/Mg–Al layered double hydroxide nanocomposite: thermal and flammability properties. *Polymer Degradation and Stability* **2007**, *92* (10), 1813–1823.
56. Ye, L.; Wu, Q. Effects of an intercalating agent on the morphology and thermal and flame-retardant properties of low-density polyethylene/layered double hydroxide nanocomposites prepared by melt intercalation. *Journal of Applied Polymer Science* **2012**, *123* (1), 316–323.
57. Matusinovic, Z.; Lu, H.; Wilkie, C. A. The role of dispersion of LDH in fire retardancy: The effect of dispersion on fire retardant properties of polystyrene/Ca– Al layered double hydroxide nanocomposites. *Polymer Degradation and Stability* **2012**, *97* (9), 1563–1568.
58. la Rosa-Guzmán, D.; Ángel, M.; Guzmán-Vargas, A.; Cayetano-Castro, N.; Del Río, J. M.; Corea, M.; Martínez-Ortiz, M. d. J. Thermal stability evaluation of polystyrene-Mg/Zn/Al LDH nanocomposites. *Nanomaterials* **2019**, *9* (11), 1528.
59. Zhang, G.; Ding, P.; Zhang, M.; Qu, B. Synergistic effects of layered double hydroxide with hyperfine magnesium hydroxide in halogen-free flame retardant EVA/HFMH/LDH nanocomposites. *Polymer Degradation and Stability* **2007**, *92* (9), 1715–1720.
60. Du, J.-Z.; Jin, L.; Zeng, H.-Y.; Shi, X.-k.; Zhou, E.-g.; Feng, B.; Sheng, X. Flame retardancy of organic-anion-intercalated layered double hydroxides in ethylene vinyl acetate copolymer. *Applied Clay Science* **2019**, *180*, 105193.
61. Xu, S.; Zhang, L.; Lin, Y.; Li, R.; Zhang, F. Layered double hydroxides used as flame retardant for engineering plastic acrylonitrile–butadiene–styrene (ABS). *Journal of Physics and Chemistry of Solids* **2012**, *73* (12), 1514–1517.
62. Jouyandeh, M.; Karami, Z.; Hamad, S. M.; Ganjali, M. R.; Akbari, V.; Vahabi, H.; Kim, S.-J.; Zarrintaj, P.; Saeb, M. R. Nonisothermal cure kinetics of epoxy/ZnxFe3-xO4 nanocomposites. *Progress in Organic Coatings* **2019**, *136*, 105290.
63. Jouyandeh, M.; Ali, J. A.; Aghazadeh, M.; Formela, K.; Saeb, M. R.; Ranjbar, Z.; Ganjali, M. R. Curing epoxy with electrochemically synthesized ZnxFe3-xO4 magnetic nanoparticles. *Progress in Organic Coatings* **2019**, *136*, 105246.
64. Jouyandeh, M.; Ganjali, M. R.; Seidi, F.; Xiao, H.; Saeb, M. R. Nonisothermal Cure Kinetics of Epoxy/Polyvinylpyrrolidone Functionalized Superparamagnetic Nano-Fe3O4 Composites: Effect of Zn and Mn Doping. *Journal of Composites Science* **2020**, *4* (2), 55.
65. Rajaei, M.; Kim, N. K.; Bickerton, S.; Bhattacharyya, D. A comparative study on effects of natural and synthesised nano-clays on the fire and mechanical properties of epoxy composites. *Composites Part B: Engineering* **2019**, *165*, 65–74.
66. Jouyandeh, M.; Tikhani, F.; Hampp, N.; Akbarzadeh Yazdi, D.; Zarrintaj, P.; Reza Ganjali, M.; Reza Saeb, M. Highly curable self-healing vitrimer-like cellulose-modified halloysite nanotube/epoxy nanocomposite coatings. *Chemical Engineering Journal* **2020**, *396*, 125196.

67. Paran, S. M. R.; Vahabi, H.; Jouyandeh, M.; Ducos, F.; Formela, K.; Saeb, M. R. Thermal decomposition kinetics of dynamically vulcanized polyamide 6–acrylonitrile butadiene rubber–halloysite nanotube nanocomposites. *Journal of Applied Polymer Science* **2019**, *136* (20), 47483.
68. Jouyandeh, M.; Jazani, O. M.; Navarchian, A. H.; Shabanian, M.; Vahabi, H.; Saeb, M. R. Surface engineering of nanoparticles with macromolecules for epoxy curing: Development of super-reactive nitrogen-rich nanosilica through surface chemistry manipulation. *Applied Surface Science* **2018**, *447*, 152–164.
69. M. de Souza, F.; Choi, J.; Bhoyate, S.; Kahol, P. K.; Gupta, R. K. Expendable Graphite as an Efficient Flame-Retardant for Novel Partial Bio-Based Rigid Polyurethane Foams. *C—Journal of Carbon Research.* **2020**, *6* (2), 27.
70. Ranaweera, C.; Ionescu, M.; Bilic, N.; Wan, X.; Kahol, P.; Gupta, R. K. Biobased polyols using thiol-ene chemistry for rigid polyurethane foams with enhanced flame-retardant properties. *Journal of Renewable Materials* **2017**, *5* (1), 1–12.
71. Bhoyate, S.; Ionescu, M.; Kahol, P.; Gupta, R. K. Sustainable flame-retardant polyurethanes using renewable resources. *Industrial Crops and Products* **2018**, *123*, 480–488.
72. Vahabi, H.; Rastin, H.; Movahedifar, E.; Antoun, K.; Brosse, N.; Saeb, M. R. Flame Retardancy of Bio-Based Polyurethanes: Opportunities and Challenges. *Polymers.* **2020**, *12* (6), 1234.
73. Bhoyate, S.; Ionescu, M.; Kahol, P.; Chen, J.; Mishra, S.; Gupta, R. K. Highly flame-retardant polyurethane foam based on reactive phosphorus polyol and limonene-based polyol. *Journal of Applied Polymer Science* **2018**, *135* (21), 46224.
74. Benício, L. P. F.; Silva, R. A.; Lopes, J. A.; Eulálio, D.; Santos, R. M. M. d; Aquino, L. A. d.; Vergütz, L.; Novais, R. F.; Costa, L. M. d.; Pinto, F. G. Layered double hydroxides: nanomaterials for applications in agriculture. *Revista Brasileira de Ciência do Solo.* **2015**, *39* (1), 1–13.
75. Mallakpour, S.; Dinari, M.; Hatami, M. Modification of Mg/Al-layered double hydroxide with L-aspartic acid containing dicarboxylic acid and its application in the enhancement of the thermal stability of chiral poly (amide-imide). *RSC Advances* **2014**, *4* (79), 42114–42121.
76. Manzi-Nshuti, C.; Hossenlopp, J. M.; Wilkie, C. A. Fire retardancy of melamine and zinc aluminum layered double hydroxide in poly(methyl methacrylate). *Polymer Degradation and Stability* **2008**, *93* (10), 1855–1863.
77. Xu, W.; Xu, B.; Li, A.; Wang, X.; Wang, G. Flame retardancy and smoke suppression of MgAl layered double hydroxides containing P and Si in polyurethane elastomer. *Industrial & Engineering Chemistry Research* **2016**, *55* (42), 11175–11185.
78. Mohammadi, A.; Wang, D.-Y.; Hosseini, A. S.; De La Vega, J. Effect of intercalation of layered double hydroxides with sulfonate-containing calix[4]arenes on the flame retardancy of castor oil-based flexible polyurethane foams. *Polymer Testing* **2019**, *79*, 106055.
79. Xu, W. Z.; Wang, S. Q.; Liu, L.; Hu, Y. Synthesis of heptamolybdate-intercalated MgAl LDHs and its application in polyurethane elastomer. *Polymers for Advanced Technologies.* **2016**, *27* (2), 250–257.
80. Xu, W.; Zhang, B.; Xu, B.; Li, A. The flame retardancy and smoke suppression effect of heptaheptamolybdate modified reduced graphene oxide/layered double hydroxide hybrids on

polyurethane elastomer. *Composites Part A: Applied Science and Manufacturing.* **2016**, *91*, 30–40.

81. Qian, Y.; Qiao, P.; Li, L.; Han, H.; Zhang, H.; Chang, G. Hydrothermal synthesis of lanthanum-doped MgAl-layered double hydroxide/graphene oxide hybrid and its application as flame retardant for thermoplastic polyurethane. *Advances in Polymer Technology* **2020**, 2020. https://doi.org/10.1155/2020/1018093.
82. Gao, L.; Zheng, G.; Zhou, Y.; Hu, L.; Feng, G.; Zhang, M. Synergistic effect of expandable graphite, diethyl ethylphosphonate and organically-modified layered double hydroxide on flame retardancy and fire behavior of polyisocyanurate-polyurethane foam nanocomposite. *Polymer Degradation and Stability* **2014**, *101*, 92–101.
83. Huang, S.-C.; Deng, C.; Wang, S.-X.; Wei, W.-C.; Chen, H.; Wang, Y.-Z. Electrostatic action induced interfacial accumulation of layered double hydroxides towards highly efficient flame retardance and mechanical enhancement of thermoplastic polyurethane/ammonium polyphosphate. *Polymer Degradation and Stability* **2019**, *165*, 126–136.
84. Kotal, M.; Srivastava, S. K.; Manu, S. K.; Saxena, A. K.; Pandey, K. N. Preparation and properties of in-situ polymerized polyurethane/stearate intercalated layer double hydroxide nanocomposites. *Polymer International* **2013**, *62* (5), 728–735.
85. Huang, S.-C.; Deng, C.; Chen, H.; Li, Y.-M.; Zhao, Z.-Y.; Wang, S.-X.; Wang, Y.-Z. Novel ultrathin layered double hydroxide nanosheets with in situ formed oxidized phosphorus as anions for simultaneous fire resistance and mechanical enhancement of thermoplastic polyurethane. *ACS Applied Polymer Materials* **2019**, *1* (8), 1979–1990.
86. Huang, S.-C.; Deng, C.; Zhao, Z.-Y.; Chen, H.; Gao, Y.-Y.; Wang, Y.-Z. Phosphorus-containing organic-inorganic hybrid nanoparticles for the smoke suppression and flame retardancy of thermoplastic polyurethane. *Polymer Degradation and Stability* **2020**, *178*, 109179.
87. Xie, H.; Ye, Q.; Si, J.; Yang, W.; Lu, H.; Zhang, Q. Synthesis of a carbon nanotubes/ZnAl-layered double hydroxide composite as a novel flame retardant for flexible polyurethane foams. *Polymers for Advanced Technologies* **2016**, 27 (5), 651–656.
88. Gao, L.; Zheng, G.; Zhou, Y.; Hu, L.; Feng, G.; Xie, Y. Synergistic effect of expandable graphite, melamine polyphosphate and layered double hydroxide on improving the fire behavior of rosin-based rigid polyurethane foam. *Industrial Crops and Products* **2013**, *50*, 638–647.
89. Gómez-Fernández, S.; Günther, M.; Schartel, B.; Corcuera, M. A.; Eceiza, A. Impact of the combined use of layered double hydroxides, lignin and phosphorous polyol on the fire behavior of flexible polyurethane foams. *Industrial Crops and Products* **2018**, *125*, 346–359.
90. Lim, K.-S.; Bee, S.-T.; Sin, L. T.; Tee, T.-T.; Ratnam, C. T.; Hui, D.; Rahmat, A. R. A review of application of ammonium polyphosphate as intumescent flame retardant in thermoplastic composites. *Composites Part B: Engineering* **2016**, *84*, 155–174.
91. Liu, Y.; Gao, Y.; Zhang, Z.; Wang, Q. Preparation of ammonium polyphosphate and dye co-intercalated LDH/polypropylene composites with enhanced flame retardant and UV resistance properties. *Chemosphere* **2021**, *277*, 130370.
92. Nyambo, C.; Kandare, E.; Wang, D.; Wilkie, C. A. Flame-retarded polystyrene: Investigating chemical interactions between ammonium polyphosphate and MgAl layered double hydroxide. *Polymer Degradation and Stability* **2008**, *93* (9), 1656–1663.

93. Khanal, S.; Lu, Y.; Jin, D.; Xu, S. Effects of layered double hydroxides on the thermal and flame retardant properties of intumescent flame retardant high density polyethylene composites. *Fire and Materials* **2021**, 1–10.
94. Qian, Y.; Qiao, P.; Li, L.; Han, H.; Zhao, R.; Zhang, H. Preparation of pillared layered antimony hydroxide and its flame retardancy in thermoplastic polyurethane. *Journal of Thermal Analysis and Calorimetry* **2019**, 1–11.
95. Yang, Y. H.; Li, Y. C.; Shields, J.; Davis, R. D. Layer double hydroxide and sodium montmorillonite multilayer coatings for the flammability reduction of flexible polyurethane foams. *Journal of Applied Polymer Science* **2015**, *132* (14) https://doi.org/10.1002/app.41767.
96. Liu, L.; Wang, W.; Hu, Y. Layered double hydroxide-decorated flexible polyurethane foam: significantly improved toxic effluent elimination. *RSC Advances* **2015**, 5 (118), 97458–97466.

Chapter 10

Phosphorus-Based Flame Retardants for Polyurethanes: Synthesis and Mechanistic Studies

Xin Wang,[1] Lei Song,[1] and Yuan Hu[*,1]

[1]State Key Laboratory of Fire Science, University of Science and Technology of China, Hefei, Anhui 230026, P.R. China

[*]Email: yuanhu@ustc.edu.cn

Polyurethanes (PUs) are a group of versatile synthetic polymer materials that have been extensively applied in the field of rubbers, flexible or rigid foams, elastomers, adhesives, and coatings. However, like most synthetic polymer materials, PUs suffer from high flammability, which potentially threatens the safety of people and property. To eliminate this potential threat, flame-retardant (FR) modification is required for PUs. Among a wide variety of FRs, phosphorus-containing compounds have attracted considerable interest because of their relatively high FR efficiency and environmental friendliness. Over the past few decades, diverse phosphorus-containing compounds have been added into PUs, either as additives or reactive FRs that can effectively improve the FR property of PUs. However, the incorporation of phosphorus-containing compounds will cause some adverse effect on the physical and mechanical properties and thermal stability. Therefore, phosphorus-containing compounds have been combined with synergistic FR agents to improve the FR efficiency and obtain exceptional comprehensive properties of PUs. The FR mechanism of phosphorus-containing compounds is discussed for both the vapor and condensed phases.

Introduction

Polyurethanes (PUs) are one kind of versatile polymeric materials that are produced from the reaction between diisocyanates or polyisocyanates and diols or polyols (*1–3*). The properties of PUs can be facilely adjusted by tuning the types of diisocyanates or polyisocyanates and diols or polyols, the ratio of –NCO/–OH, and the types of catalyst. Owing to their versatile and adjustable properties, PUs have been extensively applied as rubbers, flexible or rigid foams, elastomers, adhesives, and coatings. For example, thermoplastic PU (TPU) is widely used for wire and cable sheath, flexible PU foam (FPUF) is the main component of upholstered furniture and mattresses, and rigid PU foam (RPUF) is mostly applied as thermal-insulating material in construction. However, PUs are highly flammable plastics because of their abundant components of carbon, hydrogen, and oxygen. After ignition, PUs show a rapid fire propagation rate and high heat release, indicating a very high fire risk. To eliminate this potential threat, a flame-retardant (FR) treatment is required for PUs.

The FRs used for PUs can be mainly classified into two categories: inorganic and organic compounds. The former category includes expanded graphite (EG), aluminum trihydrate, magnesium hydroxide, ammonium polyphosphate (APP), and various nanomaterials such as layered double hydroxide, graphene, molybdenum disulfide, MXene, silica, and so on. The latter category can be subdivided into halogen-, phosphorus-, nitrogen-, and silicon-containing compounds. Among these FRs, phosphorus-containing compounds have attracted considerable interest in both academic and industrial communities because of their relatively high FR efficiency and environmental friendliness.

In this chapter, the advances in phosphorus-based FRs for PUs are reviewed and summarized. The combustion behavior of PUs is introduced first to fundamentally understand the FR mechanism. Then, the main part of this chapter is a synthesis of diverse phosphorus-based FRs for PUs and their mechanisms, in which the additive-type and reactive-type phosphorus-based FRs for PUs are discussed in detail. Finally, a summary of the opportunities and challenges for phosphorus-based FRs for PUs is provided to guide the future development of this promising class of FR plastics.

Combustion Behavior of PUs

PUs are highly flammable polymeric materials. The limiting oxygen index (LOI) of PU foams (PUFs) is less than 18% (*4*). When heated to a certain temperature, PUs begin to decompose by means of the cleavage of the carbamate bond, generating various oligomeric and small molecules, including isocyanates, polyols, substituted amines, and carbon dioxide. With the further increase of temperature, the polyol segments start to decompose to form small molecules, such as alcohols. These decomposition products evaporate to the surface of the polymer and mix with oxygen. When the accumulated heat radiation is strong enough, the mixture ignites. The combustion generates more heat radiation, which aggravates the thermal decomposition of PU, leading to more combustible fragments evolving, thereby supporting continued combustion (Figure 1) (*5*). During combustion, the scission of the macromolecular network of PU results in the formation of a burning liquid such that a pool fire occurs (*6*), suggesting an extremely low melt viscosity. Accompanying the fierce heat release, the melt dripping phenomenon is also observed. Because most PUs are converted to liquid and further pyrolytic volatiles, the char yield is very low after burning. Therefore, PUs are polymers with poor carbonization capacity.

The flammability of PUs is closely related to the molecular structure of their polyols and isocyanates. PUF made from aromatic-based polyols exhibits a better FR property than PUF made from aliphatic-based polyols (*7, 8*). In another study, PUF made from polyurea-based polyol displayed a superior FR behavior over its counterparts made from styrene–acrylonitrile-grafted or polyisocyanurate-based polyols (*9*). The density of PUFs is also a crucial factor for the fire propagation rate (*10*).

Diverse Phosphorus-Based FRs for PUs and Their Mechanisms

In this section, diverse phosphorus-based FRs for PUs are discussed in terms of their comprehensive properties, especially FR performance. According to the chemical structure of FRs in the substrates, phosphorus-based FRs for PUs are primarily divided into two categories: additive and reactive. The additive-type approach involves the physical mixing of the FR fillers with PUs during the preparation process. Although the additive-type approach is regarded as an affordable, inexpensive, and feasible strategy, the interaction between the FR fillers and the PU matrices is relatively weak, leading to several undesirable side effects, such as migration and leaching of the FR fillers. To overcome this shortcoming, the reactive-type approach has been employed to produce FR-PUs. The reactive-type approach mainly involves the covalent linkage of the FR groups into the molecular chains of PUs. The merits and demerits of these two strategies are summarized in Table 1.

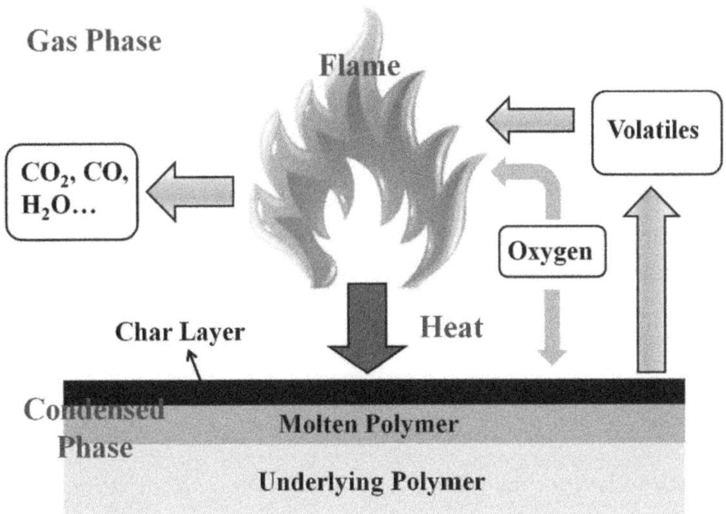

Figure 1. Diagrammatical illustration of the combustion process of polymers. Reproduced with permission from reference (5). Copyright 2017 Elsevier.

Table 1. The Merits and Demerits of the Additive-Type and Reactive-Type FR Strategies

Strategies	*Merits*	*Demerits*
Additive type	Easy operation, cost effective, suitable for industrial production	Migration and leaching of FR fillers, poor dispersion of FR fillers
Reactive type	Strong interaction between FR fillers and PU matrices, permanent and intrinsic FR effect	Increased complexity of the preparation reaction

Additive-Type Phosphorus-Based FRs for PUs

Red Phosphorus

Red phosphorus (RP) is an FR additive known for its high phosphorus content (100%). Despite its high FR efficacy, RP has some disadvantages, such as poor water resistance and poor compatibility with matrices. To overcome these shortcomings, microencapsulation technology has been employed to modify RP. Cao et al. compared the FR and smoke-suppressant effects between pristine RP and melamine–formaldehyde resin encapsulated RP (MFeRP) on the RPUFs (*11*). When the loading of RP or MFeRP is 80 parts per hundred polyols (pphp) or higher, the resulting RPUFs show a high LOI (> 25%) and a UL-94 V-0 rating. In addition, RPUF containing 100 pphp of MFeRP displays much lower heat release and smoke emission than the control RPUF and RPUF containing 100 pphp of RP, indicating better FR and smoke-suppressant efficiency. Based on the results from the pyrolysis–gas chromatography–mass spectrometry and Fourier transform infrared spectrometry, P_4 was detected in the gas phase during the thermal decomposition process of both RP and MFeRP, whereas the phosphorous content in the char of the RPUF containing MFeRP was much higher than that of the RP-containing counterpart. These results indicate that the gas phase mechanism plays a predominant role in the case of RP, whereas both the gas phase and condensed phase mechanisms work for MFeRP. This is because the wrap effect of melamine-formaldehyde resin inhibits the volatilization of RP into the gas phase. Therefore, the surface modification of RP not only enhances its compatibility with the matrices but also changes the mode of FR action.

APP

APP is one of the most widely used commercial phosphorus-based FRs because of its high FR efficiency and low cost. APP is produced by the reaction between ammonia and polyphosphoric acid (PPA); its molecular structure is shown in Figure 2. When heated, APP will decompose to generate ammonia and PPA. Ammonia is a nonflammable gas that can dilute the combustible gases during the combustion process, whereas PPA serves as a catalyst that promotes the dehydration, cyclization, and carbonization of polymer substrates to form a protective char layer. However, APP suffers from poor water tolerance and poor compatibility with polymer matrices, which obviously deteriorates the mechanical properties and environmental resistance of the polymers. To overcome these shortcomings, APP has been modified by the microencapsulation method. Tang et al. prepared microencapsulated APP by melamine–formaldehyde resin (*12*) and silica (*13*). The results of the LOI and UL-94 vertical burning test indicated that microencapsulated APP showed higher FR efficiency than unmodified APP. In addition, the FR RPUFs containing the microencapsulated APP also exhibited much better water tolerance and compressive strength than those containing unmodified APP. This is because the shell materials in microencapsulated APP not only enhance the water resistance and compatibility with polymer matrices, but they also create an FR synergism with APP. Hao's group reported a modified APP with β-cyclodextrin (β-CD) and incorporated it into RPUF (*14*). Compared with RPUF with unmodified APP, the RPUF containing the β-CD–modified APP showed a lower flame spreading rate and peak heat release rate (PHRR). In addition, the RPUF containing the β-CD–modified APP also displayed higher glass-transition temperature, cross-linking density, and thermal stability than its counterpart with unmodified APP. The authors attributed this superior FR effect of the β-CD–modified APP over the unmodified APP to the formation of an intumescent FR system. In this intumescent FR system, APP serves as an acid source and as a gas source and β-CD works as a carbon source, thus generating a prominent charring ability in the condensed phase. In addition to the heat release hazard, PU also emits a large amount of toxic gases, such as carbon monoxide and hydrogen cyanide, which usually cause casualties during fire accidents. Therefore, the smoke suppression effect is another issue of concern for the FR-PUs. Chen et al. utilized the combination of APP and ferrite yellow to prepare FR and smoke-suppressant TPU composites (*15*). The incorporation of ferrite yellow could not only decrease the PHRR and total heat release (THR) but also reduce the amount of smoke production and total smoke release. The decomposition products from ferrite yellow could stabilize the char residue to form an intact and dense char layer that inhibits the escape of the volatile products to feed the flame and convert them into smoke particles.

To explore the FR mechanism of APP on PU plastics, Grassie and Perdomo Mendoza investigated the influence of APP on the thermal decomposition behavior of the TPUs made from methylenediphenyl 4,4′-diisocyanate and polyethylene glycol (PEG) (*16*). It was found that APP could promote the elimination of carbon dioxide by the absence of the scission of the PU chains (Figure 3a). In addition, APP could catalyze the reaction between an acid species of polyphosphate and urethane hydrogen, resulting in the evolution of water and the cross-linking structure (Figure 3b). The subsequent reaction is the primary decomposition stage of the PU substrates. The macromolecular chains of the PU substrates split into methylenediphenyl 4,4′-diisocyanate and PEG under the acid-catalyzed depolymerization reaction by APP (Figure 3c). The isocyanate groups might then react with PPA, resulting in the formation of amines and carbon dioxide (Figure 3d). In addition, PEG undergoes dehydration and cyclization under the catalysis of APP, leading to the generation of 1,4-dioxane and water (Figure 3e). Thus, the PU substrates decompose into water and carbon dioxide during the thermal degradation process and can dilute the flammable volatile gases. It has also been found that most of the phosphorus is retained in the char, despite some being converted to high-boiling point volatiles, implying that APP causes action in the condensed phase (*17*).

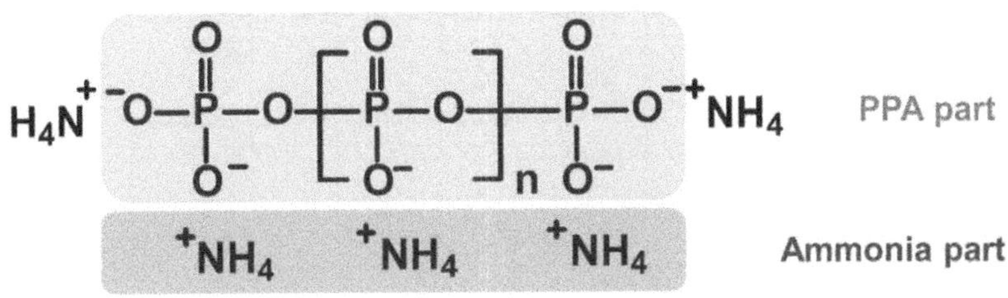

Figure 2. The molecular structure of APP.

Melamine–Phosphate Salts

Melamine polyphosphate (MPP) is another one of the most widely used commercial phosphorus-based FRs because of its high FR efficacy and low expense. MPP is manufactured from the reaction between melamine and PPA. The modes of FR action of MPP consist of two aspects: (1) The melamine decomposes to generate the nonflammable gases ammonia and nitrogen, which dilute the flammable volatiles; and (2) the PPA catalyzes the charring process of the polymer substrates to form an insulating char layer, which shields the heat irradiation. However, like APP, MPP has the disadvantages of poor water tolerance and poor compatibility with polymer matrices owing to its strong polarity. Therefore, the encapsulation strategy has been utilized for the surface modification of MPP. Zhang's group prepared microencapsulated MPP (MCMPP) by cross-linking carboxymethyl chitosan (Figure 4) (*18*). Subsequently, the MCMPP was melt blended with TPU plastics. With the addition of 15 wt % MCMPP, the PHRR, THR, total smoke release, and carbon monoxide production of the resulting TPU composite declined by 65.9, 24.1, 87.3, and 60.6%, respectively, all of which were more significant than those of the control sample containing the equivalent MPP. In addition, after immersion in hot water (70 °C) for 3 days, the TPU containing 15 wt % MPP burned out without any classification in the UL-94 vertical test, whereas the counterpart containing 15 wt % MCMPP retained the UL-94 V-0 classification, demonstrating the better water tolerance of MCMPP. This is because the surface encapsulation modification could prevent the inner MPP from leaching during the immersion. Li and collaborators synthesized an analog of MPP, dimelamine pyrophosphate (DMPY) (*19*). With the incorporation of 15 wt % DMPY, the TPU sample showed an LOI value of 27.3% and a UL-94 V-0 classification. After immersion in hot water (70 °C) for 1 week, the LOI and UL-94 performances of the TPU/DMPY exhibited almost no change, demonstrating the excellent water resistance of DMPY. It was noted that the pyrolytic gases from the decomposition of DMPY formed a high-speed jet to blow out the flame during the UL-94 burning test. The presence of DMPY was also conducive to forming a dense and intumescent char layer that played a crucial role in oxygen and heat insulation during the combustion process. The blowing-out effect in combination with the superior charring ability accounts for the dramatically improved FR characteristics of the TPU plastics.

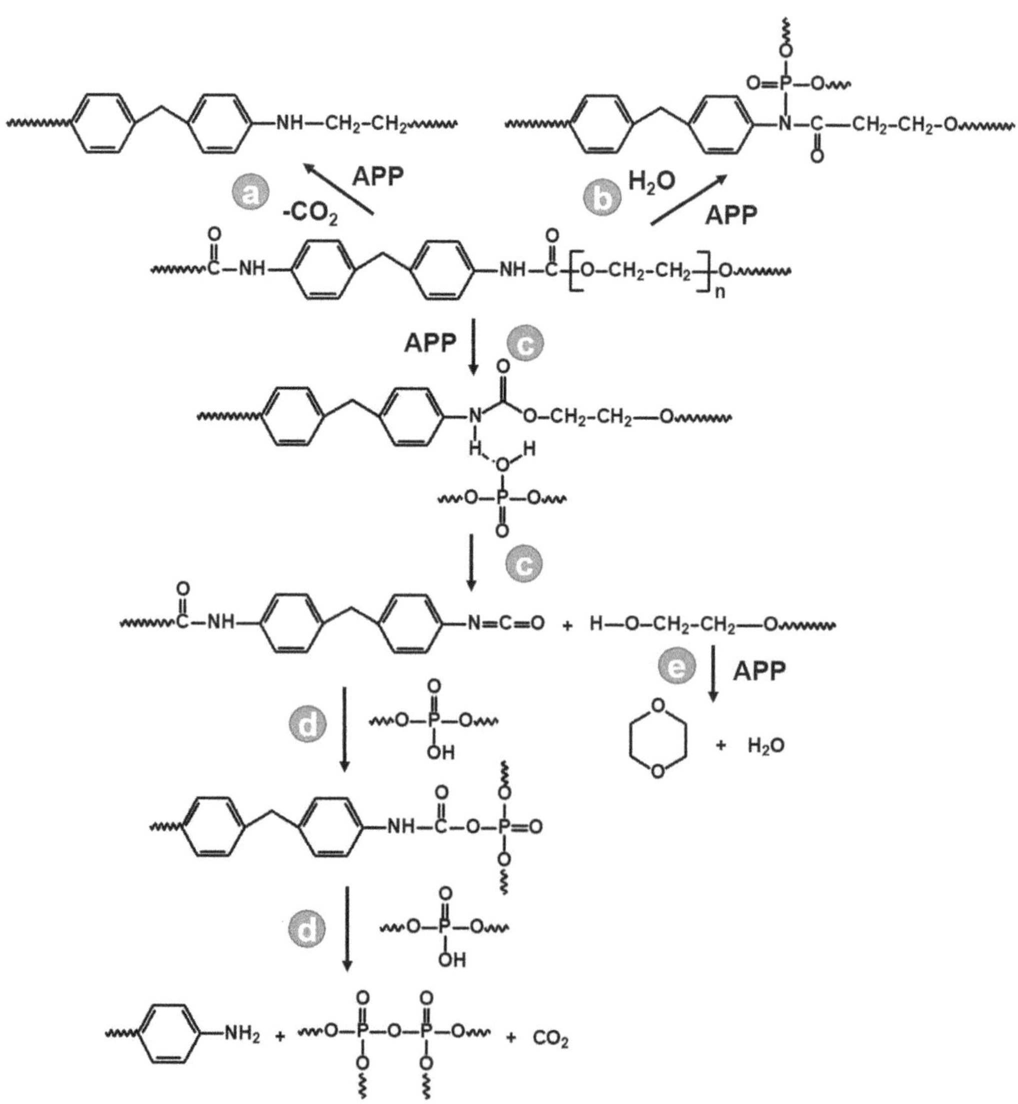

Figure 3. The thermal decomposition pathway of the TPUs in the presence of APP: (a) elimination of carbon dioxide; (b) evolution of water; (c) acid-catalyzed depolymerization reaction by APP; (d) reaction between isocyanate groups and PPA to yield amines and carbon dioxide; and (e) dehydration and cyclization under the catalysis of APP.

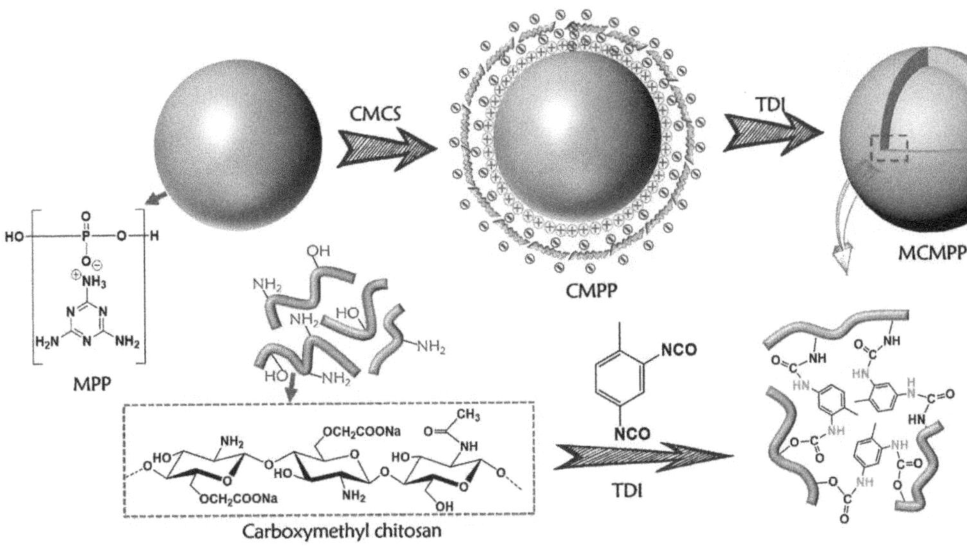

Figure 4. Schematic illustration of the preparation of MCMPP microcapsules. CMCS: carboxymethyl chitosan; CMPP: CMCS coated MPP; TDI: toluene-2, 4-diisocyanate. Reproduced with permission from reference (18). Copyright 2019 Elsevier.

Aryl Phosphates

Aryl phosphates (phosphonates) are one of the most common and effective types of FRs for PUs. Phosphonates such as triphenyl phosphate (Figure 5a) (*20*) and diphenyl cresyl phosphate (Figure 5b) (*21*) have been commercialized for PUF plastics. However, a serious disadvantage is the harm that triphenyl phosphate causes to aquatic organisms. This applies not only to triphenyl phosphate, but also to commercial triphenyl phosphate mixtures. In order to find an alternative to triphenyl phosphate or its mixtures, Lanxess invented a 9,10-dihydro-9-oxa-10-phosphaphenanthrene-10-oxide (DOPO)-derived FR additive (Figure 5c) (Levagard TP LXSS114) and applied it to produce flame-resistant PUFs with low core discoloration (*22*).

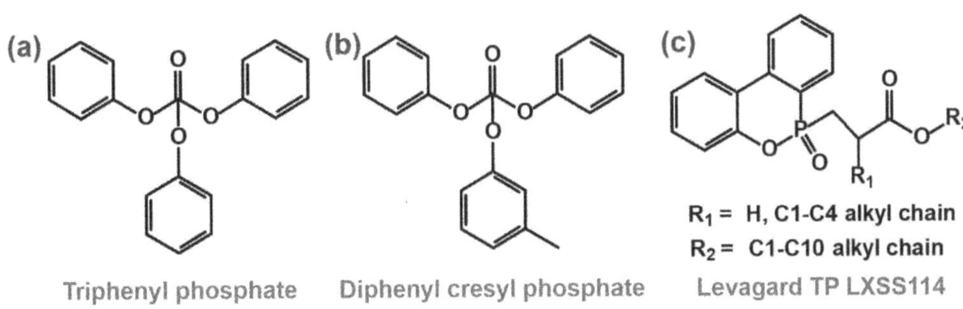

Figure 5. Molecular structure of (a) triphenyl phosphate; (b) diphenyl cresyl phosphate; and (c) Levagard TP LXSS114.

Phosphate–, Phosphonate–, and Phosphinate–Metal Salts

As one of the most widely used commercial halogen-free FRs, aluminum hypophosphite (AHP) has been used as an FR additive for various polymer materials (*23–26*). The effect of the different contents of AHP on the FR and mechanical properties of TPU was studied (*27*). It was found that the addition of 30 wt % AHP into TPU improved the LOI value to 30.2% and achieved a UL-94 V-0 rating. The condensed FR mechanism was proposed for the TPU/AHP composites, in which the presence of AHP was conducive to the formation of a more intumescent and dense char layer. However, such a high dosage of AHP caused the deterioration in tensile strength and elongation at the break of the TPU/AHP composites. In order to improve the FR efficacy, some synergistic agents are combined with AHP. For instance, the TPU composite containing 0.0625 wt % ionic liquid and 19.9375 wt % AHP exhibited a high LOI value of 35.75% and a UL-94 V-0 rating (*28*). The combination of APP and AHP was also reported for the FR-PU sealants (*29*). With the addition of 41.25 wt % FR additives (APP:AHP = 5:1), the PU sealant displayed a maximum LOI value of 35.1%, much higher than the counterparts containing only APP or AHP, suggesting the formation of a good synergism between APP and AHP. The synergistic mechanism between APP and AHP was attributed to the combined mechanism of both vapor and condensed phases. In the vapor phase, the thermal decomposition of APP and AHP induced the emission of the nonflammable volatiles (ammonia and water) that had a fuel dilution effect. Moreover, some phosphorus-containing radicals (PO·, O_2P·, etc.) were generated to trap the HO· and O· radicals, thus quenching the chain reaction of combustion. In the condensed phase, APP and AHP decomposed to generate the PPA that then catalyzed the formation of a dense and insulating char barrier. Furthermore, aluminum salts were converted into thermally stable aluminum oxide, strengthening the thermal oxidative resistance of the char layer.

In addition to inorganic AHP, the effect of some organophosphorus (OP)–metal salts on the FR property of PU materials have also been investigated. Hu et al. studied the influence of aluminum diethylphosphinate (ADP) on the thermal stability and flame retardancy of FPUFs (*30*). With the dosage of 20 wt % ADP, the PHRR and THR of the resulting FPUF declined by 35 and 33%, respectively. However, there was no classification in the UL-94 vertical burning test for the FPUF containing 20 wt % ADP. Therefore, ADP has been combined with other FR additives to improve the FR effect. The coaddition of 10 pphp of ADP and 20 pphp of EG improved the LOI of the RPUF composite to 25.9% and achieved a UL-94 V-1 rating (*31*). In another study, the TPU composite with the addition of 40 wt % OP–metal salt (consisting of aluminum OP, MPP, and zinc borate) achieved a high LOI value of 33.5% and a UL-94 V-0 rating (*32*). The FR mechanism of the TPU containing 40 wt % OP–metal salt was attributed to the condensed phase charring action by generating a thicker and more compact char layer during combustion (Figure 6). Modesti and collaborators carried out a comparative study on the FR efficacy of aluminum phosphinate (inorganic) and aluminum diethylphosphinate (organic) in PUFs (*33*). Both inorganic and organic phosphinates could effectively improve the FR property of PUFs through flame quenching by the phosphorus-containing radicals in the vapor phase. However, better behaviors in the cone calorimeter test were observed for the inorganic phosphinate, and they could be associated with the better quality of the char residue in the condensed phase.

Based on the previous description, it can be found that a high loading of AHP or ADP is required to obtain a satisfactory FR effect for PU materials. In order to find an alternative to AHP or ADP, other OP–metal salts have been synthesized. Han et al. synthesized a dicyclohexyl aluminum hypophosphate (DAH) (Figure 7) and compared its FR effect with that of ADP in the RPUFs (*34*). The average combustion time of the RPUF containing 15 wt % ADP was 12.5 s, whereas that of the RPUF containing the equivalent DAH was significantly reduced to 4.4 s owing to the higher charring ability of DAH. In another study, three kinds of salen-based OP–metal (nickel, zinc, and manganese) salts (Figure 7) were synthesized and used as FR additives for TPU (*35*). The PHRR value of the TPU composites containing 3 wt % salen-OP-Ni, salen-OP-Zn, and salen-OP-Mn was declined by

53.8, 41.5, and 45.6%, respectively, compared with that of the pristine TPU. In addition, the carbon monoxide emission of the TPU composites containing 3 wt % salen-OP-Ni and salen-OP-Zn was also decreased by 78.2 and 80.0%, respectively. The FR mechanism of the salen-based OP–metal salts could be attributed to the fuel dilution effect caused by releasing more nonflammable carbon dioxide as well as to the formation of high-quality char that consisted of thermally stable metal oxide and had a more graphited structure.

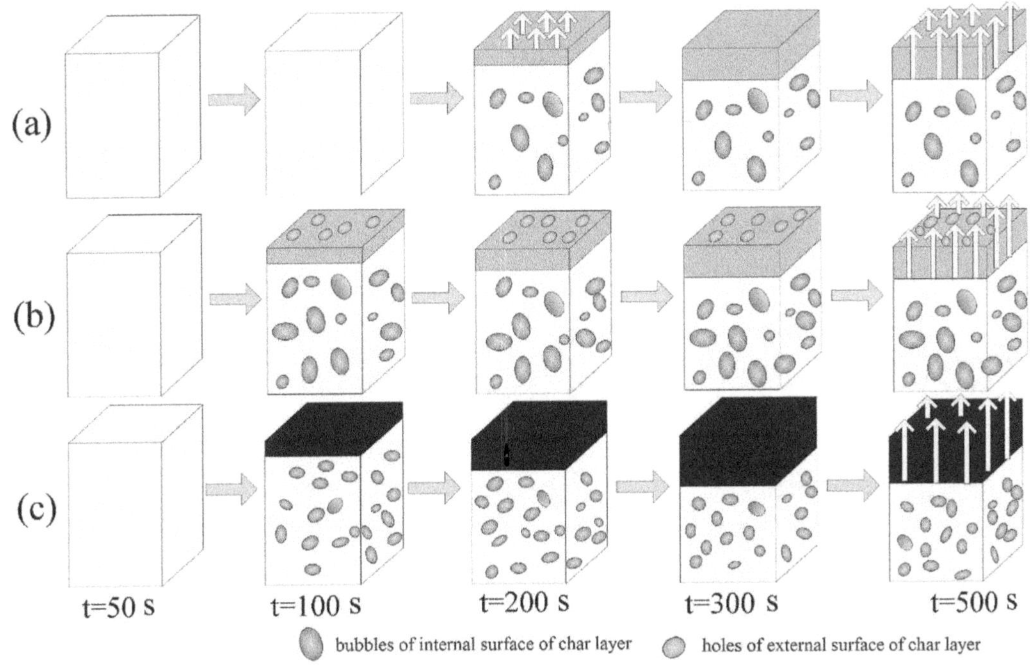

Figure 6. Diagrammatical illustration of the formation of chars during combustion: (a) TPU; (b) TPU containing 20 wt % OP–metal salt; and (c) TPU containing 40 wt % OP–metal salt. Reproduced with permission from reference (32). Copyright 2015 Elsevier.

Structural Influence of OP Compounds on the FR Property of PUs

As previously discussed, the phosphorous-containing FR additives include phosphates, phosphonates, and phosphinates. These phosphorous-containing compounds differ in their oxidation state of phosphorus. In order to clarify the influence of the phosphorus valency on the FR property of PUs, Modesti et al. studied the thermal decomposition behavior of the RPUFs containing AHP, dimethylpropanphosphonate, triethylphosphate, and APP (*36*). It was found that the vapor and condensed phase action occurred simultaneously at a lower oxidation state of phosphorus, whereas only the condensed phase action occurred at a higher oxidation state of phosphorus (Figure 8) (*37*). Gaan's group also investigated the structural influence of OP compounds (phosphoramidates, phosphonates, and phosphates) (Figure 9) on the FR property of FPUFs (*38*). The FPUFs containing phosphonates and phosphoramidates demonstrated better FR performance than those with phosphates, and the allyl-containing phosphonate and phosphoramidate demonstrated better FR performance than the propyl-containing analogs. Based on gas chromatography–mass spectrometry analysis, it was found that the FPUFs containing phosphonate and phosphoramidate, especially allyl phosphonate and allyl phosphoramidate, were more effective in releasing PO· and O_2P· radicals in the vapor phase than the phosphate-containing analog. This was responsible for the better FR performance in the FPUFs.

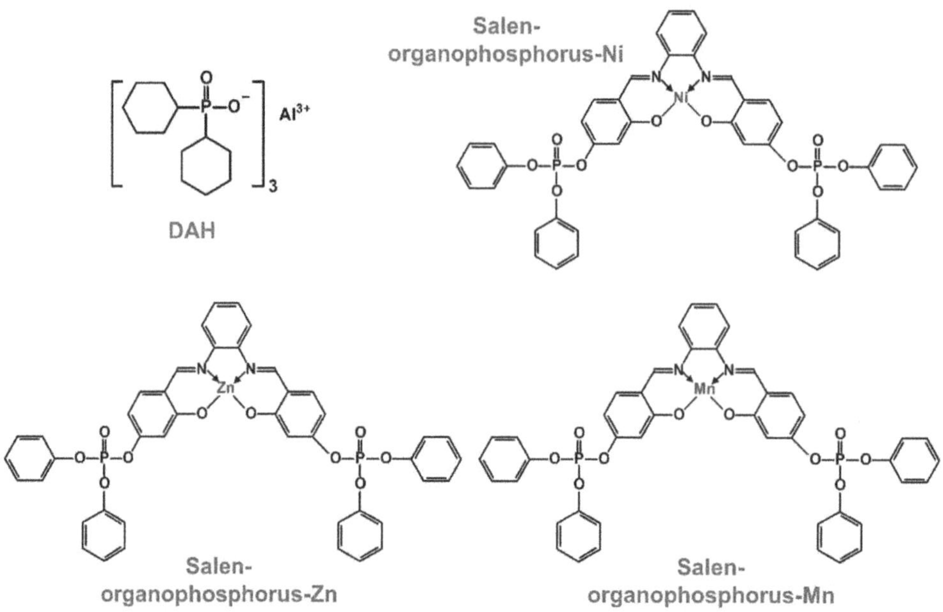

Figure 7. Molecular structures of DAH and salen-based OP–metal salts.

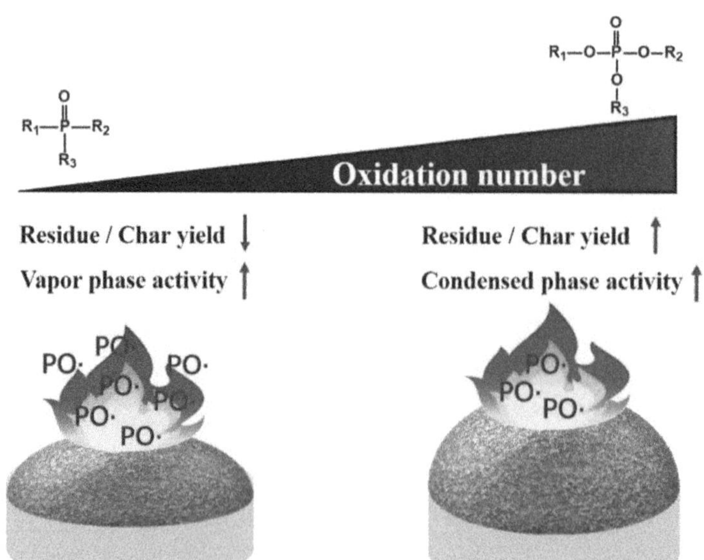

Figure 8. The influence of the oxidation state of phosphorus on the FR mode of action.

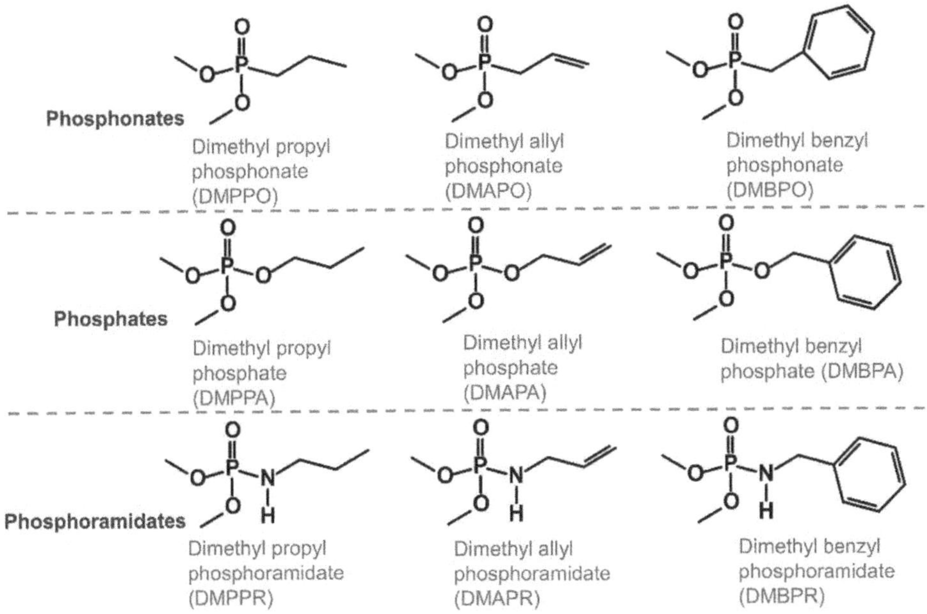

Figure 9. Molecular structure of phosphoramidates, phosphonates, and phosphates.

OP Polymeric FR Additives

Compared with low-molecular-weight phosphorus-containing FR additives, OP polymeric FR additives possess many advantages, including nonmigration, richer aromatic groups, and higher efficiency (*39–42*). Wang et al. synthesized a Schiff-base polyphosphate ester (SPE) for FR TPU (*43*). With only 5 wt % SPE, the resulting TPU composite achieved a UL-94 V-0 classification and an LOI of 29%. Table 2 compares the FR TPU performance between SPE and other low-molecular-weight FR additives previously reported. It can be seen that a much lower dosage of SPE was required to obtain a satisfactory FR level compared with other low-molecular-weight FR additives. The high efficiency of the SPE could be attributed to its FR mode of action in both the condensed and vapor phases.

Table 2. Comparison of the FR TPU Performance Between SPE and Other Low-Molecular-Weight FR Additives

Type of FRs	*Loading (wt %)*	*UL-94 Rating*	*LOI (%)*	*Reduction in PHRR (%)*	*Reference*
SPE	5	V-0	29.0	61.7	(*43*)
MPP	15	V-2	26.1	57.8	(*18*)
MCMPP	15	V-0	29.4	65.9	(*18*)
DMPY	15	V-0	27.3	88.7	(*19*)
AHP	20	V-1	32.5	91.6	(*28*)
AHP	30	V-0	30.2	38.8	(*27*)
OP	40	V-0	33.5	48.3	(*32*)
Salen-OP-Ni	3	/	/	53.8	(*35*)

Table 2. (Continued). Comparison of the FR TPU Performance Between SPE and Other Low-Molecular-Weight FR Additives

Type of FRs	*Loading (wt %)*	*UL-94 Rating*	*LOI (%)*	*Reduction in PHRR (%)*	*Reference*
Salen-OP-Zn	3	/	/	41.5	(35)
Salen-OP-Mn	3	/	/	45.6	(35)

Reactive-Type Phosphorus-Based FRs for PUs

The reactive-type FR method refers to the inclusion of FR groups into the molecular chains of polymers by a formation of covalent bonding. Given that PU materials are produced from the reaction between diisocyanates or polyisocyanates and diols or polyols, the reactive-type phosphorus-based FRs are mostly selected from those that could react with diisocyanates or polyisocyanates. The reactions between the phosphorus-based FRs and diisocyanates or polyisocyanates are summarized in Figure 10. According to the molecular structures, the reactive-type phosphorus-based FRs are divided into diols and polyols, primary and secondary amines, and epoxy-based compounds.

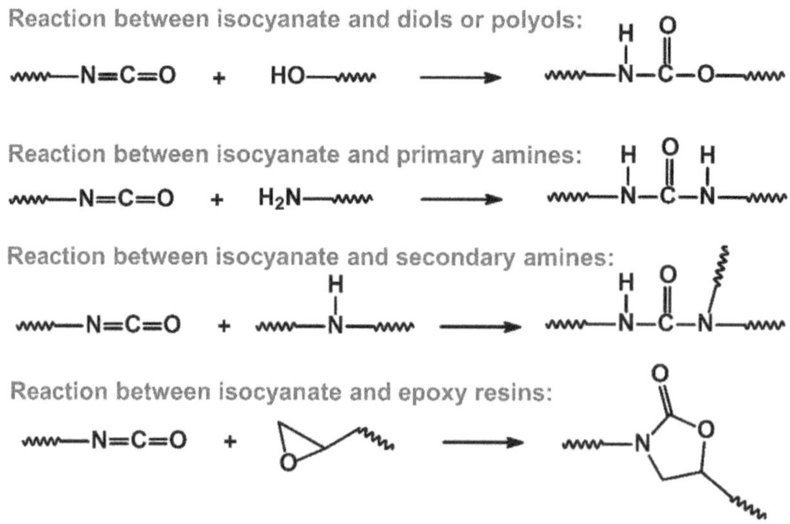

Figure 10. The reaction mechanisms of reactive-type phosphorus-based FRs.

Phosphorus-Containing Alcohols

The use of phosphorus-containing alcohols is the most common method used to prepare the reactive-type FR-PUs because of their relatively high FR efficiency. A phosphorus-containing diol, 2-ethyl-2-(2-oxo-5,5-dimethyl-1,3,2-dioxaphosphorinanyl-2-methylene)-1,3-propanediol (EPPD), was synthesized and used as a chain extender to obtain a series of FR waterborne PU (FRWPU) (*44*). The FRWPU containing 12 wt % EPPD exhibited an LOI value of 26.6% (Figure 11) and a UL-94 V-0 classification, indicating good FR effect. The catalytic formation of a rugged carbonaceous char in the condensed phase by EPPD was believed to account for the enhanced antiflammability of FRWPU. Morgan and collaborators synthesized six phosphorus-containing diols (Figure 12a) and incorporated them into PU materials (*45*). The molecular structure of the diols played an important role in the heat release reduction, with an aromatic core being somewhat more effective than an aliphatic one. Gupta et al. prepared a phosphorus-containing diol (PPA-PO) (Figure 12b) from the reaction between phenyl phosphonic acid and propylene oxide (*46*). The PUF with 18.1 wt % PPA-PO displayed a 160% increment in the compressive strength owing to the reactive nature of PPA-PO. In the cone calorimeter test, the PHRR and THR declined by 68.6 and 23.4%, respectively, for the PUF with 18.1 wt % PPA-PO; the decreases were attributed to the formation of a higher-quality char residue containing more graphitic phase. However, the antiflammable behaviors in terms of the LOI and UL-94 vertical burning test were not mentioned. In this regard, some commercial reactive-type FRs have also been invented, such as diethyl bis(2-hydroxyethyl)aminomethylphosphonate (Levagard 4090N, Lanxess) (Figure 12c), hydroxymethyl phosphonates (Figure 12d) (*47*), and 9,10-dihydro-9-oxa-10-oxide-10-[2,3-di-(2-hydroxylethoxyl)carbonylpropyl]-10-phosphaphenanthrene-10-oxide (DOPO-itaconic acid diglycol ester) (Figure 12e) (*48*), which have been applied in fabricating FR-PU coatings.

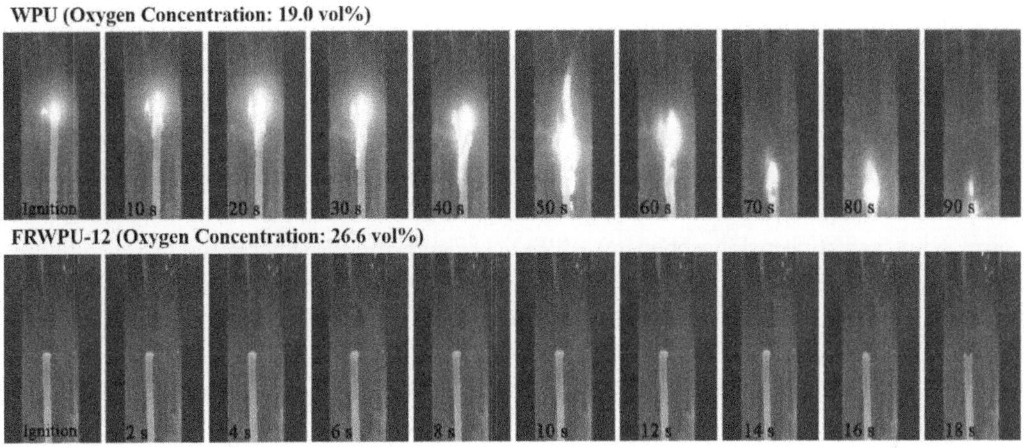

Figure 11. Combustion processes of waterborne PU (WPU) and FRWPU-12 samples at oxygen concentrations of 19.0 and 26.6%, respectively. Reproduced with permission from reference (44). Copyright 2015 Elsevier.

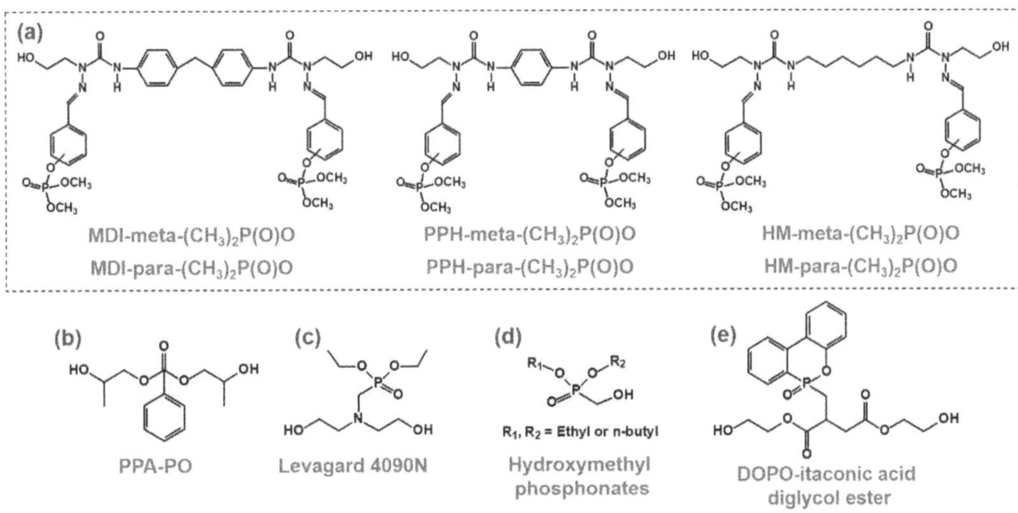

Figure 12. Chemical structures of (a) six kinds of phosphorus-containing diols; (b) PPA-PO; (c) Levagard 4090N; (d) hydroxymethyl phosphonates; and (e) DOPO-itaconic acid diglycol ester.

In addition to the low-molecular-weight phosphorus-containing diols, several phosphorus-containing polymeric diols have been reported for FR-PUs. Cheng et al. prepared a phosphorus-containing polyester diol (FRD) (Figure 13a) and incorporated it into a waterborne PU (WPU) (*49*). The resulting WPU containing 7 wt % FRD had a high LOI value of 30.5%, as well as a UL-94 V-0 grade. Because the FRD simultaneously contained phosphorus with two oxidization states, it had a quenching effect on free radicals in the vapor phase and a catalytic charring effect in the condensed phase. Wang's group synthesized phosphorus-containing polymeric diols named phenylphosphonic ethylene glycol oligomer (PDEO) (Figure 13b) (*50*) and diethanol amine methyl oligomeric phosphonate (DMOP) (Figure 13c) (*51*) as the chain extenders for the preparation of FPUF. The FR FPUF containing 10 pphp PDEO or DMOP showed self-extinguishing behaviors in the vertical burning test. It was noted that, after the thermal aging test, the FPUF sample containing PDEO or DMOP displayed a persistent FR effect, whereas the control sample containing dimethyl methyl phosphonate showed a significantly deteriorated antiflammability. The FR mechanism could be attributed to the fact that the phosphorus-containing pyrolytic products released from the decomposition of PDEO or DMOP suppressed the effective heat of combustion of the products in the vapor phase. However, the addition of PDEO or DMOP had a limited suppression influence on the PHRR and THR of the FPUF materials and even an adverse effect on the carbon monoxide yield and smoke production during the combustion. Wang's group also prepared another two kinds of phosphorus-containing polymeric diols—poly(2-butyne-1,4-diol phenylphosphonate) (PPBP) and poly(1,4-butanediol phenylphosphonate) (PPBOP) (Figure 13d,e)—for FR TPU (*52*). Although the addition of 12 wt % PPBP or PPBOP enabled the TPU composite a UL-94 V-0 material, only the TPU containing 12 wt % PPBP showed antidripping effect. This is because of the superior charring capability of PPBP over PPBOP (char yield at 700 °C under nitrogen: 45.0% vs 1.6%). The alkynyl-group played a vital role in the formation of a cross-linking structure during the thermal decomposition process, resulting in a dramatically improved charring capability (Figure 14). In another study, the effect of a commercial phosphorus-containing polymeric diol (Exolit OP560) on the FR property of aqueous PU dispersions (PUDs) was studied (*53*). As the OP560 dosage increased, the FR performance of the PUDs was improved gradually. Specifically, the PUD containing 8.7 wt % OP560 achieved an LOI value of 30.4% and a UL-94 V0 grade. The exceptional antiflammability was attributed to a combined mechanism of the fuel dilution effect in the vapor phase and the formation of a protective char in the condensed phase.

A series of phosphorus-containing polyols have been designed and successfully synthesized for FR- PUs with outstanding comprehensive properties. Sabnis et al. reported a phosphorus-containing triol (Figure 15a) that was produced from the reaction between phosphorus oxychloride and *N*-methylaminoethanol (*54*). This phosphorus-containing triol was covalently incorporated into the PU backbone in different dosages (30, 40, and 50% on an equivalent basis) to obtain a series of FR aqueous PUDs (FRPUDs). The FRPUD with 50% phosphorus-containing triol exhibited a high LOI value of 37% and UL-94 VTM-0 grade at the cost of its flexibility and impact resistance. Mhaske and coworkers synthesized a cyclic phosphorus-containing polyol (FRPOL) (Figure 15b) as a chain extender for PU coatings (*55*). The LOI value of the FR-PU coating with FRPOL was increased to 32% from 21% for the control sample, and the FR-PU coating with FRPOL also passed the UL-94 V-0 test. In addition, the higher glass-transition temperature and water resistance were also observed for the FR-PU coating with FRPOL owing to the increased cross-linking density. Patel and Mishra prepared a series of phosphorus-containing polyols (Figure 15c) containing three different molecular weights of polypropylene glycol (PPG) (*56*). It was found that the FR effect decreased with the increase of the molecular weight of PPG in the phosphorus-containing polyols owing to the relatively low phosphorus content in the higher-molecular-weight PPG-containing polyols. In contrast to the FR property, the scratch hardness and the pencil hardness of the PU coatings increased as the molecular weight of PPG increased. Therefore, the optimum comprehensive properties can be balanced by adjusting the molecular structure of the phosphorus-containing polyols. Other phosphorus-containing polyols (Figure 15d–f) have also been fabricated using triazine (*57*) and phosphazene (*58*, *59*) as core materials. These FRs have achieved a certain FR effect for the RPUFs owing to the synergistic effect of phosphorus and nitrogen on the enhanced charring ability. However, the LOI value was not very high (≤ 26%) when the dosage of the phosphorus-containing polyols was 19–20.0 wt %. The development of phosphorus-containing polyols with high FR efficacy remains a challenging task.

With the increasing concern for the depletion of crude oil resources and the emission of greenhouse gases, replacement of petroleum-based polyols by biobased ones for the fabrication of FR-PUs has received more and more attention. Various phosphorus-containing biobased polyols have been synthesized from ricinoleic acid (*60*), castor oil (*61*, *62*), sebacic acid (*63*), thioglycerol (*64*), tung oil (*65*), and gallic acid (*66*). These phosphorus-containing biobased polyols have been demonstrated to be effective in reducing the heat release rate of PU coatings and foams, but only PU coatings have achieved the UL-94 V-0 grade. This is because the porous structure of PUFs has a higher contact surface with air when burning, making it more difficult for PUFs to achieve self-extinguishing behaviors. To improve the FR effect for PUFs, conventional FR additives such as EG and triethyl phosphate (*67*) can be combined with phosphorus-containing biobased polyols to create a synergistic action.

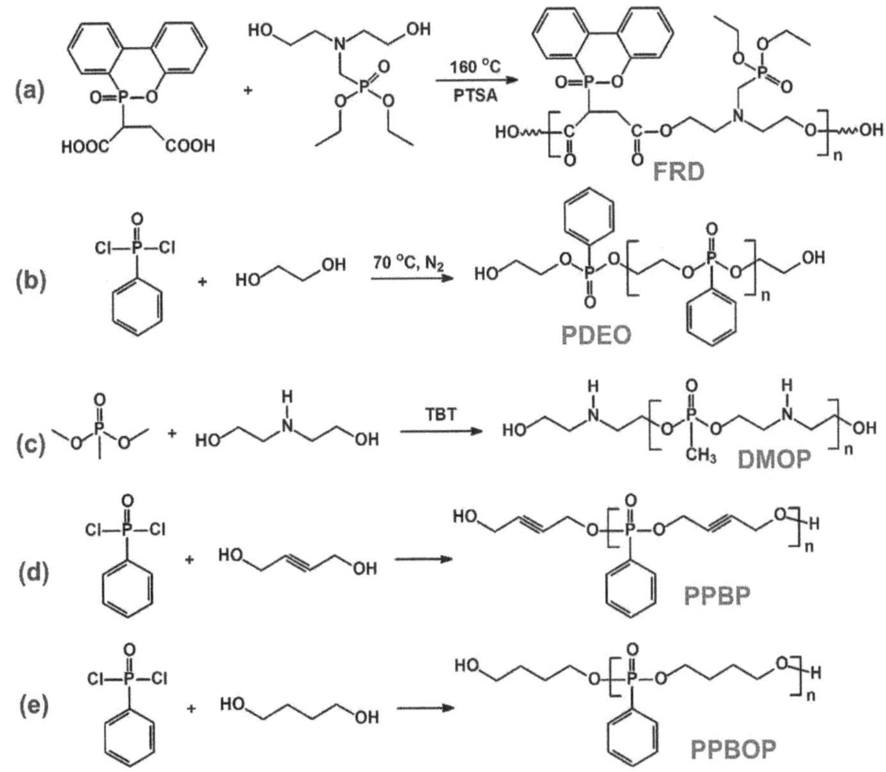

Figure 13. The synthetic route of several phosphorus-containing polymeric diols: (a) FRD; (b) PDEO; (c) DMOP; (d) PPBP; and (e) PPBOP.

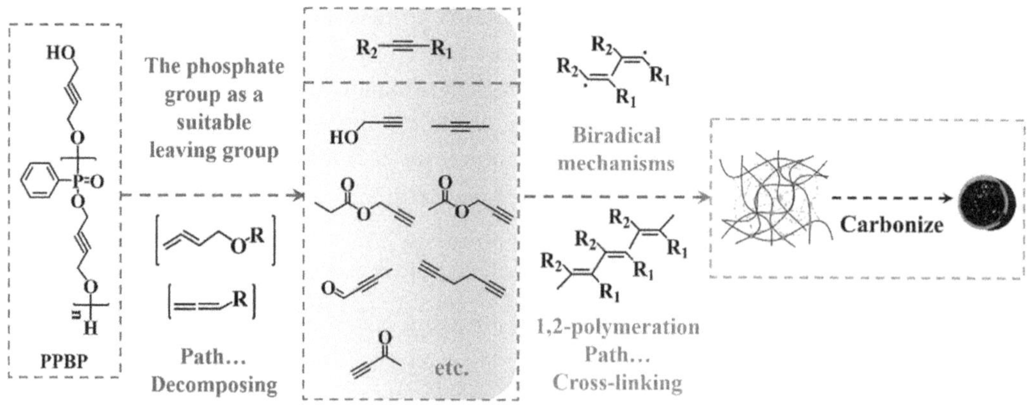

Figure 14. Schematic illustration of the charring mechanism of PPBP. Reproduced with permission from reference (52). Copyright 2020 Elsevier.

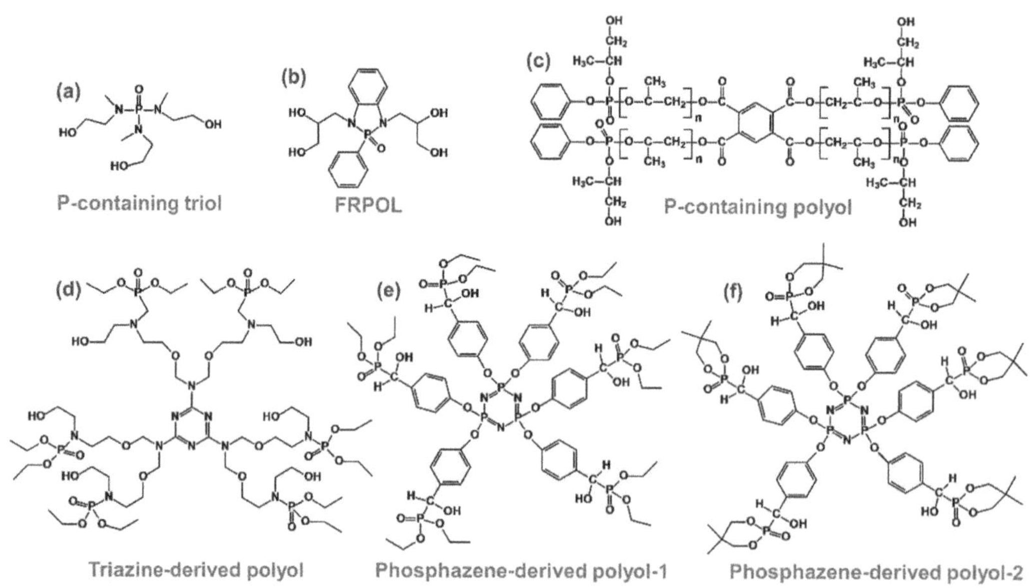

Figure 15. The molecular structure of a series of phosphorus-containing polyols: (a) P-containing triol; (b) FRPOL; (c) P-containing polyol; (d) triazine-derived polyol; (e) phosphazene-derived polyol-1; and (f) phosphazene-derived polyol-2.

Phosphorus-Containing Amines

Phosphorus-containing amines have also attracted increasing interest for FR-PUs because they have some advantages, including high reactivity with isocyanate and synergism between phosphorus and nitrogen. Wang and collaborators fabricated a multifunctional PU elastomer by covalently bonding a phosphorus-containing amine, tri(2-furyl) phosphoramide (TFP) (Figure 16a), into the molecular chain of a maleimide-terminated PU (MPU) (68). The as-prepared MPU displayed a high tensile strength (37.11 MPa) and a good self-healing efficiency (91.8%) owing to the presence of a thermoreversible cross-linking network and multiple hydrogen interactions. Although the LOI value of the MPU containing 8 wt % TFP was increased to 28.5% from 19.5% for the unmodified MPU, the reduction in the PHRR and THR was limited in the cone calorimeter test. Luo et al. synthesized a cyclic phosphorus-containing amine, octahydro-2,7-di(*N,N*-dimethylamino)-1,6,3,8,2,7-dioxadiazadiphosphecine (ODDP) (Figure 16b), as a chain extender for the preparation of an FRWPU (*69*). The WPU with 15 wt % ODDP showed a high LOI of 30.6% and a UL-94 V-0 rating. The improved antiflammability was attributed to the catalytic charring effect of ODDP, which served as a physical barrier for insulating oxygen and heat. In another work, a phosphorus-containing secondary amine, 2-((2-hydroxyphenyl)(phenylamino)methyl5,5-dimethyl-1,3,2-dioxaphosphinane 2-oxide (HAPO) (Figure 16c), was successfully synthesized and applied for FR RPUF (*70*). When the HAPO content was 10 wt %, the PHRR, THR, and total smoke production values of the resulting RPUF were decreased by 36.6, 12.3, and 38.8%, respectively, compared with those of the pristine PU. It was also noted that the RPUF containing 15 wt % HAPO could pass the UL-94 V-0 grade. The remarkable improvement in antiflammability was mainly ascribed to the formation of a thermally stable phosphorus-rich char layer that could inhibit the diffusion of heat and evolved decomposition products during combustion.

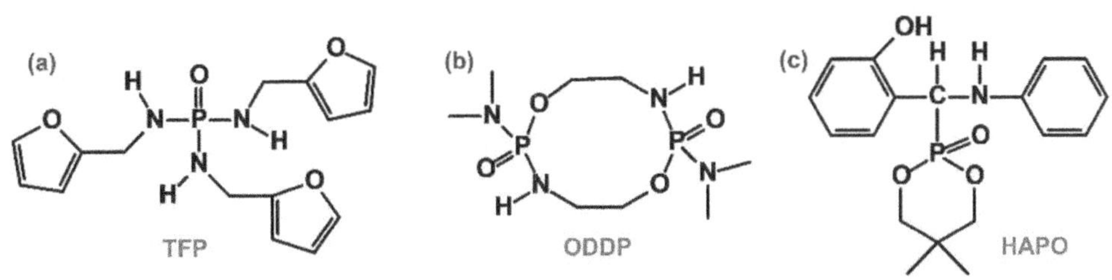

Figure 16. The molecular structure of (a) TFP, (b) ODDP, and (c) HAPO.

Among the various phosphorus-containing compounds, DOPO is one of the most commonly used candidates to manufacture phosphorus-containing amines. For example, diverse phosphorus-containing amines, such as DOPO-benzylideneaniline (*71*), 2-[anilino-(6-oxobenzo[c][2,1] benzoxaphosphinin-6-yl)methyl]phenol (*72*), DOPO-melamine (*73*), DOPO-dicyandiamide (*73*), and DOPO-urea (*73*) have been synthesized (Figure 17), and these compounds have endowed both the RPUFs and FPUFs with excellent antiflammable properties. Gaan et al. reported a series of DOPO-based phosphonamidates for manufacturing FR FPUF (*74*). The UL-94-horizontal burning test results demonstrated that the DOPO-based phosphonamidates exhibited a better antiflammable property as compared with the commercial FRs, including tris(1-chloro-2-propyl) phosphate (TCPP), DOPO, and Exolit OP 560. The FR mechanism analysis showed that these DOPO-phosphonamidates acted mainly in the vapor phase rather than the condensed phase (as evidenced by the low char yield at high temperature). In addition, among these DOPO-phosphonamidates, 6,6′-(ethane-1,2-diylbis-(azanediyl)-bis-(6H-dibenzo[c,e][1,2]-oxaphosphine-6-oxide phosphon amidate displayed the best FR behavior, which was attributable to the higher thermal stability of 6,6′-(ethane-1,2-diylbis-(azanediyl)-bis-(6H-dibenzo[c,e][1,2]-oxaphosphine-6-oxide that presumably provided more sustained vapor phase activity during the entire combustion process. Gaan et al. also synthesized a series of DOPO-based primary and secondary amines and compared their FR effects in RPUFs with conventional FRs, including TCPP and a reactive phosphorus-containing oligomeric diol (PLF140) (*75*). As shown in Table 3, the incorporation of 2 wt % phosphorus in the RPUFs achieved the LOI value of 22.2–23.2% and HF-1 grade in the UL-94 horizontal test. Furthermore, the RPUFs containing DOPO-based amines showed lower PHRR and THR values and higher char yields than those containing TCPP and PLF140. Unlike TCPP, DOPO-based amines were not only active in the vapor phase but also possessed a potent charring activity in the condensed phase, inhibiting the diffusion of more pyrolytic products to the vapor phase, thus reducing fuel and smoke sources.

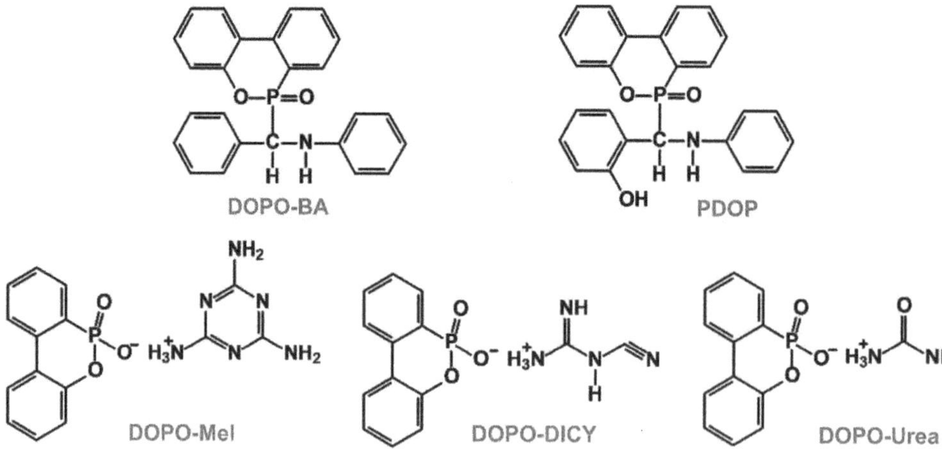

Figure 17. The molecular structure of DOPO-benzylideneaniline (BA), 2-[anilino-(6-oxobenzo[c][2,1]benzoxaphosphinin-6-yl)methyl]phenol (PDOP), DOPO-melamine (Mel), DOPO-dicyandiamide (DICY), and DOPO-urea.

Table 3. Comparison of the FR Effect of DOPO-Based Amines with Conventional TCPP and PLF140 in RPUFs[a]

Sample	*UL-94 HB*	*LOI (%)*	*PHRR (W/g)*	*THR (kJ/g)*	*Residue (wt %)*
RPUF	HBF	19.3	262.7	20.8	11.1
RPUF/TCPP	HF-1	23.2	193.4	20.2	11.2
RPUF/PLF140	HF-1	22.2	228.7	17.5	19.0
RPUF/ETA-DOPO[b]	HF-1	22.2	121.7	18.2	22.1
RPUF/DTE-DOPO[c]	HF-1	22.3	179.7	16.8	25.2
RPUF/AAM-DOPO[d]	HF-1	22.2	139.2	17.1	21.8

[a] Adapted with permission from reference (75). Copyright 2017 Elsevier. [b] ETA-DOPO: 6-(2-((6-oxido-6H-dibenzo[c,e][1,2]oxaphosphinin-6-yl)amino)ethoxy)-6H-dibenzo[c,e][1,2]oxaphosphinine 6-oxide. [c] DTE-DOPO: 6-(2-(4,6-diamino-1,3,5-triazin-2-yl)ethyl) dibenzo[c,e][1,2]oxaphosphinine 6-oxide. [d] AAM-DOPO: 3-(6-oxidodibenzo[c,e][1,2]oxaphosphinin-6-yl)propenamide.

Phosphorus-Containing Epoxy-Based Compounds

Until now, there have been few reports on the FR-PUs using phosphorus-containing epoxy-based compounds. He et al. synthesized a reactive triglycidyl phosphate (Figure 18a) from phosphorus oxychloride and glycidol and then covalently bound it to the molecular chains of an RPUF (76). Despite a 30.2% reduction being observed for the FR RPUF, the increment in the LOI was limited and the UL-94 V-0 grade could not be achieved. In another study, He et al. also synthesized a DOPO derivative containing a terminal epoxy group (e-DOPO) (Figure 18b) (77). Compared with DOPO and triethylphosphate, the FR RPUF using e-DOPO achieved lower PHRR and THR values during combustion, but its LOI value was not effectively improved. Moreover, the authors had not mentioned the UL-94 vertical burning behavior. Rolf and Rüdiger also invented a 9,10-dihydro-9-oxa-10-oxide-10-[1-propane-2,3-oxide]-10-phosphaphenanthrene-10-oxide (Figure 18c) for an FRPUD (*48*).

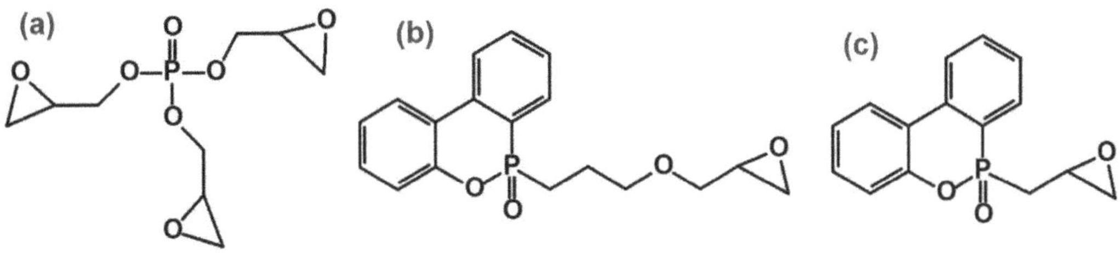

Figure 18. Chemical structure of (a) triglycidyl phosphate; (b) e-DOPO; and (c) 9,10-dihydro-9-oxa-10-oxide-10-[1-propane-2,3-oxide]-10-phosphaphenanthrene-10-oxide.

Selection of FRs for PUs

PUs are mainly applied as rigid foams, flexible foams, coatings and sealants, and elastomers. Table 4 summarizes the commercial FRs available for different applications of PUs. The selection of FRs for PUFs is affected by several crucial factors, including compatibility, volatility, smoke and toxicity, expense, and color stability. Thus, the choice of FR additives for PUFs is rather limited. Fortunately, some reactive phosphorus-containing FRs have been developed to meet the FR requirements according to the building code criteria (for RPUFs) and the automotive and furniture specifications (for FPUFs). Compared with rigid and flexible foams, PU elastomers and coatings have a wider selection of FRs. Generally, FRs seem to be more efficient in PU coatings, sealants, and elastomers. This is because there is less air contact surface and a lower specific surface ratio in these nonfoamed PUs.

Table 4. Commercial FRs for Different Applications of PUs

Type of PUs	*Additive FRs*	*Reactive FRs*
Rigid foams	Tris(2-chloroethyl) phosphate, tris(1-chloro-2-propyl) phosphate, tris(2-chloroisopropyl) phosphate, dimethyl methyl phosphonate, triphenyl phosphate, isopropylphenyl diphenyl phosphate, tricresyl phosphate, trixylenyl phosphate, APP-based intumescent system (Exolit AP 750)	Phosphorus-containing diols or polyols (e.g., Fyrol 6; Exolit OP 550)
Flexible foams	Tris(2-chloroisopropyl) phosphate, tris(1,3-dichloro-2-propyl) phosphate, oligomeric ethyl phosphate, triaryl phosphates	Phosphorus-containing diols or polyols (e.g., Exolit OP 550)
Coatings and sealants	Type-II APP, tris(chloroisopropyl) phosphate, tris(1,3-dichloro-2-propyl) phosphate, alkyl phosphates (phosphonates)	Phosphorus-containing diols or polyols (e.g., Exolit OP560; Levagard 4090N)
Elastomers	Type-II APP, RP, tris(chloroisopropyl) phosphate, tris(1,3-dichloro-2-propyl) phosphate, tris(butoxyethyl) phosphate, diphenyl octyl phosphate	Phosphorus-containing diols or polyols (e.g., Exolit OP560)

Summary

The utilization of FRs is currently the most common way for overcoming the flammability of PUs. Chlorinated FR additives are regarded as one of the most efficient FRs for PUs, especially FPUFs. However, chlorinated FR additives emit toxic and corrosive hydrogen halides during combustion that harm human health and the environment. Thus, there is an inevitable trend to develop halogen-free FRs. Among a wide variety of halogen-free FRs, phosphorus-containing compounds have attracted considerable interest because of their relatively high FR efficiency and environmental friendliness. According to the chemical structure of the FRs in the PU substrates, phosphorus-containing compounds are divided into two primary categories: additive type and reactive type. The additive-type FRs for PUs mainly include RP; APP; melamine–phosphate salts; phosphonates; phosphate–, phosphonate–, and phosphinate–metal salts; and some phosphorus-containing polymers. These additive-type FRs can be facilely mixed with PU substrates to obtain FR materials, but the high dosage is the main disadvantage because it causes an adverse effect on the mechanical property and thermal stability of PUs. The reactive-type phosphorus-containing FRs for PUs are primarily emphasized in the phosphorus-containing alcohols, phosphorus-containing amines, and phosphorus-containing epoxy-based compounds. Although the reactive-type phosphorus-containing FRs have been demonstrated to be effective for WPUs and TPUs, they usually require the addition of synergistic agents to obtain a satisfactory FR effect for RPUFs or FPUFs (*78–81*). In addition, compared with low-molecular-weight phosphorus-containing compounds, polymeric or reactive phosphorus-containing FRs avoid the migration or fogging side effect seen in FPUFs. The FR mechanism of phosphorus-containing FRs for PUs involves the quenching effect of some phosphorus-containing radicals (i.e., PO· and O_2P·) in the vapor phase or the catalytic charring effect achieved by generating PPA in the condensed phase. If one phosphorus-containing FR can be active in both the vapor and condensed phases, it would offer high FR efficacy. The phosphorus-containing compounds are one class of the most important halogen-free FRs for PUs, and the diversity of this class of FRs provides a promising solution for overcoming the fire risk of PUs.

Table 5. List of Abbreviations

Abbreviations	*Full Name*
β-CD	β-Cyclodextrin
ADP	Aluminum diethylphosphinate
AHP	Aluminum hypophosphite
APP	Ammonium polyphosphate
DAH	Dicyclohexyl aluminum hypophosphate
DMOP	Diethanol amine methyl oligomeric phosphonate
DMPY	Dimelamine pyrophosphate
DOPO	9,10-dihydro-9-oxa-10-phosphaphenanthrene-10-oxide
e-DOPO	DOPO derivative containing a terminal epoxy group
EG	Expanded graphite
EPPD	2-Ethyl-2-(2-oxo-5,5-dimethyl-1,3,2-dioxaphosphorinanyl-2-methylene)-1,3-propanediol

Table 5. (Continued). List of Abbreviations

Abbreviations	*Full Name*
FPUF	Flexible polyurethane foam
FR	Flame retardant
FRD	Phosphorus-containing polyester diol
FRPOL	Cyclic phosphorus-containing polyol
FRPUD	Flame-retardant aqueous polyurethane dispersion
FRWPU	Flame-retardant waterborne polyurethane
HAPO	Hydroxyacetophenone oxime
LOI	Limiting oxygen index
MCMPP	Microencapsulated melamine polyphosphate
MFeRP	Melamine–formaldehyde resin encapsulated red phosphorus
MPP	Melamine polyphosphate
MPU	Maleimide-terminated polyurethane
ODDP	Octahydro-2,7-di(*N*,*N*-dimethylamino)-1,6,3,8,2,7-dioxadiazadiphosphecine
OP	Organophosphorus
PDEO	Phenylphosphonic ethylene glycol oligomer
PEG	Polyethylene glycol
PHRR	Peak heat release rate
PPA	Polyphosphoric acid
PPBOP	Poly (1,4-butanediol phenylphosphonate)
PPBP	(4-Phenyl-1-(4-phenylbutyl) piperidine
PPG	Polypropylene glycol
pphp	Parts per hundred polyol
PU	Polyurethane
PUF	Polyurethane foam
PUD	Polyurethane dispersions
RP	Red phosphorus
RPUF	Rigid polyurethane foam
SPE	Schiff-base polyphosphate ester
TCPP	Tris(1-chloro-2-propyl) phosphate
TFP	Tri(2-furyl) phosphoramide
THR	Total heat release
TPU	Thermoplastic polyurethane
WPU	Waterborne polyurethane

References

1. Chattopadhyay, D. K.; Webster, D. C. Thermal stability and flame retardancy of polyurethanes. *Prog. Polym. Sci.* **2009**, *34*, 1068–1133.
2. Xie, F.; Zhang, T.; Bryant, P.; Kurusingal, V.; Colwell, J. M.; Laycock, B. Degradation and stabilization of polyurethane elastomers. *Prog. Polym. Sci.* **2019**, *90*, 211–268.
3. Joshi, M.; Adak, B.; Butola, B. S. Polyurethane nanocomposite based gas barrier films, membranes and coatings: A review on synthesis, characterization and potential applications. *Prog. Mater. Sci.* **2018**, *97*, 230–282.
4. Levchik, S. V.; Weil, E. D. Thermal decomposition, combustion and fire-retardancy of polyurethanes—a review of the recent literature. *Polym. Int.* **2004**, *53*, 1585–1610.
5. Wang, X.; Kalali, E. N.; Wan, J. T.; Wang, D. Y. Carbon-family materials for flame retardant polymeric materials. *Prog. Polym. Sci.* **2017**, *69*, 22–46.
6. Kramer, R. H.; Zammarano, M.; Linteris, G. T.; Gedde, U. W.; Gilman, J. W. Heat release and structural collapse of flexible polyurethane foam. *Polym. Degrad. Stabil.* **2010**, *95*, 1115–1122.
7. Buszard, D. L.; Dellar, R. J. The Performance of Flame Retardants in Rigid Polyurethane Foam Formulations. In *Fire and Cellular Polymers*; Buist, J. M., Grayson, S. J., Woolley, W. D. , Eds.; Springer Netherlands: Dordrecht, 1986; pp 265–277.
8. Creyf, H.; Fishbein, J. Advance of Flexible Polyurethane Foam Technology. In *Fire and Cellular Polymers*; Buist, J. M., Grayson, S. J., Woolley, W. D. , Eds.; Springer Netherlands: Dordrecht, 1986; pp 279–288.
9. Batt, A. M.; Appleyard, P. The Mechanism and Performance of Combustion Modified Flexible Foams in Small Scale Fire Tests. *J. Fire Sci.* **1989**, *7*, 338–363.
10. Schrock, A. K.; Solis, R.; Beal, G. E.; Skorpenske, R. G.; Parrish, D. P. The influence of polymer morphology on the combustion of melamine filled flexible foams. *J. Fire Sci.* **1990**, *8*, 174–193.
11. Cao, Z. J.; Dong, X.; Fu, T.; Deng, S. B.; Liao, W.; Wang, Y. Z. Coated vs. naked red phosphorus: A comparative study on their fire retardancy and smoke suppression for rigid polyurethane foams. *Polym. Degrad. Stabil.* **2017**, *136*, 103–111.
12. Tang, G.; Jiang, H. H.; Yang, Y. D.; Chen, D. P.; Liu, C. L.; Zhang, P.; Zhou, L.; Huang, X. J.; Zhang, H.; Liu, X. Y. Preparation of melamine-formaldehyde resin-microencapsulated ammonium polyphosphate and its application in flame retardant rigid polyurethane foam composites. *J. Polym. Res.* **2020**, *27*, 375.
13. Yang, Y. D.; Chen, W.; Liu, M. R.; Zhu, Q. M.; Liu, X. L.; Zhang, B.; Chen, D. P.; Liu, X. Y.; Zhang, K.; Tang, G. Flame retarded rigid polyurethane foam composites based on gel-silica microencapsulated ammonium polyphosphate. *J. Sol-Gel Sci. Techn.* **2021**, *98*, 212–223.
14. Li, Q.; Wang, J.; Chen, L.; Shi, H.; Hao, J. Ammonium polyphosphate modified with β-cyclodextrin crosslinking rigid polyurethane foam: Enhancing thermal stability and suppressing flame spread. *Polym. Degrad. Stabil.* **2019**, *161*, 166–174.
15. Chen, X. L.; Jiang, Y. F.; Jiao, C. M. Smoke suppression properties of ferrite yellow on flame retardant thermoplastic polyurethane based on ammonium polyphosphate. *J. Hazard. Mater.* **2014**, *266*, 114–121.
16. Grassie, N.; Mendoza, G. A. P. Thermal degradation of polyether-urethanes: Part 4—Effect of ammonium polyphosphate on the thermal degradation of polyether-urethanes prepared from

methylene bis(4-phenylisocyanate) and low molecular weight poly(ethylene glycols). *Polym. Degrad. Stabil.* **1985**, *11*, 145–166.

17. Grassie, N.; Perdomo Mendoza, G. A. Thermal degradation of polyether-urethanes: 5. Polyether-urethanes prepared from methylene bis(4-phenylisocyanate) and high molecular weight poly(ethylene glycols) and the effect of ammonium polyphosphate. *Polym. Degrad. Stabil.* **1985**, *11*, 359–379.
18. Liu, X. D.; Sun, J.; Zhang, S.; Guo, J.; Tang, W. F.; Li, H. F.; Gu, X. Y. Effects of carboxymethyl chitosan microencapsulated melamine polyphosphate on the flame retardancy and water resistance of thermoplastic polyurethane. *Polym. Degrad. Stabil.* **2019**, *160*, 168–176.
19. Liu, L. B.; Xu, Y.; He, Y. T.; Xu, M. J.; Shi, Z. X.; Hu, H. C.; Yang, Z. C.; Li, B. An effective mono-component intumescent flame retardant for the enhancement of water resistance and fire safety of thermoplastic polyurethane composites. *Polym. Degrad. Stabil.* **2019**, *167*, 146–156.
20. Kenji, S.; Shigeaki, K.; Toru, O.; Take, S. *Production of polyurethane foam*; European Patent EP0170206-A1, 1984-07-26.
21. Norbert, A.; Werner, D.; Manfred, G.; Reinhard, K.; Rolf, W. *Process for the preparation of rigid polyurethane foams*. Europena Patent EP0308733-A1, 1989-03-29.
22. Jan-Gerd, H.; Heiko, T. *Flame-resistant polyurethane foams with low core discolouration*. European Patent EP2985336-B1, 2016-02-17.
23. Tian, S. H.; He, H.; Wang, D. K.; Yu, P.; Jia, Y. C.; Luo, Y. F. Study of using aluminum hypophosphite as a flame retardant for low-density polyethylene. *Fire Mater.* **2017**, *41*, 983–992.
24. Xu, M. J.; Wang, J.; Ding, Y. H.; Li, B. Synergistic effects of aluminum hypophosphite on intumescent flame retardant polypropylene system. *Chinese J. Polym. Sci.* **2015**, *33*, 318–328.
25. Tang, G.; Wang, X.; Xing, W.; Zhang, P.; Wang, B.; Hong, N.; Yang, W.; Hu, Y.; Song, L. Thermal degradation and flame retardance of biobased polylactide composites based on aluminum hypophosphite. *Ind. Eng. Chem. Res.* **2012**, *51*, 12009–12016.
26. Ge, H.; Tang, G.; Hu, W. Z.; Wang, B. B.; Pan, Y.; Song, L.; Hu, Y. Aluminum hypophosphite microencapsulated to improve its safety and application to flame retardant polyamide 6. *J. Hazard. Mater.* **2015**, *294*, 186–194.
27. Xiao, S. S.; Chen, M. J.; Dong, L. P.; Deng, C.; Chen, L.; Wang, Y. Z. Thermal degradation, flame retardance and mechanical properties of thermoplastic polyurethane composites based on aluminum hypophosphite. *Chinese J. Polym. Sci.* **2014**, *32*, 98–107.
28. Chen, X. L.; Ma, C. Y.; Jiao, C. M. Synergistic effects between [Emim]PF6 and aluminum hypophosphite on flame retardant thermoplastic polyurethane. *RSC Adv.* **2016**, *6*, 67409–67417.
29. Ding, H. Y.; Huang, K.; Li, S. H.; Xu, L. N.; Xia, J. L.; Li, M. Flame retardancy and thermal degradation of halogen-free flame-retardant biobased polyurethane composites based on ammonium polyphosphate and aluminium hypophosphite. *Polym. Test* **2017**, *62*, 325–334.
30. Wang, X.; Zhang, P.; Huang, Z.; Xing, W.; Song, L.; Hu, Y. Effect of aluminum diethylphosphinate on the thermal stability and flame retardancy of flexible polyurethane foams. *Fire Safety J.* **2019**, *106*, 72–79.

31. Hu, Y. X.; Zhou, Z. J.; Li, S. S.; Yang, D.; Zhang, S.; Hou, Y. K. Flame retarded rigid polyurethane foams composites modified by aluminum diethylphosphinate and expanded graphite. *Front. Mater.* **2021**, 7, 629284.

32. Li, H. X.; Ning, N. Y.; Zhang, L. Q.; Wang, Y. X.; Liang, W. L.; Tian, M.; Chan, T. W. Effect of content of organophosphorus on flame retardancy mode of thermoplastic polyurethane. *Polymer* **2015**, *67*, 1–11.

33. Lorenzetti, A.; Modesti, M.; Gallo, E.; Schartel, B.; Besco, S.; Roso, M. Synthesis of phosphinated polyurethane foams with improved fire behaviour. *Polym. Degrad. Stabil.* **2012**, *97*, 2364–2369.

34. Zhang, G.; Lin, X.; Zhang, Q.; Jiang, K.; Chen, W.; Han, D. Anti-flammability, mechanical and thermal properties of bio-based rigid polyurethane foams with the addition of flame retardants. *RSC Adv.* **2020**, *10*, 32156–32161.

35. Zhang, Y. B.; Cui, J. F.; Wang, L. R.; Liu, H.; Yang, B. P.; Guo, J. H.; Mu, B.; Tian, L. Phosphorus-containing Salen-metal complexes investigated for enhancing the fire safety of thermoplastic polyurethane (TPU). *Polym. Advan. Technol.* **2020**, *31*, 1150–1163.

36. Lorenzetti, A.; Modesti, M.; Besco, S.; Hrelja, D.; Donadi, S. Influence of phosphorus valency on thermal behaviour of flame retarded polyurethane foams. *Polym. Degrad. Stabil.* **2011**, *96*, 1455–1461.

37. Velencoso, M. M.; Battig, A.; Markwart, J. C.; Schartel, B.; Wurm, F. R. Molecular firefighting-how modern phosphorus chemistry can help solve the challenge of flame retardancy. *Angew. Chem. Int. Edit.* **2018**, *57*, 10450–10467.

38. Liang, S. Y.; Neisius, M.; Mispreuve, H.; Naescher, R.; Gaan, S. Flame retardancy and thermal decomposition of flexible polyurethane foams: Structural influence of organophosphorus compounds. *Polym. Degrad. Stabil.* **2012**, *97*, 2428–2440.

39. Wang, X.; Song, L.; Xing, W.; Lu, H.; Hu, Y. A effective flame retardant for epoxy resins based on poly(DOPO substituted dihydroxyl phenyl pentaerythritol diphosphonate). *Mater. Chem. Phys.* **2011**, *125*, 536–541.

40. Wang, X.; Hu, Y.; Song, L.; Xing, W.; Lu, H.; Lv, P.; Jie, G. Flame retardancy and thermal degradation mechanism of epoxy resin composites based on a DOPO substituted organophosphorus oligomer. *Polymer* **2010**, *51*, 2435–2445.

41. Wang, X.; Hu, Y.; Song, L.; Xing, W.; Lu, H. Thermal degradation mechanism of flame retarded epoxy resins with a DOPO-substituted organophosphorus oligomer by TG-FTIR and DP-MS. *J. Anal. Appl. Pyrol.* **2011**, *92*, 164–170.

42. Wang, X.; Hu, Y.; Song, L.; Yang, H.; Xing, W.; Lu, H. Synthesis and characterization of a DOPO-substituted organophosphorus oligomer and its application in flame retardant epoxy resins. *Prog. Org. Coat.* **2011**, *71*, 72–82.

43. Yang, A. H.; Deng, C.; Chen, H.; Wei, Y. X.; Wang, Y. Z. A novel Schiff-base polyphosphate ester: Highly-efficient flame retardant for polyurethane elastomer. *Polym. Degrad. Stabil.* **2017**, *144*, 70–82.

44. Zhang, P. K.; Tian, S. Q.; Fan, H. J.; Chen, Y.; Yan, J. Flame retardancy and hydrolysis resistance of waterborne polyurethane bearing organophosphate moieties lateral chain. *Prog. Org. Coat.* **2015**, *89*, 170–180.

45. Wang, K.; Morgan, A. B.; Benin, V. Preparation and studies of new phosphorus-containing diols as potential flame retardants. *Fire Mater.* **2017**, *41*, 973–982.
46. Bhoyate, S.; Ionescu, M.; Kahol, P. K.; Chen, J.; Mishra, S. R.; Gupta, R. K. Highly flame-retardant polyurethane foam based on reactive phosphorus polyol and limonene-based polyol. *J. Appl. Polym. Sci.* **2018**, *135*, 46224.
47. Rolf, A.; Stephan, S.; Horst-Josef, S.; Hans-Peter, H.; Andre, N.; Nicole, W.; Catherine, L. *Methods for producing flame-retardant PUR/PIR foam materials*. World Patent WO2020201187-A1, Germany Patent EP3719047-A1, Oct 8, 2020.
48. Rolf, S.; Rüdiger, S. *Aqueous dispersions of pre-polymers containing isocyanate-terminated phosphorus and coatings made from same and use*. European Patent EP3150646-A1, Germany Patent DE102015116528-A1, April 5, 2017.
49. Wang, S.; Du, X. S.; Fu, X. H.; Du, Z. L.; Wang, H. B.; Cheng, X. Highly effective flame-retarded polyester diol with synergistic effects for waterborne polyurethane application. *J. Appl. Polym. Sci.* **2020**, *137*, 48444.
50. Rao, W. H.; Zhu, Z. M.; Wang, S. X.; Wang, T.; Tan, Y.; Liao, W.; Zhao, H. B.; Wang, Y. Z. A reactive phosphorus-containing polyol incorporated into flexible polyurethane foam: Self-extinguishing behavior and mechanism. *Polym. Degrad. Stabil.* **2018**, *153*, 192–200.
51. Rao, W. H.; Xu, H. X.; Xu, Y. J.; Qi, M.; Liao, W.; Xu, S. M.; Wang, Y. Z. Persistently flame-retardant flexible polyurethane foams by a novel phosphorus-containing polyol. *Chem. Eng. J.* **2018**, *343*, 198–206.
52. Chen, H.; Deng, C.; Zhao, Z. Y.; Huang, S. C.; Wei, Y. X.; Wang, Y. Z. Novel alkynyl-containing phosphonate ester oligomer with high charring capability as flame retardant additive for thermoplastic polyurethane. *Compos. Part B-Eng.* **2020**, *199*, 108315.
53. Tabatabaee, F.; Khorasani, M.; Ebrahimi, M.; Gonzalez, A.; Irusta, L.; Sardon, H. Synthesis and comprehensive study on industrially relevant flame retardant waterborne polyurethanes based on phosphorus chemistry. *Prog. Org. Coat.* **2019**, *131*, 397–406.
54. Wazarkar, K.; Kathalewar, M.; Sabnis, A. Improvement in flame retardancy of polyurethane dispersions by newer reactive flame retardant. *Prog. Org. Coat.* **2015**, *87*, 75–82.
55. Arora, S.; Mestry, S.; Naik, D.; Mhaske, S. T. o-Phenylenediamine-derived phosphorus-based cyclic flame retardant for epoxy and polyurethane systems. *Polym. Bull.* **2020**, *77*, 3185–3205.
56. Mishra, V. K.; Patel, R. H. Processing and characterizations: Effect of PPG molecular weight on properties of phosphate based polyurethanes. *Prog. Org. Coat.* **2020**, *147*, 105868.
57. Jia, D. K.; Yang, J. X.; He, J. Y.; Li, X. M.; Yang, R. J. Melamine-based polyol containing phosphonate and alkynyl groups and its application in rigid polyurethane foam. *J. Mater. Sci.* **2021**, *56*, 870–885.
58. Yang, R.; Hu, W. T.; Xu, L.; Song, Y.; Li, J. C. Synthesis, mechanical properties and fire behaviors of rigid polyurethane foam with a reactive flame retardant containing phosphazene and phosphate. *Polym. Degrad. Stabil.* **2015**, *122*, 102–109.
59. Yang, R.; Wang, B.; Han, X. F.; Ma, B. B.; Li, J. C. Synthesis and characterization of flame retardant rigid polyurethane foam based on a reactive flame retardant containing phosphazene and cyclophosphonate. *Polym. Degrad. Stabil.* **2017**, *144*, 62–69.

60. Ding, H. Y.; Wang, J. F.; Wang, C. P.; Chu, F. X. Synthesis of a novel phosphorus and nitrogen-containing bio-based polyols and its application in flame retardant polyurethane sealant. *Polym. Degrad. Stabil.* **2016**, *124*, 43–50.
61. Bhoyate, S.; Ionescu, M.; Kahol, P. K.; Gupta, R. K. Castor-oil derived nonhalogenated reactive flame-retardant-based polyurethane foams with significant reduced heat release rate. *J. Appl. Polym. Sci.* **2019**, *136*, 47276.
62. Patel, R. H.; Shah, M. D.; Patel, H. B. Synthesis and Characterization of Structurally Modified Polyurethanes Based on Castor Oil and Phosphorus-Containing Polyol for Flame-Retardant Coatings. *Int. J. Polym. Anal. Ch.* **2011**, *16*, 107–117.
63. Patel, R.; Kapatel, P. Waterborne polyurethanes: A three step synthetic approach towards environmental friendly flame retardant coatings. *Prog. Org. Coat.* **2018**, *125*, 186–194.
64. Bhoyate, S.; Ionescu, M.; Kahol, P. K.; Gupta, R. K. Sustainable flame-retardant polyurethanes using renewable resources. *Ind. Crop Prod.* **2018**, *123*, 480–488.
65. Zhou, W.; Jia, P. Y.; Zhou, Y. H.; Zhang, M. Preparation and characterization of tung oil-based flame retardant polyols. *Chinese J. Chem. Eng.* **2018**, *26*, 2664–2671.
66. Patel, M.; Mestry, S.; Khuntia, S. P.; Mhaske, S. Gallic acid-derived phosphorus-based flame-retardant multifunctional crosslinking agent for PU coating. *J. Coat. Technol. Res.* **2020**, *17*, 293–303.
67. Zhang, L. Q.; Zhang, M.; Zhou, Y. H.; Hu, L. H. The study of mechanical behavior and flame retardancy of castor oil phosphate-based rigid polyurethane foam composites containing expanded graphite and triethyl phosphate. *Polym. Degrad. Stabil.* **2013**, *98*, 2784–2794.
68. Yang, S. W.; Wang, S.; Du, X. S.; Du, Z. L.; Cheng, X.; Wang, H. B. Mechanically robust self-healing and recyclable flame-retarded polyurethane elastomer based on thermoreversible crosslinking network and multiple hydrogen bonds. *Chem. Eng. J.* **2020**, *391*, 123544.
69. Gu, L. M.; Ge, Z.; Huang, M. H.; Luo, Y. J. Halogen-free flame-retardant waterborne polyurethane with a novel cyclic structure of phosphorus-nitrogen synergistic flame retardant. *J. Appl. Polym. Sci.* **2015**, *132*, 41288.
70. Lin, Z. J.; Zhao, Q. Q.; Fan, R. L.; Yuan, X. X.; Tian, F. L. Flame retardancy and thermal properties of rigid polyurethane foam conjugated with a phosphorus-nitrogen halogen-free intumescent flame retardant. *J. Fire Sci.* **2020**, *38*, 235–252.
71. Zhang, M.; Luo, Z. Y.; Zhang, J. W.; Chen, S. G.; Zhou, Y. H. Effects of a novel phosphorus-nitrogen flame retardant on rosin-based rigid polyurethane foams. *Polym. Degrad. Stabil.* **2015**, *120*, 427–434.
72. Zhao, Q. Q.; Chen, C. Y.; Fan, R. L.; Yuan, Y.; Xing, Y. L.; Ma, X. Halogen-free flame-retardant rigid polyurethane foam with a nitrogen-phosphorus flame retardant. *J. Fire Sci.* **2017**, *35*, 99–117.
73. Ma, S. C.; Xiao, Y. L.; Zhou, F.; Schartel, B.; Chan, Y. Y.; Korobeinichev, O. P.; Trubachev, S. A.; Hu, W. Z.; Ma, C.; Hu, Y. Effects of novel phosphorus-nitrogen-containing DOPO derivative salts on mechanical properties, thermal stability and flame retardancy of flexible polyurethane foam. *Polym. Degrad. Stabil.* **2020**, *177*, 109160.
74. Gaan, S.; Liang, S. Y.; Mispreuve, H.; Perler, H.; Naescher, R.; Neisius, M. Flame retardant flexible polyurethane foams from novel DOPO-phosphonamidate additives. *Polym. Degrad. Stabil.* **2015**, *113*, 180–188.

75. Liu, X.; Salmeia, K. A.; Rentsch, D.; Hao, J. W.; Gaan, S. Thermal decomposition and flammability of rigid PU foams containing some DOPO derivatives and other phosphorus compounds. *J. Anal. Appl. Pyrol.* **2017**, *124*, 219–229.
76. Jia, D. K.; Guo, X. Y.; He, J. Y.; Yang, R. J. An anti-melt dripping, high char yield and flame-retardant polyether rigid polyurethane foam. *Polym. Degrad. Stabil.* **2019**, *167*, 189–200.
77. Jia, D. K.; He, J. Y.; Yang, R. J. The synthesis and characterization of a DOPO derivative bearing an active terminal epoxy group: e-DOPO and its application in rigid polyurethane foam. *J. Fire Sci.* **2019**, *37*, 47–66.
78. Xi, W.; Qian, L.; Chen, Y.; Wang, J.; Liu, X. Addition flame-retardant behaviors of expandable graphite and [bis(2-hydroxyethyl)amino]-methyl-phosphonic acid dimethyl ester in rigid polyurethane foams. *Polym. Degrad. Stabil.* **2015**, *122*, 36–43.
79. Xi, W.; Qian, L.; Huang, Z.; Cao, Y.; Li, L. Continuous flame-retardant actions of two phosphate esters with expandable graphite in rigid polyurethane foams. *Polym. Degrad. Stabil.* **2016**, *130*, 97–102.
80. Rao, W. H.; Liao, W.; Wang, H.; Zhao, H. B.; Wang, Y. Z. Flame-retardant and smoke-suppressant flexible polyurethane foams based on reactive phosphorus-containing polyol and expandable graphite. *J. Hazard. Mater.* **2018**, *360*, 651–660.
81. Qian, L. J.; Li, L. J.; Chen, Y. J.; Xu, B.; Qiu, Y. Quickly self-extinguishing flame retardant behavior of rigid polyurethane foams linked with phosphaphenanthrene groups. *Compos. Part B-Eng.* **2019**, *175*, 107186.

Chapter 11

Recent Developments in Nitrogen- and Phosphorous-Based Flame Retardants for Polyurethanes

Anil M. Palve,*,[1] Jagruti S. Suroshe,[2] and Ram K. Gupta*,[3]

[1]**Department of Chemistry, Mahatma Phule Arts, Science, and Commerce College, Panvel, Navi Mumbai, MH 410206, India**

[2]**Department of Chemistry, DSPM's K.V. Pendharkar College, Dombivli, Thane, MH 421203, India**

[3]**Department of Chemistry, Kansas Polymer Research Center, Pittsburg State University, Pittsburg, Kansas 66762, United States**

***Email: palve_anil@yahoo.com**

***Email: ramguptamsu@gmail.com**

The high flammability of polyurethanes is a major concern for their commercial applications. Blending polyurethanes with flame retardants (FRs) can significantly reduce their flammability. Nonhalogen FRs are very popular because of the increasing demand for environmentally friendly materials for flame inhibition. The broad view of the type of FRs such as phosphorus- and nitrogen-based compounds is elaborated in this chapter. Nitrogen-based FRs are often used as FRs in many polymeric materials because of their eco-friendly nature and high flame-inhibition properties. In this chapter, different types of nitrogen-based FRs and their uses in different polymer systems are discussed. Along with well-known ammonia- and melamine-based FRs, some unique compounds of this family—such as azoalkanes, phosphazenes, phospham, and *N*-alkoxy amines—are also discussed.

Introduction

Polyurethanes (PUs) are synthesized by the chemical interaction between a di/polyisocyanate and a diol or polyol, giving repeated urethane groups. This is often in the existence of a catalyst, a lengthened chain, and other additives. The presence of two basic elements (i.e., a polyol and an isocyanate) decide the properties of PUs. By changing a polyol and an isocyanate, a drastic change in the characteristics of PUs can be achieved. The polyol and isocyanate from separate segments in PU are known as a soft segment and a hard segment, respectively. Soft segments controlled by polyols tend to have extended chain lengths, which impart excellent mobility and therefore provide good flexibility to PUs. Meanwhile, hard compartments are controlled by isocyanates and chain extenders that are incorporated. These short-length chain molecules make PU very hard and nonflexible. The presence of both segments enhances the characteristics of PUs, making them very suitable for a range of applications (*1*, *2*). For example, PUs with good adhesion, excellent drying ability, chemical resistivity, abrasion resistance, and low-temperature flexibility make them suitable for coating purposes. PU-based sealants come up with exceptionally tight seals, whereas adhesive-based PUs provide excellent bonding strength. PUs are very useful in the automotive industry. Because of their low density, they can be used for the front bumpers and body of the car, interior panels in aircrafts, ceiling sections, and so forth. PUs can replace metals or some other materials in automobiles so that automobiles are less heavy and the fuel economy rises. The use of PUs in car seats and pillows provides good relaxation to the passengers. Because of excellent mechanical properties, toughness, cost-effectiveness, biocompatibility, and availability, PUs are also used for medical purposes such as tubing, blood-contacting materials, surgical drapes, coating for cardiac pacemaker leads, breast implants, catheters, hospital bedclothes, vascular devices, wound dressings, and many other applications (*3*).

Many industrial PUs are synthesized using petrochemicals. However, the current trend in PU research is to find alternative and eco-friendly materials and processes. For this purpose, many materials from bioresources and green approaches are being used by researchers (*4*, *5*). During burning, PUs foams show less thermal resistance, show high flammability, and burn with a sooty flame. PU foams (PUFs) are highly flammable and are vulnerable to degradation when exposed to elevated temperatures during a fire accident. Porous or spongy PUs are known for their fast spread of flames. Highly porous lightweight combustible foams are likely to have fast flame spread and high thermal emission. This limits their uses, and many studies are devoted to flame retardancy (*6–10*). Research has been focused on upgrading the fire behavior and thermal stability of PUFs. During the 1970s, the use of flame-retardant (FR) materials gained worldwide attention, and they were applied inflammable materials to suppress the combustion process (*11*). There are major five categories of FRs: halogen FRs, inorganic FRs, nitrogen FRs, intumescent FRs (IFRs), and phosphorus-based FRs. Among all FRs, nitrogen-based FRs (NFRs) have certain advantages such as low toxicity, absence of dioxins, and low smoke release during a fire incident. Nitrogen FRs can be recycled by reextrusion, as they have high decomposition temperatures. They are a fast developing group of FRs that are environmentally friendly, are less toxic, have good recyclability, and have good efficiency (*12*, *13*). In this chapter, we are mainly focusing on NFRs because of their various advantages, as mentioned previously.

Types of FRs

FRs are introduced in materials to reduce their flammability. They can be utilized during the manufacturing process as a chemical moderation technique, or they can be introduced during the processing of polymeric materials. FRs are broadly categorized into reactive FRs (chemically incorporated into PUs) and additives (physically incorporated into PUs). How they are categorized most often depends on their "active" chemical component—the major group products are based on halogens (usually Br and Cl), P, N, Si, minerals such as Al and Mg, and, to a lesser extent B, Zn, and C (*14*). With the rapid development of nanotechnology, nanomaterials are used as effective and environment-friendly FRs. Nanoclay and multiwalled carbon nanotubes have been used as FRs. Polymer/layered inorganic nanocomposites—including montmorillonite, layered double hydroxides, and carbon-based materials—have been extensively investigated to improve the thermal stability and flame retardancy of polymers (*15*, *16*). Major categories of FRs are discussed as follows. Figure 1 indicates various types of FRs. In the following subsections, some of the FRs and their properties are discussed.

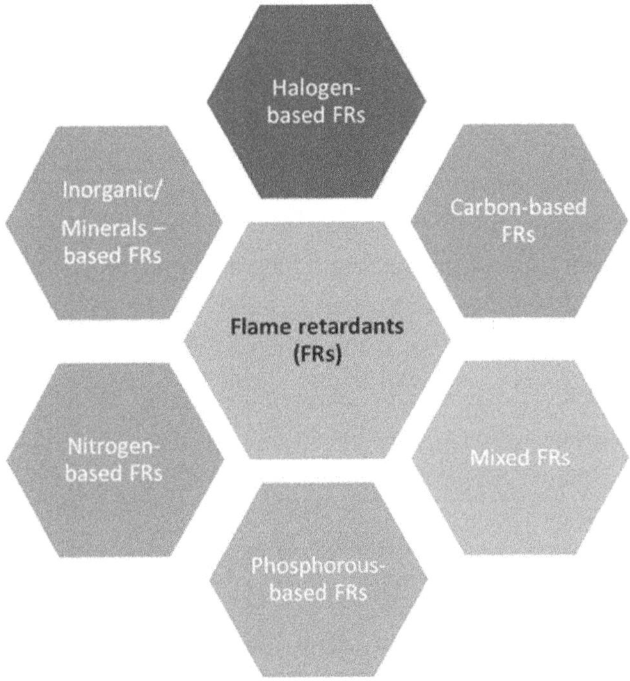

Figure 1. Types of FRs.

Halogen-Based FRs

Depending upon the chemical compositions, halogen FRs can be additionally classified into two main categories: brominated and chlorinated FRs. Brominated FRs are the most regularly applicable ones, and they dominate over other halogen FRs because of their lower decomposition temperature, excellent efficiency of trapping free radicals, and low cost (*17*, *18*). Halogen-based FRs are very efficient in preventing and spread of flames. The emphasis of the action of halogen is on quickly scavenging H● and OH●. The effective smoke suppression process yields thermostable volatile products that present the mass transfer from these radicals back to the polymer, as well as dilute oxygen in the media.

Brominated bisphenols, decabrominated diphenyl ethers, and cyclododecanes are some of the three major classes of brominated FRs that represent larger production volumes. Brominated FRs can be used as additives or reactive components in polymers such as polystyrene, epoxy resins, and so forth. These have applications in commercial products such as furniture foams, computers, thermal insulators, and electronic equipment. In the early 1970s, some brominated FRs were rejected from the market after an accidental poisoning turned into the loss of livestock because of the ingestion of polybrominated biphenyl-contaminated animal feed. This illustrated the harmfulness of this brominated FR class. Tris(2,3-dibromopropyl) phosphate is a brominated FR that was banned from clothing for children (such as pajamas) because of its mutagenic and nephrotoxic effects (*19*, *20*).

Inorganic FRs

Most of the inorganic metal oxides serve as FRs or as additives within the FR system. To achieve the desired results, these inorganic compounds must be utilized in high concentrations. To be very effective, they must be used in conjunction with another type of FR. Antimony trioxide, antimony pentoxide, and antimony–metal compounds are used as FRs. Antimony oxides themselves do not have any FR properties, but when they are used with bromine- or chlorine-based FRs, they serve as a synergists (*21*). Boron compounds that can be used as FRs include zinc borate, barium metaborate, boric acid, sodium borate, and ammonium fluoroborate. These boron and antimony compounds can be used as FRs in rigid PUFs. Boron-based buffers remarkably decrease the gravity of thermal degradation and also serve as additives in FR actions (*22*, *23*). A more commonly used aluminum FR compound is alumina trihydrate. It is used as a secondary FR and as a smoke controller for flexible poly(vinyl chloride) plastics. It can combine with antimony and halogen FRs. Some inorganic FRs that can be employed individually consist of aluminum and magnesium hydroxides. Inorganic FRs help to reduce the ignition of PUs by means of liberating nonreactive gases (such as water vapors), forming defensive char coatings, or energy absorption. Tang et al. explored the synergistic effect of three inorganic boron materials—zinc borate, boron phosphate, and boron oxide—blended with phosphaphenanthrene epoxy resins (*24*). Unlu et al. demonstrated a relative study of boron material and aluminum trihydroxide as FR additives in epoxy resins (*25*).

NFRs

NFRs are very efficient and can be considered as an individual class of FRs. Based on their mode of action, they can be further divided into ammonia-based FRs and melamine-based FRs. Ammonium polyphosphate (APP) and other ammonium salts such as ammonium pentaborate (APB) and ammonium sulfamate can also be used as FRs. APB solutions can be applied to paper through sprinkling to create an FR property. For cellulosic materials, they can be employed as constituents in flame-proofing materials and used in urethane foam, epoxies, thermoplastic PUs (TPUs), and so forth (*26*).

Melamine and its products are often used as FRs. They can be used with phosphorus-based FRs for a variety of applications. In the condensed phase, melamine changes into cross-linked structures and it gives char, which prohibits the delivery of oxygen. This is a model of a solid-phase char FR. Melamine, or its derivatives, are introduced to intumescent varnishes or paints. Melamine is added into flexible PU cellular plastics for the synthesis of combustion-modified high resilient foams. Melamine cyanurate (MCA) is used for unreinforced nylons. Compared with halogen FRs, NFRs have higher decomposition temperatures, so they do not react with dangerous products like dioxins. Thermoplastic materials with added NFRs may be re-ejected many times without the loss of FRs or mechanical possessions. This prevents injury to the operators by liberating dangerous gases (*12*, *27*, *28*).

Intumescent Coatings

Intumescent coatings are made up of FR additives such as APP (an acid source), pentaerythritol (a carbon source), and a blowing agent with a binder (*29*, *30*). The main objective of the coatings is to protect the materials from fire. Intumescent coatings are highly approved as passive fire protection in construction, as they can save both humans and assets. They can be applied to paints and are suitable for construction materials, such as steel beams or wood. When they come into contact with heat, the coatings form an intumescence char layer, which serves as a thermal barrier that effectively holds the substrate against fire. This can prevent structural damage (*31*, *32*). FR coatings have many advantages, as they do not alter the intrinsic properties of the materials, are easy to process, and are used on a variety of materials such as polymers (*11*), textiles (*33*), wood (*34*), and metallic materials (*29*). Various studies have illustrated that only a few polymers can take part in the intumescent phenomenon (*35*, *36*).

Phosphorus-Based FRs

Halogen flame retarding can be replaced by phosphorus-based FRs because of their better fire properties and good environmental stability (*37*). Phosphorus-based FRs have many advantages, such as fire reduction, char enrichment, intumescence, and so forth (*38–41*). It has been reported that in several polymers, aluminum diethyl phosphinate exhibits better action in the gas phase (*42–45*). A small amount left in the condensed phase is used for char or residue formation (*46*). The esterification and dehydration processes are mainly involved in phosphorus-based FRs (*47*, *48*). Although they are released in the gas phase, they are good for a fire-dropping mechanism (*49*). Depending on the decomposition temperature and mechanism, the materials—such as bisphenol-A bis(diphenyl phosphate), triphenyl phosphate, and resorcinol bis(diphenyl phosphate)—can cause fire inhibition (*50*).

General Mechanism of FRs

Finding the fire-inhibiting effect of volatile halogen derivatives is an eventual stepping-stone to the radical trap-gas-phase mechanism. The performance of the FRs can be studied with the current mechanisms. It is a little complicated to understand which mechanism plays an important role, as the main process involves various mechanistic principles simultaneously. FRs synthesized via a condensed-phase chemical mechanism alter the decomposition way, which lowers the number of gaseous combustibles (*51*). It is important to design the flame-retarding agent and quantity, which help in decreasing the release of heat. The decomposition of the PUs slows down with less transfer of heat on the surface because of FRs used. FR additives used should reach the flame in the gaseous form quickly, must be volatile, and must create pyrolysis to provide the active part of the molecule with the FRs in the gaseous phase.

Gas-Phase Mechanism

A brief discussion about the gas-phase mechanism of the FRs is given in this section of the chapter (*52*). Polymers on combustion produce hydrogen radicals that further react to oxygen, giving oxygen, hydrogen, and hydroxyl radicals. This scheme further generates the fuel ignition by the branching reactions (eqs 1 and 2):

$$H^{\cdot} + O_2 \rightarrow OH^{\cdot} + O^{\cdot} \quad (1)$$

$$H_2 + O^{\cdot} \rightarrow OH^{\cdot} + H^{\cdot} \quad (2)$$

For the maintenance of the flame, a large amount of energy is needed. This is supplied by a major exothermic reaction (eq 3), which is shown as follows:

$$OH^{\cdot} + CO \rightarrow CO_2^{\cdot} + H^{\cdot} \quad (3)$$

To terminate the process of pyrolysis, it is important to block propagation (eqs 1 and 2). The prohibiting outcomes of bromine and chlorine compounds are investigated to work through the gas-phase mechanism. In the first case, this effect is observed either by the release of a halogen-provided FR molecule that does not consist of hydrogen or by liberating a hydrogen halide (eqs 4 and 5):

$$MX \rightarrow M^{\cdot} + X^{\cdot} \tag{4}$$

$$MX + H^{\cdot} \rightarrow M^{\cdot} + HX \tag{5}$$

In the previously mentioned reactions, $M^{\cdot}$ is known as the precipitate of the FR's molecule. Hydrogen halide is synthesized by the reaction of halogen with fuel (eq 6):

$$RH + X^{\cdot} \rightarrow R^{\cdot} + HX \tag{6}$$

The hydrogen halide impacts the chain branching and is therefore known to be a real flame inhibitor (eqs 7 and 8):

$$HX + H^{\cdot} \rightarrow X^{\cdot} + H_2 \tag{7}$$

$$HX + OH^{\cdot} \rightarrow X^{\cdot} + H_2O \tag{8}$$

The rate of eq 7 was found to be about twice the rate of eq 8. Equation 7 is the major inhibiting reaction in the mechanism (53). It was noted that the comparison between eqs 7 and 1 explains the inhibiting effect. For every hydrogen atom absorbed, eq 1 gives two free radicals, although eq 1 gives one halogen radical. They recombine to form the comparatively stable halogen molecule.

Condensed-Phase Mechanism

In this type of mechanism, there will be a chemical relationship between a polymer and an FR agent at lower decomposition temperatures, where the FR agent is added in an inconsiderable amount. Dehydration and cross-linking are two principal modes suggested for this type of interaction.

Major Approaches for the Condensed-Phase Mechanism

Char Formation and Dehydration

The performance of various phosphorus-based materials in various polymers depends upon polymer sensitivity toward dehydration and char formation. The mechanism demonstrates that with a decrease in oxygen content of the polymer, there will be a decrease in its activity. Although cellulosic materials are sufficiently FR with approximately 2% phosphorus, 4–5% of it is required for polyolefins. Polymers that do not consist of hydroxyl groups interact slowly with phosphorous derivatives. Therefore, they must undergo the process of oxidation. During this process, the main phosphorus compounds possibly evaporate in the form of phosphorus pentoxide because of the decomposition of the phosphorus derivatives (54).

In cellulose for the condensed phase, two different mechanisms have been proposed: acid helps in the removal of water from cellulosic material, and acid-forming agents of phosphorus and sulfur form char. Equation 9 is shown as follows (where Z = acyl radical of acid), representing the decomposition of the esters, and eq 10 represents carbonium ion catalysis:

$$R_2CH - CHR'OH + ZOH\ (acid) \rightarrow R_2CH - CHR'OZ + H_2O \rightarrow R_2CH = CHR' + ZOH \quad (9)$$

$$R_2CH - CHR'OH \rightarrow R_2CH - CHR'OH_2^+ \rightarrow R_2CH - C^+HR' + H_2O \quad (10)$$

The results obtained from the oxygen index (OI) and differential thermal analysis indicate that phosphorus-based compounds decrease the flammability of cellulosic material first by the mechanism shown in eq 9. This process is relatively slow, which may be because of the excellent polymer structure. Less-ordered regions can be pyrolyzed at relatively low temperatures compared with those of the high crystalline part of the compound. This leads to a decrease in the effectiveness of flame retardancy and requires a large quantity of phosphorus, attained by the sulfation process, with ammonium sulfamate. This gives sulfated celluloses, which further dehydrate by the carbonium ion that is disproportioned (eq 10) and display activity of strong acid. This quickly hydrolyzes the crystalline part of the compound.

Cross-Linking and Char Formation

Side chains in compounds help char production during the decomposition of celluloses (*55*, *56*), but they also enhance the OI of phenolics and do not appreciably change the flammability of epoxides (*57*). Cross-linked polystyrene yields 47% char (*58*). Adding together organometallic compounds and polymer increases side chain and char formation (*59*). Meanwhile, a low degree of cross-linking leads to a decrease in thermal stability. It has been noticed that the OI of cotton is enhanced nominally with an increase in formaldehyde cross-linking, whereas rayon significantly reduces the OI (*60*).

In cellulose, the char production is commenced by rapid autocross-linking. This autocross-linking is proved by quick weight loss in the first stage of pyrolysis (i.e., at a temperature of 251 °C) because of water molecule loss, and it is equivalent to the amount of char. It has been proposed that cross-linking may lead to a rise in the stickiness of the softened polymer in the decomposition chamber and therefore lower the transport kinetics of the decomposed yield to the flame. NFRs have various advantages over other types of FRs. Therefore, In this chapter, our discussion will be focused on NFRs.

NFRs for PUs

PUs are used for many purposes—such as in furniture, automobiles, covers, coatings, and so forth—but they can easily catch fire when coming into contact with a flame. An excess amount of heat and hazardous gases evolve during the burning of PUs. In the 1970s, because of flexible PUs, domestic fire issues were reported, and the United Kingdom banned non-FR-PUFs in furnishing applications (*27*). As a result, many PUFs now consist of FRs. Phosphorus-based FRs can be among the important substitutes for halogen FRs. More commonly, phosphorus-based FRs are used with NFRs. NFRs have several advantages over other types of FRs. In NFRs, gases or vapors evolved during combustion have low toxicity. Like halogens, NFRs do not exhibit antagonism with hindered amine light stabilizers. Therefore, it is possible to manufacture plastics that burn at low rates with long-lasting reliability (*12*). Because of high decomposition temperatures, they can be recycled by reextrusion. In the cable industry, the use of halogen FRs leads to their dissociation in hydrogen chloride and hydrogen bromide, which leads ultimately to corrosion of wires, switches, and plugs. If halogen FRs are replaced by NFRs, hydrogen cyanide is formed with a dissociation constant of 7 × 10 (*10*), and we can avoid that corrosion of those materials (*12*, *13*).

Ammonia-Based FRs

APP

Among various ammonium salts, APP is the most important NFR. The chemical formula of APP is $(NH_4PO_3)_n$. APP is a phosphorus-based FR that has a nitrogen–phosphorus combination and is applicable in thermosets, thermoplastics, foams, and coatings (*61*). APP exists in various crystalline forms. Form I of APP can be manufactured by simple heating of a combination of ammonium orthophosphate and urea. Heating a combination of ammonium orthophosphate, diphosphorus pentoxide, and urea will give us form II APP (*62, 63*). APP with a small particle size can be used as an FR for organic polymer materials of thin films. Watanabe et al. demonstrated the preparation of APP forms II, V, VI and also studied the flame retardation effect using the organic polymers polyethylene and polypropylene (PP) (*64*). To have a high retardation effect, FRs must have good dispersibility, which is more dependent on particle size. The FR performance of APP can be appreciably improved by the introduction of harmonious compounds such as carbohydrates, foaming agents, pentaerythritol derivatives, and so forth. The decomposition mechanism of APP is shown in Figure 2.

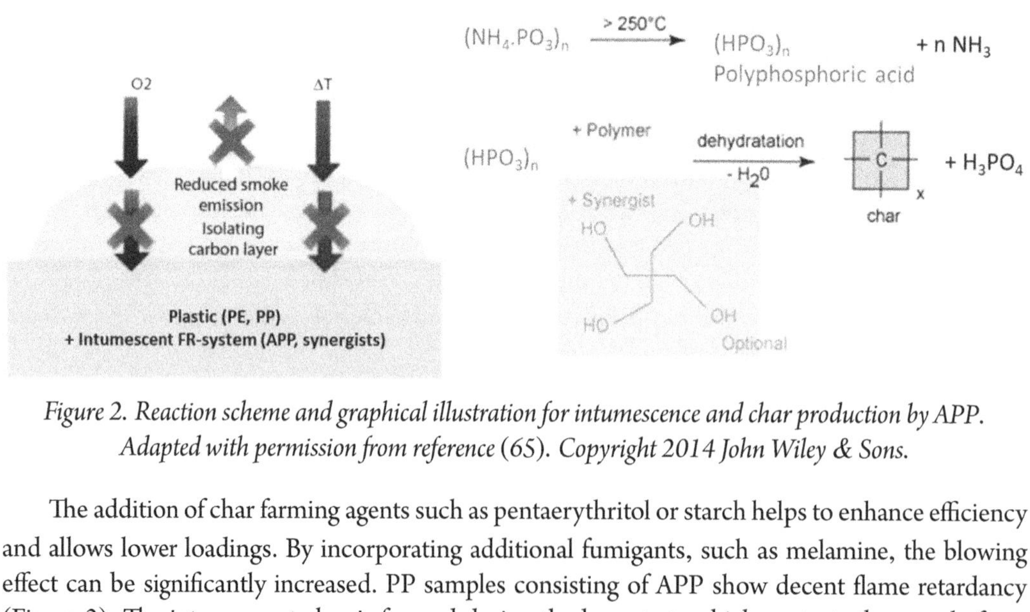

Figure 2. Reaction scheme and graphical illustration for intumescence and char production by APP. Adapted with permission from reference (65). Copyright 2014 John Wiley & Sons.

The addition of char farming agents such as pentaerythritol or starch helps to enhance efficiency and allows lower loadings. By incorporating additional fumigants, such as melamine, the blowing effect can be significantly increased. PP samples consisting of APP show decent flame retardancy (Figure 3). The intumescent char is formed during the burn test, which protects the sample from further burning. The dependency of limiting OI (LOI) on the loading level of a usual synergistic combination (Exolit AP 760) in PP is demonstrated in Figure 4. It is observed that with an increase in loading of FR and thickness of the material, LOI also increases.

Yao et al. investigated the effect of the addition of soluble APP (SAPP) in PUF (SPUF) (*66*). The FR was synthesized using water-blown SPUF and APP. The study was focused on the effect of SAPP on the flame-resistant, thermal degradation, and mechanical properties of the SPUF. Different quantities of SAPP (5, 7, 9, 11, 13, 15, 17, and 19 wt %) were added in order to understand the FR properties of PU. The mixture of additives and monomers was then stirred for 60 s, using an electric stirrer to get a homogenous mixture, followed by the addition of isocyanate. The mixture was then poured into a foaming box. The foam was heated for 24 h at 50 °C in an oven, and the sample measurement was performed. The results are summarized in Table 1. It was concluded that with increasing APP, the LOI increased, whereas the rate of horizontal burning decreased. Observations from the horizontal burning test suggested that plain SPUF did not exhibit a reduction in burning properties until the percentage of the APP extended to 7 wt %. The highest LOI of 24.5% was recorded for SPUF/APP19. The results showed that SAPP effectively enhanced the FR properties of SPUF. However, it also increased the viscosity. Therefore, SAPP should not exceed 19% to maintain a working viscosity. A genuine decrease in burning characteristics was observed after the addition of SPUF. With the combined effects of nitrogen and phosphorus, APP serves as an acid and gas source for foam combustion. It also encourages the development of porous carbonaceous chars and prevents droplet formation.

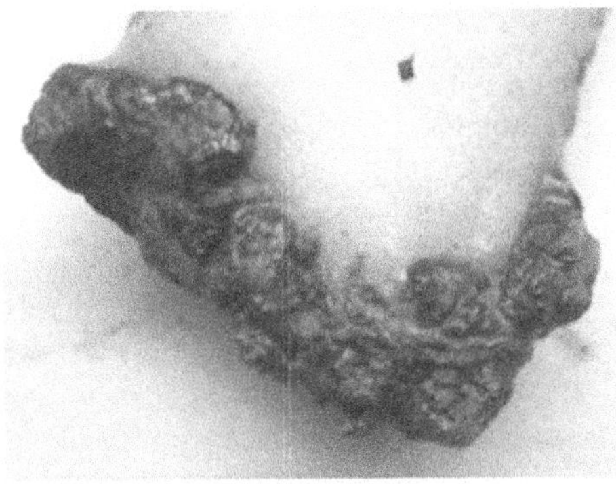

Figure 3. APP FR PP by UL-94 testing. Adapted with permission from reference (65). Copyright 2014 John Wiley & Sons.

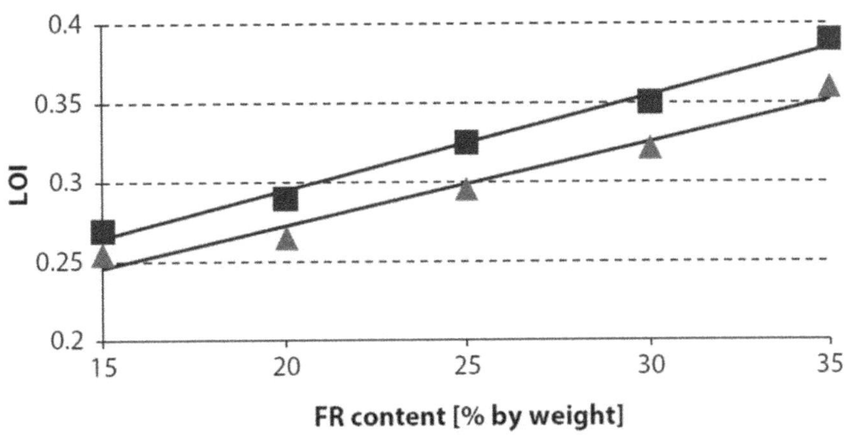

Figure 4. Graphical presentation to illustrate the dependency of LOI on the loading level of Exolit AP 760 in PP; thickness of the triangle is 1.6 mm and of the square is 3.2 mm. Adapted with permission from reference 65. Copyright 2014 John Wiley & Sons.

In comparison with halogen-based FRs, APP is environmentally friendly and does not produce a large amount of smoke (67). The smoke density of PP in the presence and absence of APP (Exolit AP 760) is in consideration of two halogens consisting of FRs—45% chlorinated cycloaliphate + antimony(III) oxide + zinc borate and 33% decabromodiphenylether + antimony(III) oxide. It has been observed that halogen FRs generate a large amount of smoke. PP in the absence of FRs (i.e., APP) also produces a sufficient amount of smoke, whereas PP in the presence of FR (APP) does not generate a large amount. This highlights the importance of APP as an FR.

Table 1. Observations Recorded for SPUF from the Burning Test with Separate Contents of APP[a]

Samples	*APP wt %*	*LOI %*	*Horizontal Burning Rate (mm/min)/ Self-Extinguished Time (seconds)*	*UL-94*
SPUF	0	18.9	49.01	/
SPUF/APP5	5	22.2	36.61	HBF
SPUF/APP7	7	22.4	172.81 SET[b]	HBF
SPUF/APP9	9	22.7	69.66 SET	HBF
SPUF/APP11	11	22.8	34.52 SET	HBF
SPUF/APP13	13	22.9	31.52 SET	HBF
SPUF/APP15	15	23.2	29.94 SET	HF-1
SPUF/APP17	17	24.1	28.59 SET	HF-1
SPUF/APP19	19	24.5	27.25 SET	HF-1

 [b] SET: Self-Extinguished Time.

Leftover Ammonia Salts as FRs

Other than APP, various materials have also been used as FRs, such as APB, ammonium sulfamate, ammonium sulfate, and so forth. APB is a colorless crystalline basic salt that can be synthesized via a reaction of ammonia, boric acid, and water. Because of easy availability and low density, APB is used as an IFR for various polymers like TPUs, epoxies, urethane foams, and so forth (*27*). Levchik et al. studied the effect of APB on the disintegration of polyamide 6 (*28*). APB reduces the thermal stability of polyamide 6 and on combustion increases the formation of the carbonated residue. Arinze et al. synthesized FR paints by incorporating ammonium sulfamate into emulsion and gloss paints (*68*). Lewin et al. proposed the role of ammonium sulfamate to obtain a large strength of flame retardancy of polyamide 6 (*69*). Mostashari et al. explained the influence of ammonium sulfate on the flammability of pure cotton cloth (*70*).

Melamine-Based FRs

Melamine (2, 4, 6-triamine-1, 3, 5 triazine) is a white crystal-like powder with a specific gravity of 1.573 g/cc and a melting point of ~354 °C. Melamine undergoes sublimation at a temperature of about 200 °C, and its sublimation energy is around 29 kcal/mol. Melamine has very low solubility in cold water but shows higher solubility in hot water. Because of good stability and rich nitrogen content, melamine is generally used in PUFs to enhance the fire resistance of foams. The use of melamine leads to a decrease in smoke density during the burning process. Price et al. explored the heat and smoke discharge kinetics of the PUF, PUF containing melamine, and polyol (*27*). They reported that the heat release rate (HRR) of PUF containing melamine was decreased by an element of two. Both PUFs got ignited early on, introducing a flow of heat, whereas polyol showed a remarkably delayed ignition (~ 25 s). The time required to touch the peak heat discharge kinetic for polyol was 120 s, which was much higher compared with PU (~ 50 s). This suggested that quick ignition of PU was not because of polyol but rather because of some other component of PU (i.e., isocyanate). The authors proposed the mechanism to investigate the reaction between melamine and isocyanate, and they mentioned that the interaction between them was also an important part of decreasing the HRR and smoke release rate. This interaction was explained by a mechanism that is shown in Figure 5. At higher temperatures (more than 250 °C), the $-NH_2$ group of melamine attacks on the isocyanate group and forms urea, which is nonvolatile and therefore cannot pass through the gas phase and fuel the fire. This helps to decrease the amount of heat release. Because of the emission of highly aromatic isocyanate during combustion, reduction of soot is also observed. This compound helps in protecting the remaining part of the foam. Therefore, it has been concluded that the reactive amino group of melamine plays a vital role in flame retardation (*27*).

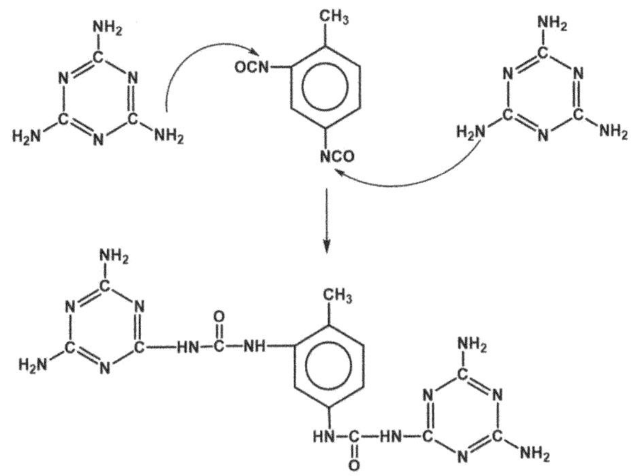

Figure 5. A reaction of melamine with toluene diisocyanate. Adapted with permission from reference (27). Copyright 2002 John Wiley & Sons.

Melamine Cyanurate-Based FRs

MCA is an important FR for many polymeric materials (*71*). Upon sublimation, it decomposes to form gaseous ammonia and carbon dioxide. MCA is also called the melamine–cyanuric acid complex. A complex of MCA is crystalline; it is synthesized by taking a 1:1 ratio of cyanuric acid and melamine. Figure 6 shows the structure of MCA. The complex is held together by a two-dimensional interconnection of hydrogen bonds between two compounds. The triangular shape of each molecule produces sheets, as shown in Figure 7 (*72*).

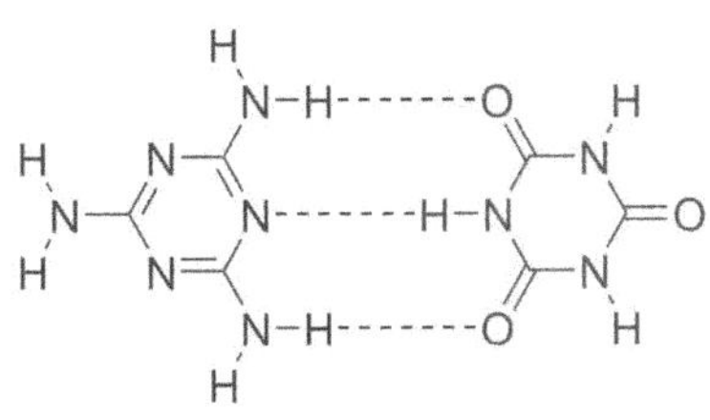

Figure 6. Structure of MCA.

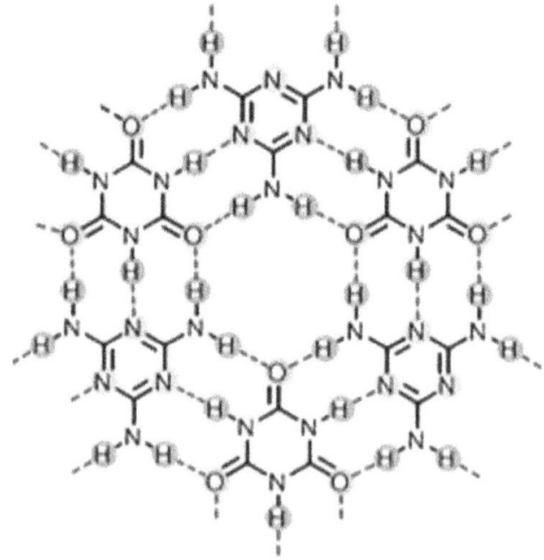

Figure 7. Hydrogen bonding within melamine–cyanuric acid. Adapted with permission from reference (73). Copyright 2018 Elsevier.

Yang et al. used APP/MCA as a synergistic FR for papermaking (74). While warming the handsheets, APP/MCA played a vital role in withdrawing high temperatures. Initially, MCA got pyrolyzed to form ammonia and carbon dioxide that further diluted the oxygen from the air and separated heat from the fire. After the APP played a dominating role in synergistic FR systems, it decomposed into phosphoric acid. Then, a compact carbide layer was formed on the fiber surface, which separated oxygen and prohibited the process of pyrolysis. These physicochemical changes improved the thermal stability of handsheets and also significantly increased the FR efficiency. MCA was one of the effective FRs for polyamides. Around 6–10% of MCA was sufficient to bring out flammability in this compound. When polyamide containing MCA was introduced to the flame, MCA was quickly decomposed, as shown in Figure 8. Decomposition of MCA took place with the formation of small bubbles, which indicated an intumescent effect. Under these conditions, no char formation was observed. The entire specimen formed a spit, which further dropped down in the next flaming time. The formed drops were very small in size, so they stopped burning before reaching the bottom because of the heat sink/diluent effect of the MCA, resulting in a UL-94 classification of V-0 (65).

Figure 8. Demonstration of UL-94 V-0 test of polyamide consisting of 6% MCA throughout the first and second and after the second application of the flame. Adapted with permission from reference (65). Copyright 2014 John Wiley & Sons.

Tabuani et al. studied fire-resistant TPU formulations consisting of different amounts of nonhalogen FRs such as MCA and nanoclays (75). They noticed that TPU consisting of both MCA and nanoclays exhibited a considerable improvement in fire properties and enhanced mechanical properties. The use of nanoclay provided a better rigidity to the matrix without compromising the failure strain. This appreciable increase in flame retardancy was attributed to the synergic effect of the two FR additives. Braun et al. proposed the flame retardancy mechanism of aluminum diethyl phosphinate with melamine polyphosphate and zinc borate in glass fiber reinforced poly(1,4-butylene terephthalate) (42). The authors concluded that the major flame retardancy ability was because of the active phosphinate in the gaseous phase, whereas MCA came up with the heat sink/ dilution effect.

Melamine Polyphosphate-Based FRs

Melamine polyphosphate (MPP) can be used as a reactive and additive FR for thermally resistant PUFs (76). MPP is a salt of melamine and phosphoric acid. At more than 200 °C, it forms melamine pyrophosphate, which with further heating (more than 260 °C) forms MPP. This process involves the formation of water, which leads to a heat sink effect. The structure of MPP is shown in Figure 9. The intumescent charring effect for polyamide with MPP during the UL-94 V-test is shown in Figure 10. The intumescent char is immediately formed during the first flaming period, and it increases throughout the second flaming period; dripping is not observed.

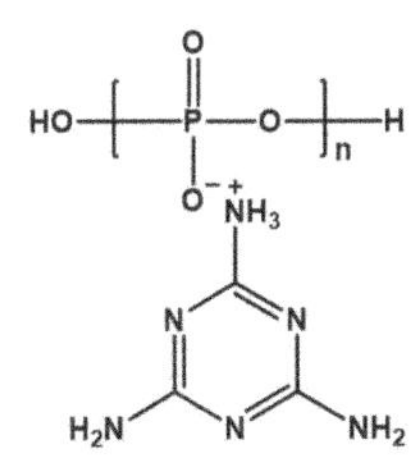

Figure 9. Structure of MPP.

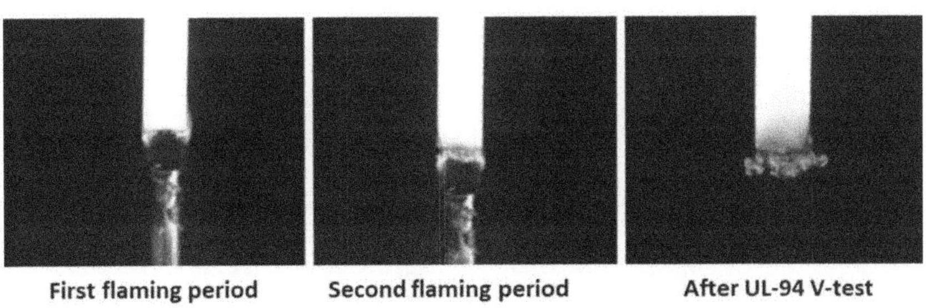

Figure 10. UL-94 V-test of a polyamide compound with MPP. Adapted with permission from reference (65). Copyright 2014 John Wiley & Sons.

Lai et al. investigated the effect of the addition of FR in propylene (77). In this work, they synthesized caged bicyclic phosphate 2,6,7-trioxa-1-phosphabicyclo[2.2.2]octane-4-methanol 1-oxide, named pentaerythritol phosphate (PEPA), and mixed it with melamine pyrophosphate to prepare FR PP. The effect of this composite was well tested using different techniques. When the 6.7 wt % of melamine pyrophosphate and 13.3 wt % of PEPA were added, the LOI value of 33.0% was observed for a PP/melamine pyrophosphate/PEPA composite with a V-0 rating (using UL-94). It was also noticed that total heat release, peak HRR, and average mass loss rate were appreciably decreased. From the structural and thermal analysis, it was concluded that while burning melamine pyrophosphate/PEPA, phosphorus-based and triazine oligomer radicals could form, which prohibited the deterioration and ignition of the polymer. Additionally, cyclization in C=C bonds with phosphorus-based material took place and formed a cross-linked char to prevent degradation and burning of the substrate. The probable mechanism of FR and thermal pyrolysis of melamine pyrophosphate/PEPA for PP is illustrated in Figure 11. At a temperature of more than 250 °C, the P–O–C bond is formed because of the formation of ester and water removal between melamine pyrophosphate and PEPA. Generally, PEPA is broken at higher temperatures to give *P*-based material with alkenes. During this reaction, triazine oligomer was obtained from melamine pyrophosphate. These radicals could squish the free radicals such as −CO• and −COO• obtained from the deterioration of PP. This led to enhancing the thermostability of PP. After this, the layer of char was obtained. Gases such as ammonia, carbon dioxide, and water, which are not capable of being burned, were evolved, and a char layer was formed. After moderate heating, melamine pyrophosphate/PEPA formed a bond between phosphorous and nitrogen and helped to provide char surrounding it. This layer further protected the polymers from degradation by reducing the transfer of heat and combustible gases (77).

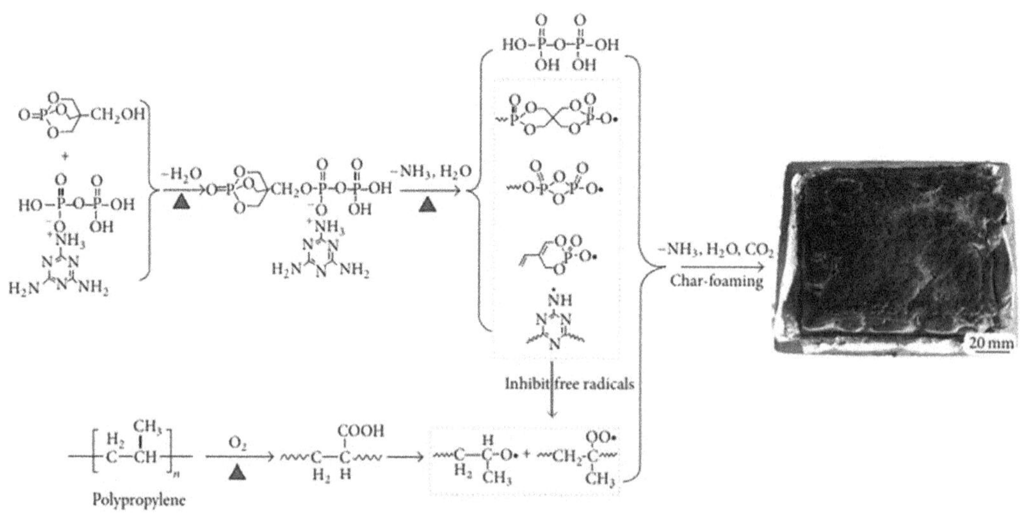

Figure 11. Proposed mechanism of FR and thermal degradation of melamine pyrophosphate/PEPA for PP. Adapted with permission from reference (77). Copyright Lai et al., some rights reserved; exclusive licensee Hindawi. Distributed under a Creative Commons Attribution License 3.0 (CC BY) https://creativecommons.org/licenses/by/3.0.

Melem and Melemium Salts-Based FRs

Melamine is used as a precursor for the synthesis of melem. Figure 12 explains the generation of melem using melamine and cyanamide (78). At about 300 °C, melamine decomposes into cyanamide, and then the combination of cyanamide and melamine, which leads to the formation of melem with the loss of ammonia (79). Lotsch et al. proposed the mechanism for the generation of melem using melamine and dicyandiamide, which is shown in Figure 13 (*80*). As shown in Figure 13, dicyandiamide may form parallel to the cyanamide monomer by thermal decomposition of melamine. Melem and its salts are employed as FRs for thermoplastics (mainly polyamide).

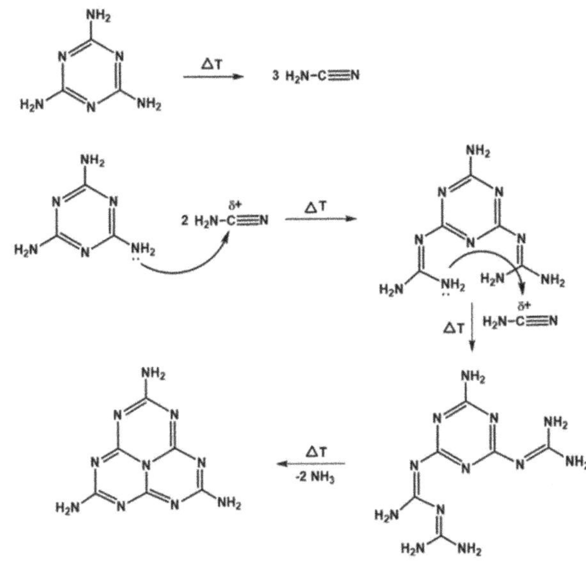

Figure 12. Synthesis of melem using melamine and cyanamide. Adapted with permission from reference (78). Copyright 2013 Elsevier.

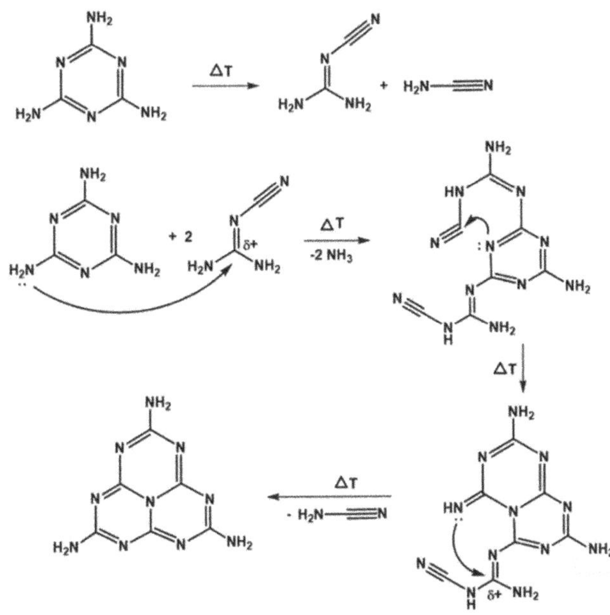

Figure 13. Preparation of melem using melamine. Adapted with permission from reference (78). Copyright 2013 Elsevier.

Nitrogen-Based Radical Generators

Nitrogen-based radical generators are current types of NFRs that have specific applications in polyolefins. Large loss and removal of the polyolefins have been observed during this process. Among various chemical structures, *N*-alkoxy hindered amines and azoalkanes are very common (*81*, *82*). Cao et al. synthesized *N* alkoxy hindered amine containing silanes (Si-NORs) by a sol–gel reaction as multifunctional FR synergists and tested them as IFRs for PPs (*83*). They investigated whether Si-NORs can replace part of IFRs and provide better flame retardancy for IFR–PP composites.

The LOI value for IFR–PP composites (with 25 wt % IFR) increased from 30 to 32, with 1 wt % loading of Si-NORs. The rating for UL-94 was V-0, which is shown in Figure 14. The possible synergistic mechanism of the IFR–PP composite with Si-NORs is given in Figure 15. Si-NORs are used as radical scavengers and for the arrest of free radicals in the gas phase. Si- NORs also help with the increase in the FRs in IFR–PP composites. IFR–PP and Si-NORs increase the UV stability and mechanical properties. The synergistic mechanism of Si-NORs in the IFR–PP composite is explained in the gas and condensed phases. During thermal decomposition, free radicals are formed, which helps the degradation of PP (*84*). Herein, nitroxyl radicals act as radical scavengers, which are produced during the pyrolysis of Si-NORs. These nitroxyl radicals help to convert free radicals into stable alcohols and ketones, and they form the cross-linking network (*85*, *86*). This network helps to stabilize cell growth, prevent the merging of cells, and give char with excellent quality surrounding it.

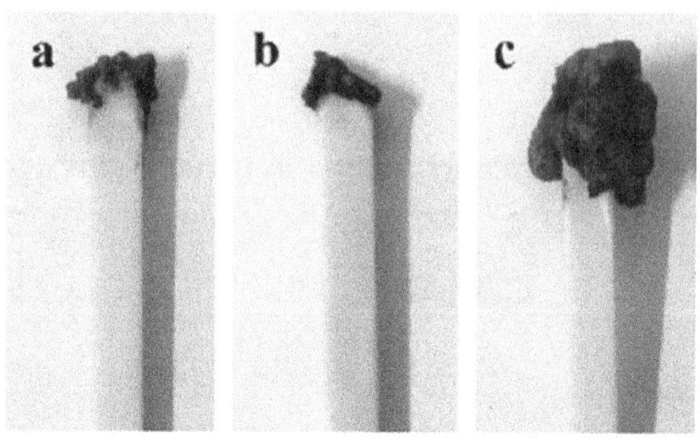

Figure 14. Digital images after LOI test for PP75/IFR25/Si-NORs composite. Oxygen was (a) 30, (b) 32, and (c) 34%. Adapted with permission from reference (83). Copyright 2013 American Chemical Society.

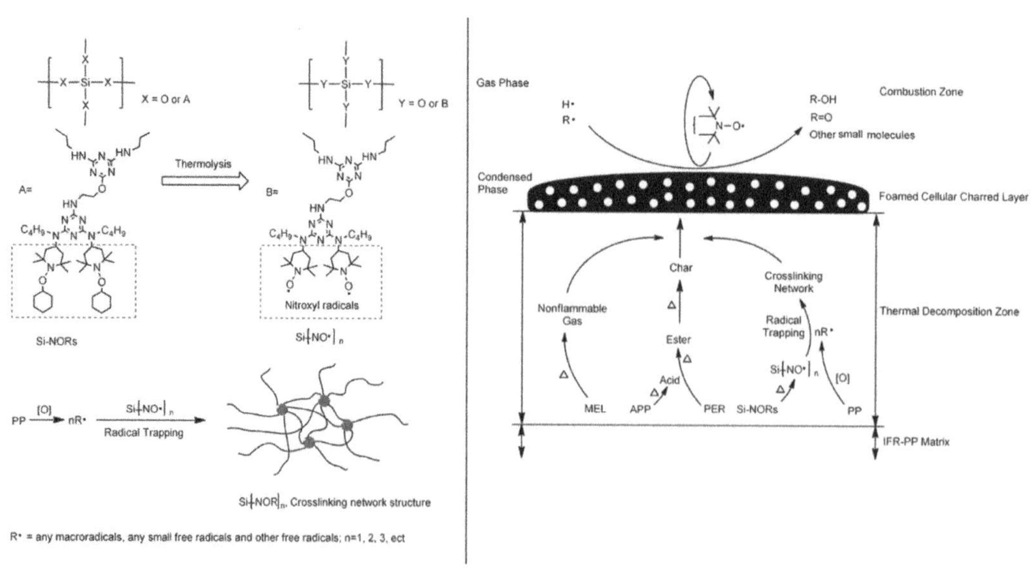

Figure 15. Proposed FR mechanism of Si-NORs in an IFR–PP composite. Adapted with permission from reference (83). Copyright 2013 American Chemical Society.

Aubert et al. discussed the use of azoalkanes as novel FRs and highlighted their structure–property relationship (*82*). They observed that azoalkanes and their compounds displayed a considerable degree of flame retardancy of PP thin films. They also acted as effective FRs in thick PP moldings. A series of compounds of azoalkanes was discussed in this article and tested for its performance. The best performance was recorded for the symmetrical high molecular weight cyclic azo compound 4,4′-bis(cyclohexylazocyclohexyl) methane (*82*).

Phospham-, Phosphazene-, and Phosphoroxynitride-Based FRs

Phospham, phosphazene, and phosphoroxynitride are compounds in which phosphorus is covalently bonded to a nitrogen atom. The structures of different polyphosphazenes are shown in Figure 16. Polyphosphazene is an organic–inorganic hybrid polymer with a high molecular weight in which the main chain is alternately attached by single and double bonds of phosphorus and nitrogen atoms. Meanwhile, the side chain is replaced by organic groups. Linear and cross-linked polyphosphazenes are the most studied ones because of their highly efficient FR and mechanical enhancement effects. The presence of both phosphorus and nitrogen elements exhibits a synergistic FR effect and provides high thermal stability (*87*, *88*). Levchik and coauthors demonstrated the use of phospham, phosphazene, and phosphoroxynitride in polybutylene terephthalate (*89*). From thermogravimetric analysis, they concluded that phospham and phosphorus oxynitride are thermally stable FR additives. Phosphorus oxynitride was observed to be an effective charring promoter in polybutylene terephthalate. The use of phospham as an efficient FR additive for nylon-4,6 was explained by Weil and coauthors (*90*).

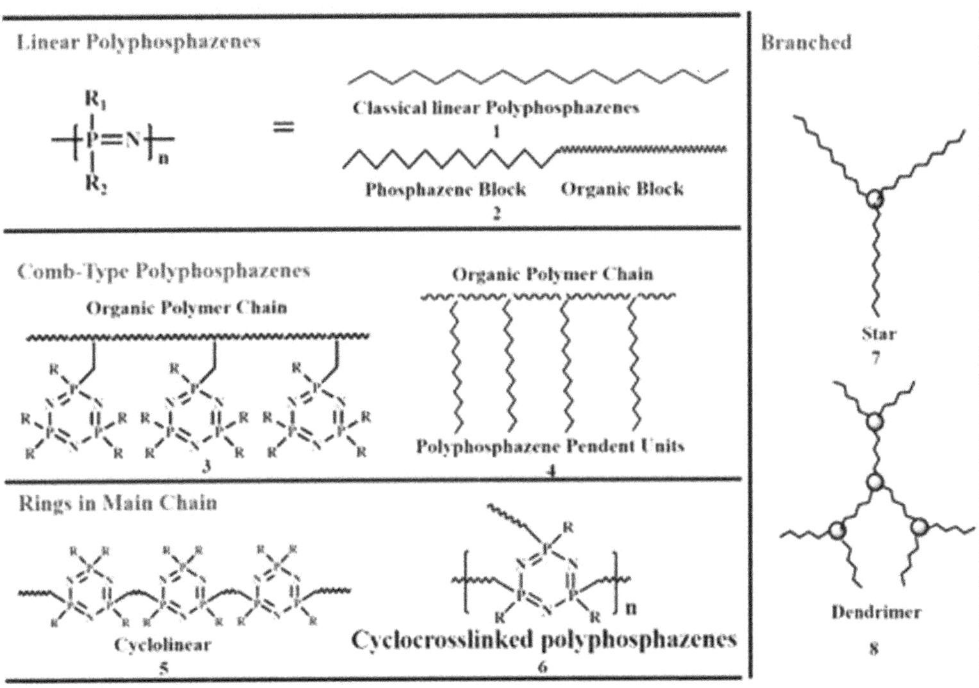

Figure 16. Structures of various types of polyphosphazenes. Adapted with permission from reference (87). Copyright 2020 Elsevier.

Cyanuric Acid-Based FRs

In the last decade, cyanuric acid-based FRs have been popular in the market. When cyanuric chloride reacts with alcohol and amine, it yields ester and amide, respectively. Morpholine reacts with bifunctional cyanuric chloride to prepare its derivatives. These derivatives undergo polymerization with bifunctional amines or esters. A polymer with morpholine and piperazine is manufactured by MCA Technologies and sold using the commercial name PPM Triazine HF (65). MCA Technologies has demonstrated the results observed from a combination of APP and PPM Triazine HF in a 3:1 ratio in PP. For pure PP, it recorded an increase in LOI from 19 to 31 for 20 wt % loading and therefore showed a UL-94 V-0 grade. Data obtained using cone calorimetry have shown a sudden decrease in temperature discharge kinetics by parts of 5 and 10 with a 10 and 20% loading, respectively. It has also been also noticed that the entire emission of smoke is appreciably decreased with 20% loading.

Conclusion and Future Outlook

NFRs can be considered as an individual class of FRs because of their variety of applications. Because of high decomposition temperature, they can be recycled by reextrusion. These compounds are more often combined with phosphorus, and their synergistic effects make them effective FRs for various polymers. NFRs act as passive supporters of the intumescent mechanism by producing foaming gases. In addition, they act as heat sinks because of endothermic decomposition. Some new nitrogen-based compounds are being introduced, and they play a highly important role in flame retardancy. Nitrogen-based radical generators—such as azoalkanes, *N*-alkoxy hindered amines, phospham, phosphazene, phosphoroxynitride, and cyanuric acid—provide a highly efficient FR effect and enhance the mechanical properties of polymers. The use of nitrogen-based synergistic effects helps to reduce the overall loading of FRs and also to develop a general trend toward sustainable FR solutions.

References

1. Šebenik, U.; Krajnc, M. Influence of the Soft Segment Length and Content on the Synthesis and Properties of Isocyanate-Terminated Urethane Prepolymers. *Int. J. Adhes. Adhes.* **2007**, 27 (7), 527–535.
2. Klinedinst, D. B.; Yilgör, I.; Yilgör, E.; Zhang, M.; Wilkes, G. L. The Effect of Varying Soft and Hard Segment Length on the Structure-Property Relationships of Segmented Polyurethanes Based on a Linear Symmetric Diisocyanate, 1,4-Butanediol and PTMO Soft Segments. *Polymer (Guildf)* **2012**, 53 (23), 5358–5366.
3. Liu, C.-G.; Wang, N.; Lin, Y.-H.; Bai, F.-W. Very High Gravity Ethanol Fermentation by Flocculating Yeast under Redox Potential-Controlled Conditions. *Biotechnol. Biofuels* **2012**, 5 (1), 61.
4. Ranaweera, C. K.; Ionescu, M.; Bilic, N.; Wan, X.; Kahol, P. K.; Gupta, R. K. Biobased Polyols Using Thiol-Ene Chemistry for Rigid Polyurethane Foams with Enhanced Flame-Retardant Properties. *J. Renew. Mater.* **2017**, 5 (1), 1–12.
5. Elbers, N.; Ranaweera, C. K.; Ionescu, M.; Wan, X.; Kahol, P. K.; Gupta, R. K. Synthesis of Novel Biobased Polyol via Thiol-Ene Chemistry for Rigid Polyurethane Foams. *J. Renew. Mater.* **2017**, 5 (1), 74–83.

6. Kirpluks, M.; Cabulis, U.; Zeltins, V.; Stiebra, L.; Avots, A. Rigid Polyurethane Foam Thermal Insulation Protected with Mineral Intumescent Mat. *Autex Res. J.* **2014**, *14* (4), 259–269.
7. Zatorski, W.; Brzozowski, Z. K.; Kolbrecki, A. New Developments in Chemical Modification of Fire-Safe Rigid Polyurethane Foams. *Polym. Degrad. Stab.* **2008**, *93* (11), 2071–2076.
8. Zhang, L.; Zhang, M.; Zhou, Y.; Hu, L. The Study of Mechanical Behavior and Flame Retardancy of Castor Oil Phosphate-Based Rigid Polyurethane Foam Composites Containing Expanded Graphite and Triethyl Phosphate. *Polym. Degrad. Stab.* **2013**, *98* (12), 2784–2794.
9. Bhoyate, S.; Ionescu, M.; Kahol, P. K.; Chen, J.; Mishra, S. R.; Gupta, R. K. Highly Flame-Retardant Polyurethane Foam Based on Reactive Phosphorus Polyol and Limonene-Based Polyol. *J. Appl. Polym. Sci.* **2018**, *135* (21), 16–19.
10. Bhoyate, S.; Ionescu, M.; Kahol, P. K.; Gupta, R. K. Sustainable Flame-Retardant Polyurethanes Using Renewable Resources. *Ind. Crops Prod.* **2018**, *123*, 480–488.
11. Bourbigot, S.; Duquesne, S. Fire Retardant Polymers: Recent Developments and Opportunities. *J. Mater. Chem.* **2007**, *17* (22), 2283–2300.
12. Horacek, H.; Grabner, R. Advantages of Flame Retardants Based on Nitrogen Compounds. *Polym. Degrad. Stab.* **1996**, *54* (2), 205–215.
13. Vahabi, H.; Rastin, H.; Movahedifar, E.; Antoun, K.; Brosse, N.; Saeb, M. R. Flame Retardancy of Bio-Based Polyurethanes: Opportunities and Challenges. *Polymers (Basel)* **2020**, *12* (6), 1234.
14. Szolnoki, B.; Toldy, A.; Konrád, P.; Szebényi, G.; Marosi, G. Comparison of Additive and Reactive Phosphorus-Based Flame Retardants in Epoxy Resins. *Period. Polytech. Chem. Eng.* **2013**, *57* (1–2), 85–91.
15. Su, D.; Tang, Z.; Xie, J.; Bian, Z.; Zhang, J.; Yuan, A.; Su, D.; Tang, Z.; Xie, J.; Bian, Z.; Zhang, J.; Yuan, A. High Performance Layered Inorganic Flame Retardants: A Review. *Gen. Chem. Chem. Chem.* **2019**, *5* (1), 180021–180021.
16. Beyer, G. Nanocomposites: A New Class of Flame Retardants for Polymers. *Plast. Addit. Compd.* **2002**, *4* (10), 22–28.
17. Alaee, M.; Arias, P.; Sjödin, A.; Bergman, Å. An Overview of Commercially Used Brominated Flame Retardants, Their Applications, Their Use Patterns in Different Countries/Regions and Possible Modes of Release. *Environ. Int.* **2003**, *29* (6), 683–689.
18. Ray, C.; Lee, S. C.; Sankar, K. V.; Jin, B.; Lee, J.; Park, J. H.; Jun, S. C. Amorphous Phosphorus-Incorporated Cobalt Molybdenum Sulfide on Carbon Cloth: An Efficient and Stable Electrocatalyst for Enhanced Overall Water Splitting over Entire PH Values. *ACS Appl. Mater. Interfaces* **2017**, *9* (43), 37739–37749.
19. Fries, G. F. Bioavailability of Soil-Borne Polybrominated Biphenyls Ingested by Farm Animals. *J. Toxicol. Environ. Health* **1985**, *16* (3–4), 565–579.
20. Söderlund, E.; Dybing, E.; Nelson, S. D. Nephrotoxicity and Hepatotoxicity of Tris(2,3-Dibromopropyl)Phosphate in the Rat. *Top. Catal.* **1980**, *56* (2), 171–181.
21. Hilado, C. J.; Kuryla, W. C.; McLaughlin, R. W.; Proops, W. R. Boran and Antimony Compounds as Flame Retardants in Rigid Polyurethane Foam. *J. Fire Flammability/Fire Retard. Chem.* **1974** (1), 175–184.

22. Nagieb, Z. A.; Nassar, M. A.; El-Meligy, M. G. Effect of Addition of Boric Acid and Borax on Fire-Retardant and Mechanical Properties of Urea Formaldehyde Saw Dust Composites. *Int. J. Carbohydr. Chem.* **2011**, *2011*, 1–6.
23. Benin, V.; Durganala, S.; Morgan, A. B. Synthesis and Flame Retardant Testing of New Boronated and Phosphonated Aromatic Compounds. *J. Mater. Chem.* **2012**, *22* (3), 1180–1190.
24. Tang, S.; Qian, L.; Qiu, Y.; Dong, Y. Synergistic Flame-Retardant Effect and Mechanisms of Boron/Phosphorus Compounds on Epoxy Resins. *Polym. Adv. Technol.* **2018**, *29* (1), 641–648.
25. Unlu, S. M.; Dogan, S. D.; Dogan, M. Comparative Study of Boron Compounds and Aluminum Trihydroxide as Flame Retardant Additives in Epoxy Resin. *Polym. Adv. Technol.* **2014**, *25* (8), 769–776.
26. Myers, R. E.; Dickens, E. D.; Licursi, E.; Evans, R. E. Ammonium Pentaborate: An Intumescent Flame Retardant for Thermoplastic Polyurethanes. *J. Fire Sci.* **1985**, *3* (6), 432–449.
27. Price, D.; Liu, Y.; Milnes, G. J.; Hull, R.; Kandola, B. K.; Horrocks, A. R. An Investigation into the Mechanism of Flame Retardancy and Smoke Suppression by Melamine in Flexible Polyurethane Foam. *Fire Mater.* **2002**, *26* (4-5), 201–206.
28. Levchik, S. V.; Balabanovich, A. I.; Levchik, G. F.; Costa, L. Effect of Melamine and Its Salts on Combustion and Thermal Decomposition of Polyamide 6. *Fire Mater.* **1997**, *21* (2), 75–83.
29. Duquesne, S.; Magnet, S.; Jama, C.; Delobel, R. Intumescent Paints: Fire Protective Coatings for Metallic Substrates. *Surf. Coatings Technol.* **2004**, *180–181*, 302–307.
30. Wladyka-Przybylak, M.; Kozlowski, R. Thermal Characteristics of Different Intumescent Coatings. *Fire Mater.* **1999**, *23* (1), 33–43.
31. Jimenez, M.; Duquesne, S.; Bourbigot, S. Characterization of the Performance of an Intumescent Fire Protective Coating. *Surf. Coatings Technol.* **2006**, *201* (3–4), 979–987.
32. Yew, M. C.; Ramli Sulong, N. H. Fire-Resistive Performance of Intumescent Flame-Retardant Coatings for Steel. *Mater. Des.* **2012**, *34*, 719–724.
33. Horrocks, A. R.; Wang, M. Y.; Hall, M. E.; Sunmonu, F.; Pearson, J. S. Flame Retardant Textile Back-Coatings. Part 2. Effectiveness of Phosphorus-Containing Flame Retardants in Textile Back-Coating Formulations. *Polym. Int.* **2000**, *49* (10), 1079–1091.
34. Koo, J. H.; Wootan, W.; Chow, W. K.; Au Yeung, H. W.; Venumbaka, S. Flammability Studies of Fire Retardant Coatings on Wood. *ACS Symp. Ser.* **2001**, *797*, 361–374.
35. Montaudo, G.; Scamporrino, E.; Vitalini, D. Intumescent Flame Retardants for Polymers. II. The Polypropylene-Ammonium Polyphosphate-Polyurea System. *J. Polym. Sci. A1.* **1983**, *21* (12), 3361–3371.
36. Zaikov, G. E. Recent Advances in Flame Retardancy of Polymeric Materials. *Int. J. Polym. Mater. Polym. Biomater.* **1997**, *36* (1–2), 33–38.
37. Levchik, S. V; Weil, E. D. A Review of Recent Progress in Phosphorus-Based Flame Retardants. *J. Fire Sci.* **2006**, *24* (5), 345–364.
38. Hastie, J. W. Molecular Basis of Flame Inhibition. *J. Res. Natl. Bur. Stand. Sect. A Phys. Chem.* **1973**, *77 A* (6), 733–754.

39. Scharte, B. Phosphorus-Based Flame Retardancy Mechanisms-Old Hat or a Starting Point for Future Development? *Materials (Basel)* **2010**, 3 (10), 4710–4745.
40. Bourbigot, S.; Le Bras, M.; Duquesne, S.; Rochery, M. Recent Advances for Intumescent Polymers. *Macromol. Mater. Eng.* **2004**, *289* (6), 499–511.
41. Nazir, R.; Gaan, S. Recent Developments in P(O/S)–N Containing Flame Retardants. *J. Appl. Polym. Sci.* **2020**, *137* (1), 47910.
42. Braun, U.; Schartel, B. Flame Retardancy Mechanisms of Aluminium Phosphinate in Combination with Melamine Cyanurate in Glass-Fibre-Reinforced Poly(1,4-butylene terephthalate). *Macromolecular Materials and Engineering* **2008**, 206–217.
43. Langfeld, K.; Wilke, A.; Sut, A.; Greiser, S.; Ulmer, B.; Andrievici, V.; Limbach, P.; Bastian, M.; Schartel, B. Halogen-Free Fire Retardant Styrene-Ethylene-Butylene-Styrene-Based Thermoplastic Elastomers Using Synergistic Aluminum Diethylphosphinate-Based Combinations. *J. Fire Sci.* **2015**, *33* (2), 157–177.
44. Gallo, E.; Schartel, B.; Braun, U.; Russo, P.; Acierno, D. Fire Retardant Synergisms between Nanometric Fe2O3 and Aluminum Phosphinate in Poly(Butylene Terephthalate). *Polym. Adv. Technol.* **2011**, *22* (12), 2382–2391.
45. Gallo, E.; Schartel, B.; Acierno, D.; Russo, P. Flame Retardant Biocomposites: Synergism between Phosphinate and Nanometric Metal Oxides. *Eur. Polym. J.* **2011**, *47* (7), 1390–1401.
46. Braun, U.; Bahr, H.; Sturm, H.; Schartel, B. Flame Retardancy Mechanisms of Metal Phosphinates and Metal Phosphinates in Combination with Melamine Cyanurate in Glass-Fiber Reinforced Poly(1,4-Butylene Terephthalate): The Influence of Metal Cation. *Polym. Adv. Technol.* **2008**, *19* (6), 680–692.
47. Braun, U.; Balabanovich, A. I.; Schartel, B.; Knoll, U.; Artner, J.; Ciesielski, M.; Döring, M.; Perez, R.; Sandler, J. K. W.; Altstädt, V.; Hoffmann, T.; Pospiech, D. Influence of the Oxidation State of Phosphorus on the Decomposition and Fire Behaviour of Flame-Retarded Epoxy Resin Composites. *Polymer (Guildf)* **2006**, *47* (26), 8495–8508.
48. Hergenrother, P. M.; Thompson, C. M.; Smith, J. G.; Connell, J. W.; Hinkley, J. A.; Lyon, R. E.; Moulton, R. Flame Retardant Aircraft Epoxy Resins Containing Phosphorus. *Polymer (Guildf)* **2005**, *46* (14), 5012–5024.
49. Pawlowski, K. H.; Schartel, B. Flame Retardancy Mechanisms of Triphenyl Phosphate, Resorcinol Bis(Diphenyl Phosphate) and Bisphenol A Bis(Diphenyl Phosphate) in Polycarbonate/Acrylonitrile-Butadiene-Styrene Blends. *Polym. Int.* **2007**, *56* (11), 1404–1414.
50. Wawrzyn, E.; Schartel, B.; Ciesielski, M.; Kretzschmar, B.; Braun, U.; Döring, M. Are Novel Aryl Phosphates Competitors for Bisphenol A Bis(Diphenyl Phosphate) in Halogen-Free Flame-Retarded Polycarbonate/Acrylonitrile-Butadiene- Styrene Blends? *Eur. Polym. J.* **2012**, *48* (9), 1561–1574.
51. Van Krevelen, D. W. Some Basic Aspects of Flame Resistance of Polymeric Materials. *Polymer (Guildf)* **1975**, *16*, 615.
52. Salmeia, K. A.; Fage, J.; Liang, S.; Gaan, S. An Overview of Mode of Action and Analytical Methods for Evaluation of Gas Phase Activities of Flame Retardants. *Polymers (Basel)* **2015**, 7 (3), 504–526.
53. Petrella, R. V. Factors Affecting the Combustion of Polystyrene and Styrene. *Flame-Retardant Polym. Mater.* **1978**, 2, 159–201.

54. Howell, B. A.; Oberdorfer, K. L.; Ostrander, E. A. Phosphorus Flame Retardants for Polymeric Materials from Gallic Acid and Other Naturally Occurring Multihydroxybenzoic Acids. *Int. J. Polym. Sci.* **2018**, *2018*, 7237236.
55. Hubbe, M. A.; Venditti, R. A.; Rojas, O. J. What Happens to Cellulosic Fibers during Papermaking and Recycling? A Review. *BioResources* **2007**, 2 (4), 739–788.
56. Hendrix, J. E.; Drake, G. L.; Barker, R. H. Pyrolysis and Combustion of Cellulose. III. Mechanistic Basis for the Synergism Involving Organic Phosphates and Nitrogenous Bases. *J. Appl. Polym. Sci.* **1972,** *16* (2), 257–274.
57. Pushpa, Bajaj Fire-Retardant Materials. *Bull. Mater. Sci.* **1992,** *15*, 67–76.
58. Akat, H.; Saltan, F. Synthesis, Characterization and Thermal Degradation of Cross-Linked Polystyrene Using the Alkyne-Functionalized Esters as a Cross-Linker Agent by Click Chemistry Method. *Polimeros* **2015,** *25* (4), 402–407.
59. Sirdesai, S. J.; Wilkie, C. A. Wilkinson's Salt: A Flame Retardant for Poly(Methyl Methacrylate). *J. Appl. Polym. Sci.* **1989,** *37* (4), 863–866.
60. Rodrig, H.; Basch, A.; Lewin, M. Crosslinking and Pyrolytic Behavior of Natural and Man-Made Cellulosic Fibers. *J. Polym. Sci .Polym. Chem. Ed.* **1975,** *13* (8), 1921–1932.
61. Liu, Y.; He, J.; Yang, R. The Effects of Aluminum Hydroxide and Ammonium Polyphosphate on the Flame Retardancy and Mechanical Property of Polyisocyanurate-Polyurethane Foams. *J. Fire Sci.* **2015,** *33* (6), 459–472.
62. Watanabe, M.; Sakurai, M.; Takahashi, M. New Preparation Processes of Phase II and V Ammonium Polyphosphate and Their Use as Flame Retardants. *Phosphorus Res. Bull.* **2003,** *16*, 39–46.
63. Watanabe, M.; Sakurai, M.; Yamaguchi, Y.; Watanabe, M. Phase Transition of APP Form I to II and Synthesis of APP Form II from the System of Ammonium Orthophosphate-Urea. *Phosphorus, Sulfur Silicon Relat. Elem.* **1999,** *144–146*, 235–238.
64. Watanabe, M.; Suzuki, H.; Kawasaki, S.; Maeda, M.; Sakurai, M. Preparation of Ammonium Polyphosphate and Its Use as a Flame Retardant. *Phosphorus Res. Bull.* **2002,** *14*, 105–110.
65. Klatt, M. Nitrogen-Based Flame Retardants. In *Non-Halogenated Flame Retardant Handbook*; Wiley Online Books; John Wiley & Sons: Hoboken, NJ, United States, 2014
66. Yao, W.; Wang, H.; Guan, D.; Fu, T.; Zhang, T.; Dou, Y. The Effect of Soluble Ammonium Polyphosphate on the Properties of Water Blown Semirigid Polyurethane Foams. *Adv. Mater. Sci. Eng.* **2017**, *2017*, 5282869.
67. Lim, K. S.; Bee, S. T.; Sin, L. T.; Tee, T. T.; Ratnam, C. T.; Hui, D.; Rahmat, A. R. A Review of Application of Ammonium Polyphosphate as Intumescent Flame Retardant in Thermoplastic Composites. *Compos. Part B Eng.* **2016,** *84*, 155–174.
68. Arinze, R. U.; Helen Okoye, N.; Uche Chris-Okafor, P.; Author, C.; Arinze, R.; Eboatu, A.; Okoye, H.; Nzekwe, P. Use of Ammonium Sulphamate as Flame Retardant Paint Additive Sustainable and Biodegradable Polymers View Project Thermal Degradation and Surface Characterization of Some Selected Tropical Timbers and Their Wood PET Composites View Project Use of Ammonium. *World Eng. Appl. Sci. J.* **2016,** 7 (2), 103–106.
69. Lewin, M.; Zhang, J.; Pearce, E.; Gilman, J. Flammability of Polyamide 6 Using the Sulfamate System and Organo-Layered Silicate. *Polym. Adv. Technol.* **2007,** *18* (9), 737–745.

70. Mostashari, S. M.; Mostashari, S. Z. Combustion Pathway of Cotton Fabrics Treated by Ammonium Sulfate as a Flame-Retardant Studied by TG. *J. Therm. Anal. Calorim.* **2008**, *91*, 437–441.
71. Gijsman, P.; Steenbakkers, R.; Fürst, C.; Kersjes, J. Differences in the Flame Retardant Mechanism of Melamine Cyanurate in Polyamide 6 and Polyamide 66. *Polym. Degrad. Stab.* **2002**, *78* (2), 219–224.
72. Prior, T. J.; Armstrong, J. A.; Benoit, D. M.; Marshall, K. L. The Structure of the Melamine–Cyanuric Acid Co-Crystal. *CrystEngComm* **2013**, *15* (29), 5838–5843.
73. Tao, W.; Li, J. Melamine Cyanurate Tailored by Base and Its Multi Effects on Flame Retardancy of Polyamide 6. *Appl. Surf. Sci.* **2018**, *456*, 751–762.
74. Yang, W.; Yang, F.; Yang, R.; Wang, B. Ammonium Polyphosphate/Melamine Cyanurate Synergetic Flame Retardant System for Use in Papermaking. *BioResources* **2016**, *11* (1), 2308–2318.
75. Tabuani, D.; Bellucci, F.; Terenzi, A.; Camino, G. Flame Retarded Thermoplastic Polyurethane (TPU) for Cable Jacketing Application. *Polym. Degrad. Stab.* **2012**, *97* (12), 2594–2601.
76. Lubczak, J.; Lubczak, R. Melamine Polyphosphate - The Reactive and Additive Flame Retardant for Polyurethane Foams. *Acta Chim. Slov.* **2016**, *63* (1), 77–87.
77. Lai, X.; Qiu, J.; Li, H.; Zeng, X.; Tang, S.; Chen, Y.; Chen, Z. Flame-Retardant and Thermal Degradation Mechanism of Caged Phosphate Charring Agent with Melamine Pyrophosphate for Polypropylene. *Int. J. Polym. Sci.* **2015**, *2015*, 360274.
78. Schwarzer, A.; Saplinova, T.; Kroke, E. Tri-s-Triazines (s-Heptazines)-From a "Mystery Molecule" to Industrially Relevant Carbon Nitride Materials. *Coord. Chem. Rev.* **2013**, *257* (13–14), 2032–2062.
79. May, H. Pyrolysis of Melamine. *J. Appl. Chem.* **1959**, *9*, 340–344.
80. Lotsch, B. V.; Döblinger, M.; Sehnert, J.; Seyfarth, L.; Senker, J.; Oeckler, O.; Schnick, W. Unmasking Melon by a Complementary Approach Employing Electron Diffraction, Solid-State NMR Spectroscopy, and Theoretical Calculations - Structural Characterization of a Carbon Nitride Polymer. *Chem. - A Eur. J.* **2007**, *13* (17), 4969–4980.
81. Xie, H.; Lai, X.; Zhou, R.; Li, H.; Zhang, Y.; Zeng, X.; Guo, J. Effect and Mechanism of N-Alkoxy Hindered Amine on the Flame Retardancy, UV Aging Resistance and Thermal Degradation of Intumescent Flame Retardant Polypropylene. *Polym. Degrad. Stab.* **2015**, *118*, 167–177.
82. Aubert, M.; Nicolas, R. C.; Pawelec, W.; Wilén, C. E.; Roth, M.; Pfaendner, R. Azoalkanes-Novel Flame Retardants and Their Structure-Property Relationship. *Polym. Adv. Technol.* **2011**, *22* (11), 1529–1538.
83. Cao, K.; Wu, S. L.; Qiu, S. L.; Li, Y.; Yao, Z. Synthesis of *N*-Alkoxy Hindered Amine Containing Silane as a Multifunctional Flame Retardant Synergist and Its Application in Intumescent Flame Retardant Polypropylene. *Ind. Eng. Chem. Res.* **2013**, *52* (1), 309–317.
84. Song, P.; Zhu, Y.; Tong, L.; Fang, Z. C60 Reduces the Flammability of Polypropylene Nanocomposites by in Situ Forming a Gelled-Ball Network. *Nanotechnology* **2008**, *19* (22), 225707.

85. Pfaendner, R. Nitroxyl Radicals and Nitroxylethers beyond Stabilization: Radical Generators for Efficient Polymer Modification. *Comptes Rendus Chim.* **2006**, *9* (11–12), 1338–1344.
86. Pfaendner, R. How Will Additives Shape the Future of Plastics? *Polym. Degrad. Stab.* **2006**, *91* (9), 2249–2256.
87. Zhou, X.; Qiu, S.; Mu, X.; Zhou, M.; Cai, W.; Song, L.; Xing, W.; Hu, Y. Polyphosphazenes-Based Flame Retardants: A Review. *Compos. Part B Eng.* **2020**, *202*, 108397.
88. Zou, W.; Basharat, M.; Dar, S. U.; Zhang, S.; Abbas, Y.; Liu, W.; Wu, Z.; Zhang, T. Preparation and Performances of Novel Polyphosphazene-Based Thermally Conductive Composites. *Compos. Part A Appl. Sci. Manuf.* **2019**, *119*, 145–153.
89. Levchik, G. F.; Grigoriev, Y. V; Balabanovich, A. I.; Levchik, S. V; Klatt, M. Phosphorus-Nitrogen Containing Fire Retardants for Poly (Butylene Terephthalate). *Polym. Int.* **2000**, *49* (10), 1095–1100.
90. Weil, E. D.; Pate1, N. G. Phospham-A Stable Phosphorus-Rich Flame Retardant. *Fire Mater.* **1994**, *18*, 1–7.

Chapter 12

Synergism in Nitrogen- and Phosphorus-Based Flame Retardants

Hilal Olcay,*,[1] Cigdem Gul,[1] and E. Dilara Kocak[2]

[1]Department of Textile Engineering, Institute of Science and Technology, Marmara University, 34722 Kadıkoy/Istanbul, Turkey

[2]Department of Textile Engineering, Faculty of Technology, Marmara University, 34722 Kadıkoy/Istanbul, Turkey

***Email: olcayhilal@gmail.com**

Polyurethane foams (PUFs) are polymer materials with extremely flammable properties. In the framework of the legal regulations made for the protection of life and property safety, it is necessary to use flame retardants (FRs) in polymers. Toxic gases are formed as a result of the combustion of most of the halogenated FRs, and the use of halogenated FRs is restricted owing to the damages they cause to human health and nature. Therefore, the development of halogen-free FRs draws attention to replacements for the halogenated ones. Moreover, the inclusion of various halogen-free FRs can provide better flame retardancy to PUFs. The use of FR combinations can lead to synergistic effects, and FR systems containing multiple FR elements—such as phosphorus–nitrogen, phosphorus–silicon, and phosphorus–nitrogen–silicon systems—have been developed. Because of the good synergy between phosphorus and nitrogen, the subject of many studies has been phosphorus/nitrogen (P/N) synergy and flame retardancy. Using synergistic systems is the most effective way to reduce the halogenated FR content in polymeric materials because P/N-containing FR systems provide good thermal stability, low toxicity, and low smoke generation. In this chapter, first, information about PUFs is given briefly; next, the types and properties of PUFs are mentioned. Then, the FR mechanisms used in PUFs and P/N FRs are investigated, the effects of FRs on the morphology and mechanical properties of PUFs are explained, and, finally, the thermal degradation behavior of PUFs and the tests used in thermal degradation are clarified.

Introduction

Polyurethane (PU) is a dominant polymer group that is widely used in various fields—including those related to construction, paints, adhesives, cooling, aerospace, medical equipment, sealants, astronautics, and oil pipelines—because of its unique properties, such as lightness, high compressibility strength, low thermal conductivity, good adhesion, and low water absorption (*1–3*).

PU can be produced with a combination of various materials and different reaction conditions in order to achieve desired properties. Basically, these materials are polyols, such as polyester and polyether, and isocyanates, such as diphenylmethane diisocyanate (MDI) and toluene diisocyanate (TDI). The thermal decomposition products of a PU are closely related to its chemical composition. Urethanes begin to decompose at 150–200 °C, progressing at a measurable rate beyond this range. Urethane bonds are decomposed by the three mechanisms shown in Schemes 1, 2, and 3 (*4*).

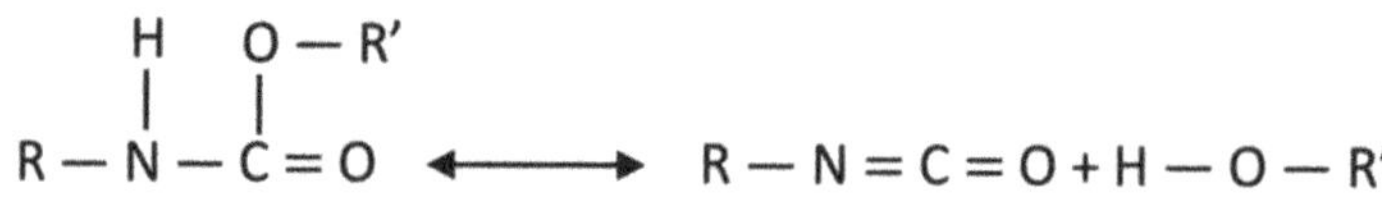

Scheme 1. Reversal to isocyanate and alcohol.

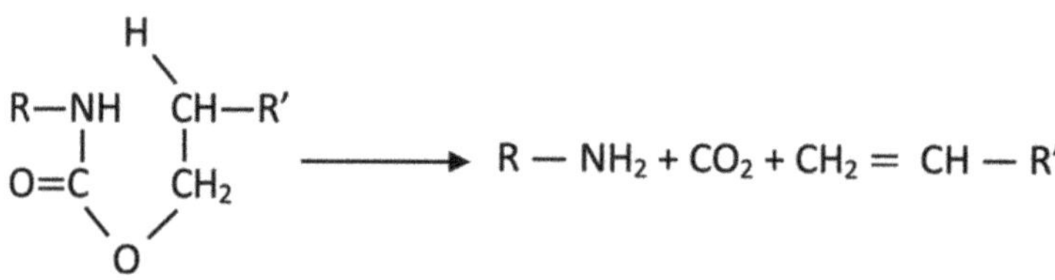

Scheme 2. Formation of the primary amine, olefin, and carbon dioxide (CO_2) from a six-membered ring.

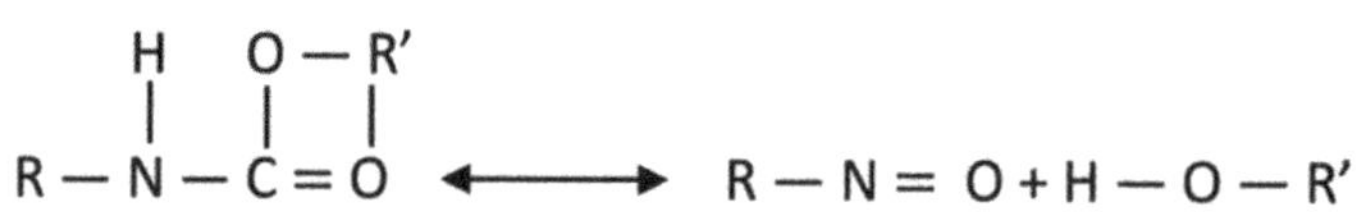

Scheme 3. Formation of the secondary amine and CO_2 from a four-membered ring.

As a result of the diversity in PU synthesis, many PU product types—such as flexible or rigid foam, elastomer, fiber, composite, and adhesive—can be obtained. The most used PU materials, especially in applications where there is a need for improvement of combustion properties, have been foams. For this reason when the literature is examined, most of the studies are related to foams. PU foams (PUFs) ignited easily and have a high combustion velocity owing to their air permeability, wide surface areas, and chemical structures. Therefore, it is very important to develop and enhance the flame retardant (FR) properties of PUFs to prevent fatal fires (*5*).

Rigid PUF (RPUF) has superior mechanical and thermal insulation properties, such as lightness, good compressive strength, ease of synthesis, low water absorption, and low thermal conductivity (*2*). Therefore, RPUF is used in many areas, including aviation, construction, sports, entertainment, rail and road transport, refrigerated transport, maritime building components, and wind turbine blades (*6*). However, RPUF's burning behavior may restrict its use. RPUFs have low thermal stability and high flammability owing to their urea, urethane, biuret, and allophanate groups. Biuret and allophanate, the most thermally unstable groups in the RPUF backbone, decompose into thermally unstable urea, urethane, and diisocyanate when exposed to fire (*7*). In addition, the porous structure of flexible PUFs (FPUFs) increases the surface area and thereby facilitates heat exposure (*8*). During combustion, RPUFs produce excess heat, toxic smoke, and gases, such as carbon monoxide (CO), nitrogen oxide, and hydrogen cyanide (HCN), which lead to a large number of deaths and serious environmental pollution. Consequently, it is important to increase their fire safety to ensure that RPUFs can be applied in a wide variety of areas (*2*).

FPUF is one of the most important types of PU materials and accounts for approximately 40% of the market (*9*). Because of its low density, high flexibility, good sound insulation, superior gas permeability, high chemical endurance, and low thermal conductivity, FPUF has a wide range of applications, including beds, furniture, shoes, construction, architectural decoration, automobile interiors, thermal insulation pipes, packaging, biomedical products, smart materials, and nanocomposites (*9–14*). However, its flammability can be particularly high because of its open-cell, porous structure, its high surface area and low density, and its limiting oxygen index (LOI) value, which is just 16–18% (*9–11, 15*). The large surface area of FPUF and its high air permeability will expedite the spread of fire during combustion, considerably limiting its usage (*4, 9*). When FPUF is inflamed, it readily decomposes and releases CO, HCN, and other toxic by-products (*11*). Hence, it is important to find or synthesize an appropriate FR to provide superior combustion behavior for FPUF (*9–11*).

Common reactions in the thermal decomposition of PUs are as follows. Flammable gases quickly react with oxygen, spreading more heat and gaseous decomposition products to the environment. Eventually, coal is produced, which reacts with oxygen, but releases heat at a much slower rate. The first stage of degradation involves the scission of the urethane bonds, separating the isocyanate and polyol groups. In the second reaction, while carbon dioxide (CO_2), carbodiimide, and isocyanurates are released, aromatic amines and CO_2 are again formed as a result of the reaction with water (*16*). Raw PUFs are highly flammable, and very porous foams cause rapid flame spread and have high heat emission (*17*). Degraded PU products drip because of their low viscosity, increasing the fire hazard (*18, 19*). This is why, for years, efforts have been focused on ways to increase the fire resistance of PU materials.

When the thermal decomposition of TDI-based FPUFs is analyzed, degradation begins with urethane bond breaks occurring at 200–300 °C, producing nonvolatile polyol components and nitrogen-rich volatiles. The latter is called "yellow smoke," and some free TDI is found along with the polymerized, or condensed, forms of TDI. At higher temperatures more degradation occurs in the polyol residue and few organic species are obtained (*4*).

For the practical use of FPUFs and RPUFs, their FR properties must meet particular demands that hinge on national or international standards and end-use application. Many studies of FRs for these foams have been and continue to be done. The flame retardancy of PUF is provided by using additive or reactive FRs: The former is added to a foam formulation as another compound, and the latter is bonded to the polymer backbone (*6*).

Several methods are used to improve the flame retardancy of PUFs: (1) including FR additives into the foam components by mechanical mixing at the composition stage; (2) chemically bonding FR compounds, including functional groups, especially hydroxyls, to the polymer chain; and (3) covering the upper surface of the combustible foam with FR materials. In the first method, additives usually cause a decline in the closed-cell ratio and strength properties and an increment in water absorption by the PUF. An outstanding decline in dimensional stability and strength properties often arises in humid conditions, especially when additives are used at greater than 15% by weight. The second method seems superior to the first given that the FR additives are involved in the foaming reaction and become part of the polymer structure. The last method is preferred solely for spray-applied foams for outdoor applications where good weather protection and low water vapor transmission can be requisitioned (*7*).

FRs and Mechanisms for PUFs

FRs are materials designed to lessen the flame spread rate and prohibit continuous combustion by avoiding or postponing blaze on the surface of combustibles (*10, 20*). They are generally divided into two groups: additive and reactive. FR additives are incorporated into foams simply by mechanical mixing at the compounding stage. Therefore, it is the most preferred method used to enhance the flame retardancy of foam. In addition, halogen (mostly chloroalkyl-phosphate), phosphorus, and nitrogen-based compounds, as well as other materials, such as silicon, expandable graphite (EG), 1,3,5-triazine-2,4,6-triamine (melamine or MEL), aluminum trihydrate, and magnesium hydroxide can be used for PUFs (*6, 9, 10, 11*). Additives with low cost and good flame retardancy require that high amounts be used to create a high FR effect but can adversely affect the mechanical and physical properties of the PUFs by increasing viscosity (*21–23*).

Phosphorus-based additives, such as phosphoramidates and 9,10-dihydro-9-oxa-10-phosphaphenanthrene-10-oxide (DOPO) are highly effective in adding flame retardancy to PUFs. Their FR effect can be correlated with the occurrence of flame active species, such as PO•. These species interfere with the combustion process of PUFs by recombining H• and OH• to prevent oxidization. These additives have restricted condensed phase activity (*22*).

MEL, an FR commonly used in PUFs, has the ability to step into the combustion period in various ways. It has a smoke-suppressing effect because it efficiently traps the volatile isocyanates released during the thermal degradation of TDI-based PUFs at temperatures greater than 250 °C, thereby decreasing the emitted toxic gas and smoke. It is also categorized as a char formation agent and is known to undergo condensing reactions, releasing ammonia and forming multi-ring structures, such as melem and melam, when heated (*5*).

Triphenyl phosphate, ammonium polyphosphate (APP), MEL, and EG, for example, are often used as intumescent FRs (IFRs) in RPUFs because of their halogen-free ingredients, low toxicity, and high efficacy. Using an intumescent system is an efficient way to enhance the flame retardancy properties of PUFs. IFRs, widely used as halogen-free additives, usually consist of a carbonization agent (e.g., pentaerythritol, sorbitol), an inorganic acid or material that yields mainly acidic species between 100 and 250 °C (e.g., APP), and a blowing agent (e.g., MEL) that decomposes at the appropriate temperature and time to allow the system to be blown. The released acids support the dehydration of the foam by producing a foaming carbon–inorganic residue due to the gases released during the decomposition period. Last, solidification takes place and a multicellular coal foam is formed. The IFR system, with its intensive expansion, creates charred layers as a physical barrier, thus preserving the underlying material from flame or flow and limiting the diffusion of flammable volatile components toward the flame and oxygen (*5, 7, 24*). APP, an efficient swelling FR for PUFs, provides its FR function by forming a carbonaceous char in the dense phase. It is converted into phosphoric acid by thermal decomposition, drying the oxygen-containing material, namely, PUF, in the dense phase and causing it to become charred (*5*). IFRs have several advantages but also have some disadvantages, such as low water solubility or poor mechanical properties due to their poor compatibility with the PUF matrix (*5, 24*). In an attempt to overcome these disadvantages, Wang et al. synthesized 2,2-diethyl-1,3-propanediol phosphoryl melamine as a phosphorus/nitrogen (P/N) single IFR for RPUFs (*24*).

Reactive FRs contain mainly FR polyols and isocyanates and are added to the rigid backbone of PU (*1, 9, 23*). There are two ways to improve the thermal resistance and combustion behavior of PUFs: through (1) the use of brominated or phosphorus-containing polyols, or, (2) for rigid foams, the inclusion of primarily isocyanurate, but also of urethidion rings or carbodiimide, into the polymer backbone, which has a more thermally resistant structure than urethane (*6*). Despite their several advantages, such as superior flame retardancy and a lower effect on physical and mechanical properties, the use of reactive FRs is restricted because of their complex structure and high cost (*25*).

The European Union has suggested limiting the usage of brominated diphenyl oxide FRs because of the highly toxic and potentially carcinogenic brominated furans and dioxins that are created during combustion (*10*). In addition, the manufacture of two main classes of bromine-based retarders, polybrominated biphenyl and polybrominated diphenyl ethers, is banned in the United States and Europe (*20*).

Several brominated polyols—such as dibromoneopentyl glycol, tetrabromophthalic anhydride-based ester, and tribromoneopentyl alcohol—can often be used together with polyether or polyester polyols (*6, 26*). In addition, some strategies can be used to obtain FR-PUs with reactive organophosphorus compounds. A widespread method is to introduce a phosphorus-containing group—such as phosphate, phosphine oxide, phosphonate, or phosphazene—into a diol and then create chain extension with a diisocyanate. Phosphorus-containing polyols and nitrogen are also effective in flame retardancy. It has been determined that polyphosphazenes, one of the cyclic compounds of phosphorus and nitrogen, are more effective in preventing combustion by increasing the charring of the polymer. The LOI of FR-PUs obtained with appropriate formulation can be higher than 30%, but the poor mechanical properties of the FR-PUs represent a disadvantage (*6*).

Another way to improve flame retardancy with a reactive method is based on isocyanate-containing reactions. Isocyanurate obtained by cyclotrimerization of isocyanate into polyisocyanurate rings with an appropriate catalyst can often be used. Under the reaction conditions, the carbodiimide bond obtained by condensation of two isocyanate groups cannot be avoided. Some polyisocyanurate foams with high thermal resistance do not have satisfactory mechanical properties. To improve this, the presence of urethane bonds is concentrated, reducing the cross-linking density and, thereby, also reducing the embrittlement (*6*).

P/N FRs

Halogen-based FRs are very effective and widely used for PUs and many other materials (*1, 10, 11, 20*). Reactive halogen compounds are usually bromine based, while additive halogen FRs are chlorine based. Various types of chlorophosphite—such as tris(chloro-isopropyl)-phosphate (TCPP), tris(chloroethyl)-phosphate, tris(1,3-dichloro-isopropyl)-phosphate (TDCP), dichloroethyl methylphosphate, and tetrakis(chloroethyl) ethylene diphosphate—have been reported in the literature as being used as FRs for both RPUFs and FPUFs (*6*). All these compounds are very efficient and versatile, but they have some disadvantages due to their chlorine content. Because, with their toxic and acidic structure, they release toxic and dense smoke that suffocates people and causes costly damage to equipment (*6, 11, 20, 27*), the use of halogen FRs has been gradually limited in architecture use. Besides, when the FR content is increased, the mechanical properties of its PUF first tend to increase and then sharply decrease (*1, 23*). Furthermore, materials containing halogenated compounds are difficult to recycle (*6*).

The development of halogen-free FRs draws attention to efforts to replace halogenated ones. In particular, systems that include phosphorus and nitrogen are of great interest (*10*). For example, halogen-free FRs, such as APP, triphenyl phosphate, modified montmorillonite, and MEL and its salts, aluminum hydroxide and aluminum hypophosphite, are widely used in PUFs because of their low cost and environmentally friendly and highly FR properties (*2, 7*).

FR systems generally serve a FR function for a limited time after ignition. Accordingly, the inclusion of various FRs can provide better flame retardancy to PUFs. In addition, the usage of FR combinations can lead to synergistic effects. The phosphorus–halogen mixture, APP with silicon or nitrogen, has been used to provide flame retardancy to PUFs (*7*). Levchik and Weil showed that MEL is generally synergistic with phosphorus additives in PUFs and is particularly useful for suppressing smoldering (*28*). Another blend with a synergistic effect is the EG/aluminum hydroxide combination. This combination has been used to obtain an eco-friendly FR system for PUFs (*5*). Singh et al. obtained a P/N additive-based RPUF with the phosphoric acid–MEL–urea–formaldehyde reaction. MEL and urea were used as the sources of nitrogen, while phosphoric acid was used as the source of phosphorus (*7*). Creating synergistic systems—such as P/N (*1, 2, 20*), phosphorus–silicon (*29*), and phosphorus–nitrogen–silicon (*30*)—is the most efficient way to decrease the required FR content in PU materials. FR activity can be considerably increased by the synergic effect between phosphorus and nitrogen. P/N FRs have received much attention and research because of their good thermal stability, low toxicity, low smoke generation, and superior performance (*1, 2, 20, 31*). P/N synergism is not a general event; it hinges on the structure of the specific nitrogen-including compound. In addition to decreasing the generation of combustible volatiles, P/N FRs can improve flame retardancy by catalyzing coal formation (*20*). In many compounds, P/N compounds are efficient radical scavengers, with FR effects in the gas and the condensed phases, as well as being efficient carbonizing agents (*1*). However, some P/N FR compounds produce only light coal residues during combustion and thus exhibit poor smoke suppression. In order to improve this feature, combining a high-efficiency carbon source, such as a P/N or EG compound, into a RPUF can be preferred (*2*).

Synthesis of P/N FRs

Many P/N FR compounds have been developed for use in PU materials. This section will focus on the synthesis of some of them.

Chen et. al. synthesized a new P/N FR, 2-carboxyethyl (phenyl) phosphinic acid melamine salt (CMA) as a result of the reaction of 2-carboxyethyl (phenyl) phosphinic acid (CEPPA) dissolved in distilled water with MEL, and then used CMA in a PUF. The $-NH_3^+$ and $-COO-$ vibration peaks observed in the Fourier transform infrared (FTIR) analysis of CMA confirmed the generation of a MEL salt between $-NH_2$ in MEL and $-COOH$ in CEPPA. In addition, CMA has been found to contain 9.18% P, 25.36% N, 42.5% C, and 5.8% H. (*10*).

In another study, Chen et al. also obtained DOPO-OH by the reaction of DOPO and a hydrogen peroxide solution. As a result of the reaction of DOPO-OH with MEL, dicyandiamide, and urea, three DOPO-derivative salts, namely, D-M, D-D, and D-U, respectively, were obtained, and their chemical structures are shown in Figure 1. The abbreviations of FPUF containing these salts are FPUF/D-M, FPUF/D-D, and FPUF/D-U, respectively (*9*).

Liu et al. synthesized 2-carboxyethyl phenylphosphinic acid glycol ester (CEPPG) by the esterification reaction of CEPPA and ethylene glycol, and the synthesis route is shown in Figure 2. Melamine phosphate (MP) and melamine phosphite (MPi) were obtained as a result of the reaction of MEL with phosphate and phosphite separately. The chemical structures of MP and MPi are shown in Figure 3. Then, six RPUF composites were obtained and the contents are given in Table 1 (*2*).

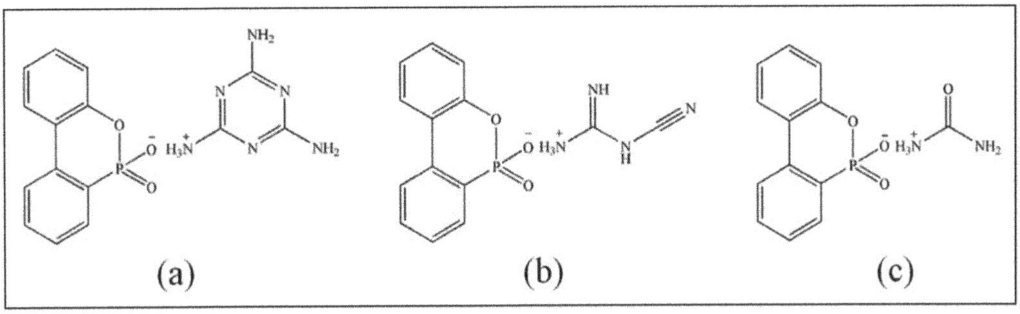

Figure 1. The chemical structure of (a) D-M, (b) D-D, and (c) D-U. Reproduced with permission from reference (9). Copyright 2020 Elsevier.

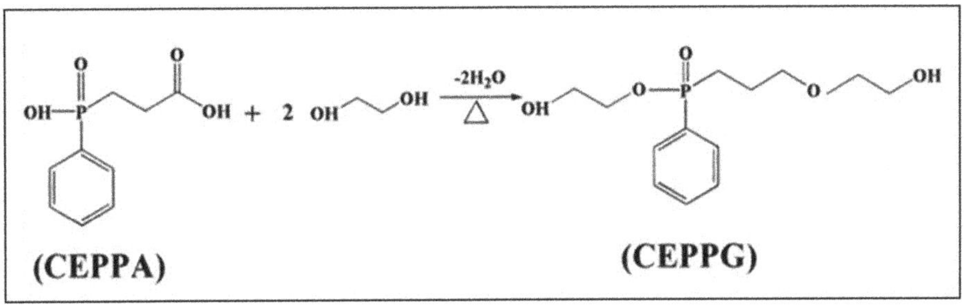

Figure 2. The synthesis of CEPPG. Adapted with permission from reference (2). Copyright, some rights reserved; exclusive licensee [MDPI]. Distributed under a Creative Commons Attribution License 4.0 (CC BY) https://creativecommons.org/licenses/by/4.0.

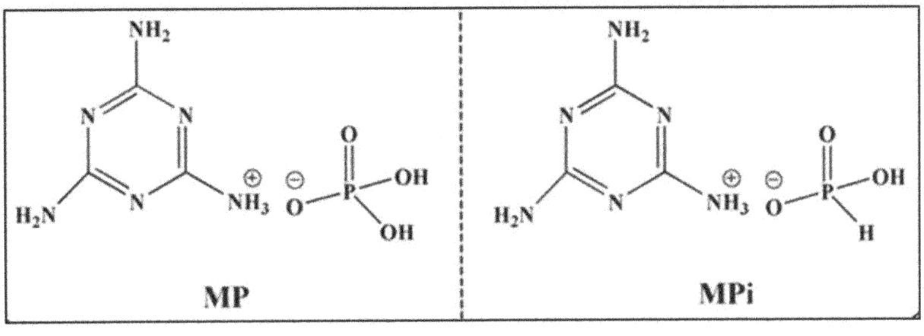

Figure 3. The chemical structures of MP and MPi. Adapted with permission from reference (2). Copyright, some rights reserved; exclusive licensee [MDPI]. Distributed under a Creative Commons Attribution License 4.0 (CC BY) https://creativecommons.org/licenses/by/4.0.

Lin et al obtained 5,5-dimethyl-1,3,2-dioxalphosphine heterocyclic-10-oxide as a result of the reaction of neopentyl glycol with phosphorus trichloride. A Schiff base was obtained by removing methanol after the reaction with aniline, anhydrous methanol, and salicylic aldehyde. 2- ((2-hydroxyphenyl) (phenylamino)methyl5,5-dimethyl-1,3,2-dioxaphosphinane 2-oxide (HAPO) was obtained as a result of the reaction between 5,5-dimethyl-1,3,2-dioxalphosphine heterocyclic-10-oxide and the Schiff base. The elementary analysis of HAPO is as follows: calculated: 62.24% C, 6.11% H, and 4.15% N; found: 62.81% C, 6.38% H, and 4.03% N. PU/HAPO samples were obtained by including HAPO at the rates of 5, 10, and 15% into PU (*1*).

Table 1. Formulations of RPUF and its composites

Sample No.	*RPUF-1*	*RPUF-2*	*RPUF-3*	*RPUF-4*	*RPUF-5*	*RPUF-6*
CEPPG (g)	0	45.06	45.06	45.06	45.06	45.06
EG (g)	0	0	7.5	7.5	8.82	8.82
MP (g)	0	0	7.5	0	8.82	0
Mpi (g)	0	0	0	7.5	0	8.82
P/N compounds wt %	0	19.2	22.4	22.4	23.0	23.0
EG wt %	0	0	3.2	3.2	3.8	3.8

Different PUF structures were obtained with phosphorus FRs, MEL-based agents, and P/N FRs in a study by Li et al. TCPP, TDCP, and tetrakis(2-chloroethyl)-2,2-di(chloromethyl)-1,3 propylene diphosphate (V6) were used as phosphorus FRs (Figure 4). Another type of FR used has been MEL, which is a triazine nitrogen-including heterocyclic compound. The phosphorus FR and MEL contents of the FPUFs obtained are given in Table 2 (*11*).

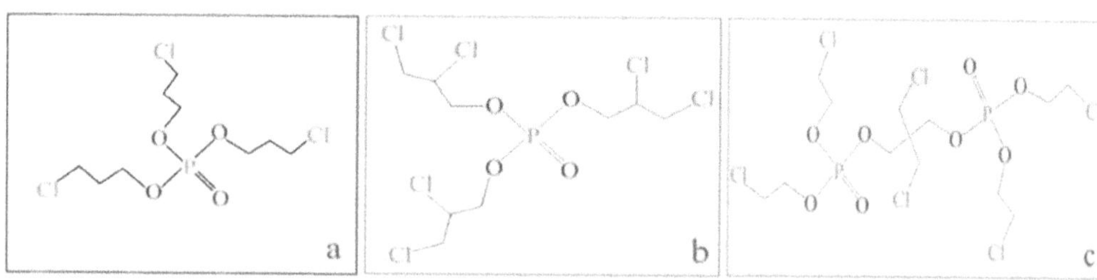

Figure 4. Chemical formulation of (a) TCPP, (b) TDCP, and (c) V6. Reproduced with permission from reference (11). Copyright 2019 Royal Society of Chemistry.

Gu et al. added triethylamine (diluted in CH_2Cl_2) drop by drop to a flask containing CH_2Cl_2, $POCl_3$, and $NH(CH_3)_2HCl$ and, as a result of the reaction, obtained triethylamine hydrochloride as a precipitate. The filtrate was obtained after removal of the precipitated triethylamine hydrochloride salt, triethylamine, and ethanolamine (dissolved in CH_2Cl_2) added dropwise. After removing the triethylamine hydrochloride precipitate, the final product, $C_8H_{22}P_2N_4O_4$ (ODDP), was obtained. The PU prepolymer was prepared by mechanical mixing of TDI, polyether polyol (OH content 99.1 mg KOH/g, molecular weight: 1000) and 2,2-Bis(hydroxy- methyl)propionic acid at 85 °C for 3 h. The P/N-synergic halogen-free FR-PU, phosphorus–nitrogen synergistic halogen- free flame-retardant waterborne PU (DPWPU), was obtained after the reaction of the PU prepolymer with the chain extender, triethylamine, and ODDP. For identification, if the ODDP content is 15% by weight, for example, the abbreviation DPWPU15 is used (*20*).

Table 2. The compositions of FPUFs

Sample code	*Control*	*V6*	*TDCP*	*TCPP*	*N-5%*	*N-10%*	*P-MEL-1%*	*P-MEL-2%*	*P-MEL-3%*	*P-MEL-5%*
Content of phosphorus FR (wt %)	0	15%	15%	15%	0	0	TCDP 15%	TCDP 15%	TCDP 15%	TCDP 15%
Content of MEL (wt %)	0	0	0	0	5%	10%	1%	2%	3%	5%

Source: Reproduced with permission from reference (*11*). Copyright 2019 Royal Society of Chemistry.

In a study by Wang et al., a new reactive FR triol 2,4,6-triphosphoric acid diethyl ester hydroxymethyl phenoxy-phosphonate (TDHTPP) was synthesized to overcome the problems encountered with FRs in RPUFs (Figure 5). In the study, the mixture of sodium hydroxide, 4-hydroxybenzaldehyde, acetone, and water was prepared, and a solution of cyanuric chloride and acetone was added to yield white solid products. This study provided an easy way to synthesize RPUFs with permanent flame retardancy and perfect thermal insulation and mechanical properties (32).

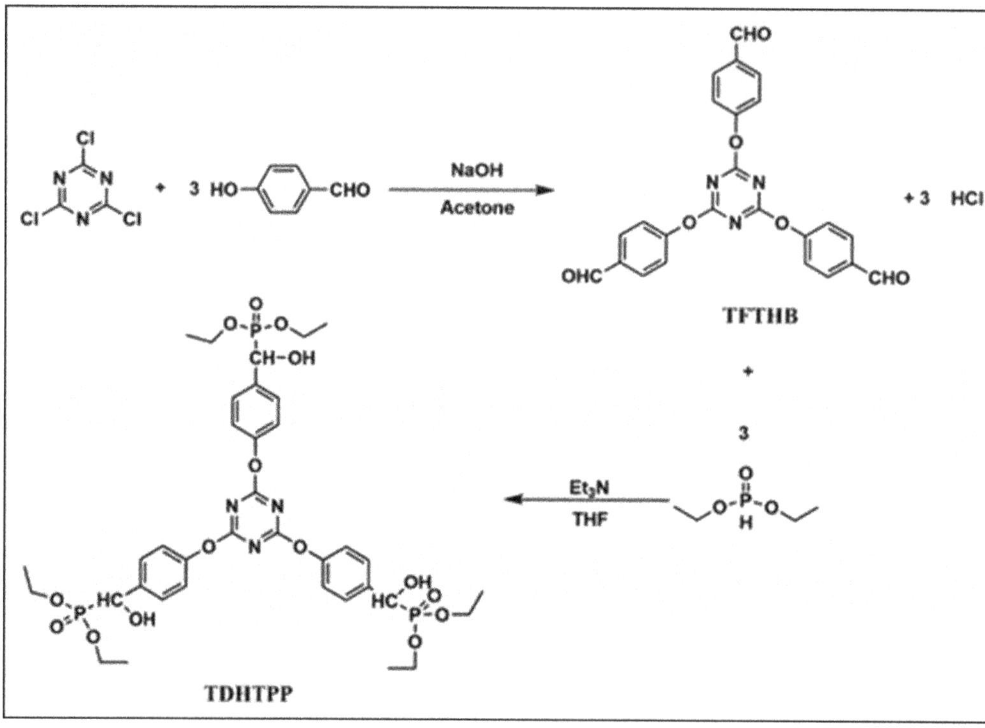

Figure 5. Synthetic routes of 2,4,6-tris(4-formylphenoxy)-1,3,5-triazine (TFTHB) and TDHTPP. Reproduced with permission from reference (32). Copyright 2018 Elsevier.

Yuan et. al. created the FR mechanism of Bis(4-hydroxybutyl) phenylphosphonate (BHPP)/ melamine-derived polyol (MADP)/EG to use in RPUF composites (Figure 6). First, the FR-RPUF degradation products are transformed into coal by the action of phosphoric acid or phosphate acids, which are formed by the decomposition of BHPP. At the same time, a chemical cross-linking reaction occurs between BHPP/MADP and PAPI and a high-quality coal residue forms on the surface as a result of the swelling effect of EG. The coal layer is able to protect the substrate from further thermal degradation (33).

Zhang et al. synthesized a new FR (DOPO-BA) benzaldehyde that contained phosphorus and nitrogen from aniline and DOPO and applied this FR RPUFs (Figure 7). The addition of a small amount of DOPO-BA enabled the formation of excellent FR-PUFs. The FR mechanism of DOPO-containing compounds suppresses fire through the release of low-molecular weight phosphorus-containing species that can scavenge H• and OH• radicals in the flame and form a high-coal layer. Thanks to the aromatic structures, DOPO and its derivatives have high thermal stability and form more effective coal layers, and DOPO-containing compounds produce lesser amounts of toxic gas and smoke than halogen-containing compounds (34).

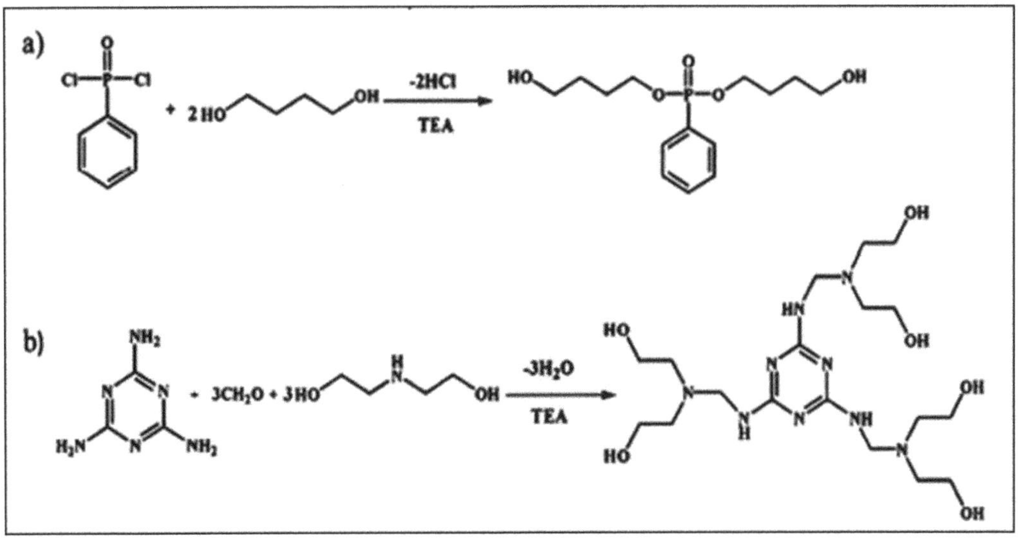

Figure 6. Synthesis of Reactive FR Polyol (a) BHPP and (b) MADP. Reproduced with permission from reference (33). Copyright 2016 American Chemical Society.

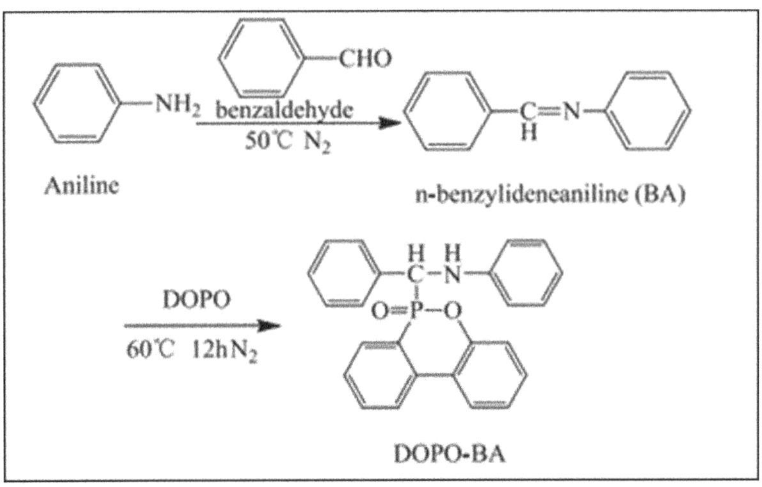

Figure 7. The schematic of DOPO-BA synthesis. Reproduced with permission from reference (34). Copyright 2015 Elsevier.

Ding et al. reacted castor oil-based fatty acid amide (CFA), hydrogen peroxide and formic acid and obtained epoxied CFA. The castor oil-based FR polyols (FR-PE) were synthesized with CFA, DOPO, and triphenyl phosphine. FR dihydric alcohol (BHAPE) was synthesized by the reaction of diethanol amine, formalin, and diethyl phosphite. The synthesis of FR-PE and BHAPE are shown in Figure 8. PUFs were synthesized in a three-step process. Initially, at room temperature, FR-PE, BHAPE, 4110 (a polyether polyol), AK8804 (a polysiloxane–polyether copolymer) as a foam stabilizer, KH560 (γ-glycidoxypropyl trimethoxysilane) as a foaming agent, dibutyltin dilaurate (DBTDL), and 141b (dichlorofluoroethane) as a foaming agent were added to a beaker and mixed with a mechanical mixer. Then the polyisocyanate PM 200 was added and mixed again. Ultimately, the viscous mixture obtained was infused into a mold and cured at 70 °C for 2 h. The other samples were synthesized using a similar process. The compositions of 4110/PUF, 4110/BHAPE/PUF, and FR-PE/BHAPE/PUF are 10 g 4110, 6 g 4110, and 4 g BHAPE; 6 g FR-PE; and 4 g BHAPE, respectively (23).

Sykam et. al. synthesized a new phosphorus- and triazole-functionalized FR monomer (PTFM) that can impart a swelling FR for PUFs. The phosphate ester, nitrogen-rich 1,2,3-triazole, and hydroxyl groups must be included in PTFM synthesis (35).

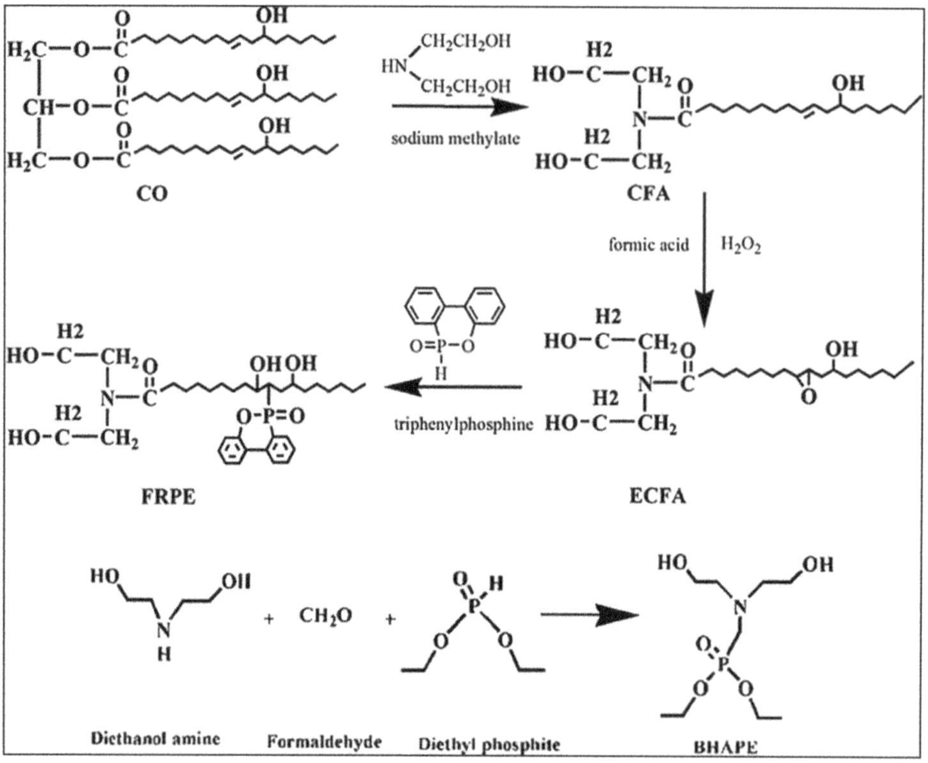

Figure 8. Synthesis of FR-PE and BHAPE. Reproduced with permission from reference (23). Copyright 2017 Elsevier.

Effect of FRs on Mechanical Properties of PUFs

FR chemistry for foams is divided into two classes: additive and reactive. Given the low FR effectiveness of additive FRs and poor compatibility with the PU matrix, the mechanical performance of PU often deteriorates rapidly, and the FR is prone to displacement and precipitation during use. To minimize the environmental impact, it is desirable to connect reactive FRs to the PU structure so that they cannot leak into the environment over time. More and more research has recently been focused on reactive FR-PU systems (36).

MEL is increasingly used as a halogen-free FR in FPUFs. Only sparse data are available in the literature regarding the application of MEL as an additive in FPUF, but the increase in the viscosity provided by MEL and its high heat capacity cause foaming growth to decrease and, thus, the height of the coal to increase with increasing MEL content. The mechanical properties, such as density and compression strength, increase with increasing MEL content. Due to the embedding of MEL in the PU matrix; however, other mechanical properties, such as tensile strength and elongation, are reduced (37).

Adding the FRs EG and APP effectively increases the flame retardancy and mechanical properties of RPUF. Compression testing and dynamic thermal mechanical analysis have showed that both compressive strength and modulus are reduced by a certain extent when an RPUF is filled with either EG or APP alone. However, the mechanical properties of these materials increases when an EG/APP mixture is added to the RPUF, that is, compression strength increases due to EG and modulus increases due to APP (*38*).

An FR system containing EG, pyrogenic silica (SiO_2), and an ionic liquid also improves the mechanical properties of RPUFs. Previous studies have shown that ionic liquids have a positive effect on the thermal stability and mechanical strength of porous PU composites that is due to the better dispersion of the filler particles in the polymer matrix (*39*).

Phosphorus- and nitrogen-containing (e.g., DOPO)–derivative salts increase the mechanical properties (tensile strength, compression set, and compression stiffness) of FPUFs while decreasing the elongation at break. DOPO-derivative salts dispersed in the FPUF can act as reinforcing particles to increase the strength of the matrix and inhibit the movement of the PU molecular chain, thereby improving the tensile strength, compression set, and compression stiffness of the FPUF. However, the presence of DOPO-derived salts in the PU matrix will cause agglomeration due to their inability to distribute properly, which will cause the foam to break more easily during stretching. Therefore, the elongation at break of the foam is reduced. Adding DOPO derivatives will increase the density of the FPUF (*9*).

Most of the phosphate esters and phosphonate FRs have P–O bonds in the main chain of the molecule, and these bonds are easily hydrolyzed. Hydrolysis of the P–O bond in the molecule will lead to the breakage of the main chain structure of a PU, which will deteriorate its mechanical performance and shorten the life of the material. On the other hand, the efficiency of phosphorus-containing FR-PU systems is limited, and it is often necessary to sacrifice the essential properties of the PU matrix (especially mechanical performance) to achieve the desired flame retardancy. Therefore, more researchers are now focusing on synergistic flame-retarding PU systems (*36*).

Toluidine spirocyclic pentaerythritol bisphosphonate (TSPB)-containing RPUFs (TSPB-RPUFs) have been synthesized using the P/N-containing IFR TSPB. TSPB has proven to have less adverse effects on the mechanical properties of PUFs. Scanning electron microscopy (SEM) and FTIR analyses have showed that the swelling coals composed of TSPB-RPUF are compact and smooth. The cell structure of the foam is known to greatly affect the mechanical properties of RPUF. The density that results from increasing the TSPB content in the foam results in decreased cell size, possibly owing to the increased TSPB particles embedded in the cell wall (*40*).

Thermal Degradation Behavior of PUFs

Thermal stability is a concept that includes thermal endurance and heat resistance. Polymers with high thermal stability have higher melting and softening points, higher thermal decomposition, less mass loss during heating at high temperatures, and higher heat deflection temperature without losing their basic properties (*16*).

PUFs are flammable because of their organic structure and improved surface structure and PUs degrade at temperatures greater than 200 °C (*41*). Thermal degradation occurs when PUFs are burned (i.e., ignited) or exposed to high temperatures. Many studies have been conducted on the thermal degradation of PUs. Their purpose has been to record thermal stability, facilitate analysis through the identification of decomposition products, and assess safety in fires by measuring the amount of smoke developed or toxic gases produced. However, the thermal degradation mechanism of PUs has not yet been fully elucidated. Examination of its behavior under fire emergency and thermally severe conditions involves considerations of pyrolysis and of combustion in oxygen-deficient atmospheres. In view of the fact that the PUF is somewhat thermally unstable and cannot be classed as a highly resistant organic polymer, extensive thermal damage would be anticipated when temperatures in excess of 250 °C are encountered (*42*).

In order to assess the effect of conditions of pyrolysis and of partial combustion the following have been studied:

1 Heat penetration rates;
2. Chemical stability;
3. Conditions for spontaneous ignition by radiating sources;
4. Extent of damage to the foam and the structure of the char remaining; and
5. Toxicological properties of products evolved during decomposition of the foam.

It is generally accepted that thermal degradation is the first step of the solid combustion process and that pyrolysis products that contain gaseous fuels support combustion. By using different types of isocyanates, polyols, or additives, PUs with different molecular structures and properties can be obtained, and the thermal degradation behavior then depends on the structure of the PU material. The majority of previous studies have focused on the effect of FRs on thermal degradation behavior. The FR mechanism of PUFs and the effectiveness of different FR additives have been investigated. The thermal degradation properties of PUF are studied through thermogravimetry and differential scanning calorimetry techniques. In situ FTIR is used to probe the characteristic functional groups of weathering residues (*43*).

The degradation of PUFs starts with the breaking of urethane bonds according to reaction (1) (Figure 9). The rupture of the hard segments occurs at a relatively low temperature, and large amounts of isocyanate are formed in the early stage degradation residues (*43*).

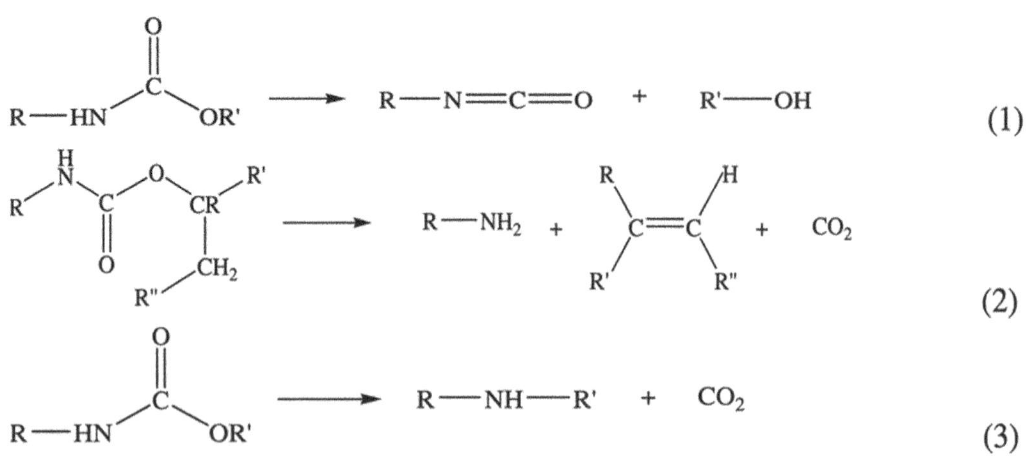

Figure 9. Thermal degradation of PUs. Reproduced with permission from reference (43). Copyright 2013 Elsevier.

Then, as the temperature increases, the polyol segments begin to decompose into some aliphatic ether alcohol species. As the temperature continues to rise, the products become more complex, and as a result of the interactions between the developing products, an epoxy compound is formed. At about 350 °C, reaction (2) and (3) dominate the process, and the materials decompose into primary amines, secondary amines, vinyl ethers, and CO_2. In addition, the nitrides HCN and NO_2 are formed in the temperature range of 300–400 °C. At temperatures greater than 500 °C, material residues continue to decompose into volatile products. Thus, branched chain aliphatic alcohols, benzene alkyls, and the small carbon molecule, CO_2, are produced. When the temperature rises to 600 °C, the entire thermal decomposition process is almost completed (*43*).

The thermal stability of PUs, including P/N FRs, can be analyzed with thermogravimetric (TG) and differential TG (DTG) curves (*2*). The thermal decomposition of PUs occurs in a two- or three-step process (*20*). In the first decomposition stage, moieties such as polyhydric alcohols, isocyanates, and amines on the PU chains are broken down to create their monomeric precursors. The latter decomposition stage is due to the decomposition of substituted urea that results from the reaction between carbodiimide and alcohol or water vapor (*2, 44*).

The thermal stability of P/N FRs obtained in some studies has also been investigated. Gu et al. examined the TG/DTG thermogram of the ODDP obtained in their study (Figure 10). The decompositions occur in four main stages, with maximum weight loss temperatures (T_{max}) at 198.7, 300.1, 415.6, and 497.8 °C, respectively. The 5% weight loss temperature ($T_{5\%}$) for the ODDP is 160.8 °C, and the residual amount at 600 °C is 19.8% (*20*).

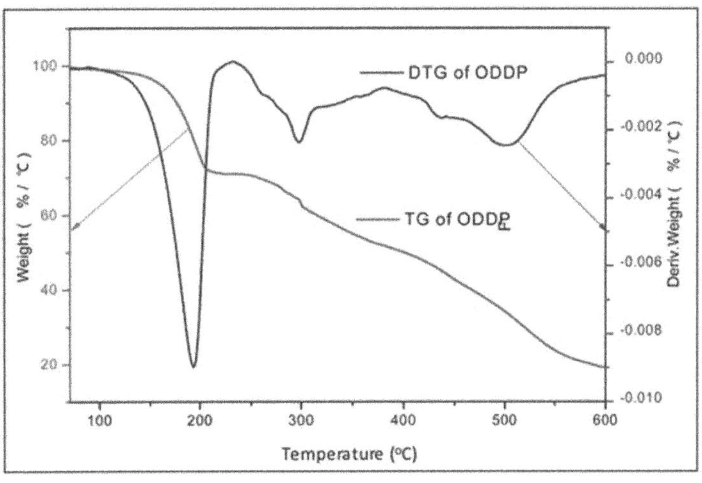

Figure 10. TG and DTG curves of ODDP. Reproduced with permission from reference (20). Copyright 2015 Wiley.

Generally, a three-stage degradation can be seen to occur in the TG graphics of PU materials that contain FRs. The first degradation stage is related to the thermal stability of the FR and is attributed to its weight loss (*20*). The other two degradation stages are compatible with the PU (*2*). Owing to the early degradation of the P–O–C and P–C bonds in P/N compounds and their poor thermal stability, a lower $T_{5\%}$ temperature can be observed for FR-PU than for pure PU (*2, 20, 23*). The addition of an FR can reduce T_{max} temperatures slightly and increase the rate of coal residue formation (*2, 20*). The decrease in $T_{5\%}$ and T_{max} temperatures is related to the decomposition of the FR and the formation of polyphosphoric acid that will speed up it (*20*). However, FR-PUFs have higher T_{max} temperatures than pure PUF. FR-PUFs exhibit an improved thermal behavior at between 400 °C and 800 °C, which has been attributed to the effect of polyphosphoric acid formed by the decomposition of the phosphorus-containing moieties on coal formation (*23, 45*).

In addition, residual amounts of coal may increase as the FR content of the material increases (*2, 20*). Gu et al. observed that the formation of coal residues increased at 480 °C with the increase of FR ODDP loading. This demonstrates the synergic effect of phosphorus and nitrogen in ODDP on coal formation, which can improve the combustion behavior of DPWPU. Moreover, ODDP supports coal formation by forming phosphoric anhydrides or related acids at higher decomposition temperatures (*20*).

Liu et. al. analyzed the TG and DTG curves of the RPUF and its composites (Figure 11). Owing to the early degradation of the P/N compounds, all the tested RPUF composites were found to have lower $T_{5\%}$ values than pure RPUF (Table 3). The T_{max} of the RPUF composites, except RPUF-2, also decreased slightly because of the addition of FRs. The increase in the T_{max} value of RPUF-2 has been associated with the excellent thermal stability of CEPPG. In addition, the formation of coal residues of all PRUF composites increased compared with those of pure RPUF. Thanks to the extraordinary catalytic carbonization of CEPPG, the highest residue rate was reached in RPUF-2. Because of the difference in *p*-value between MP and MPi, the charring effect of MP is higher. Thus, coal residues of RPUF composites are enhanced with increased FR content (*2*).

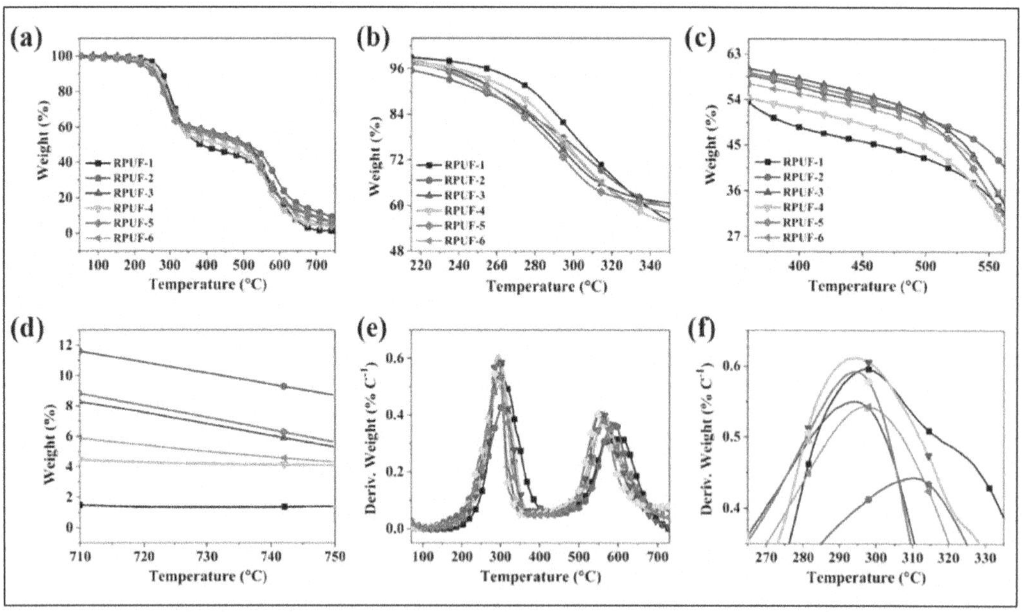

Figure 11. TG and DTG curves of RPUF and its composites. Adapted with permission from reference (2).

Son et. al. observed that TG thermograms are similar at temperatures of less than 300 °C (Figure 12). However, the degradation of products resulting from the interaction between polymer, DAP, or DAP and PUF-RH in the temperature range of 300–350 °C accelerates high thermal generation. This result can be seen in Figure 12. The degradation of products resulting from the interaction between polymer, DAP, or DAP and PUF-RH accelerated the formation of a high thermal property for the experimental TG analysis curve in the temperature range of ~350 to 650 °C. These results reveal that the interaction between DAP and PUF-RH or their decomposed products significantly improves the thermal properties in the high temperature range (*46*).

Table 3. The thermal data of RPUF and its composites

Sample No.	T_5 (°C)	T_{50} (°C)	T_{max} (°C)		*Residues at 750 (°C) (wt.%)*
			Step 1	*Step 2*	
RPUF-1	260.8	383.4	297.4	604.6	1.41
RPUF-2	220.4	504.6	310.2	590.3	8.74
RPUF-3	240.6	505.4	294.1	556.5	5.32
RPUF-4	245.1	437.9	294.4	564.4	4.18
RPUF-5	236.6	497.7	294.3	550.6	5.65
RPUF-6	236.2	486.4	297.2	565.4	4.33

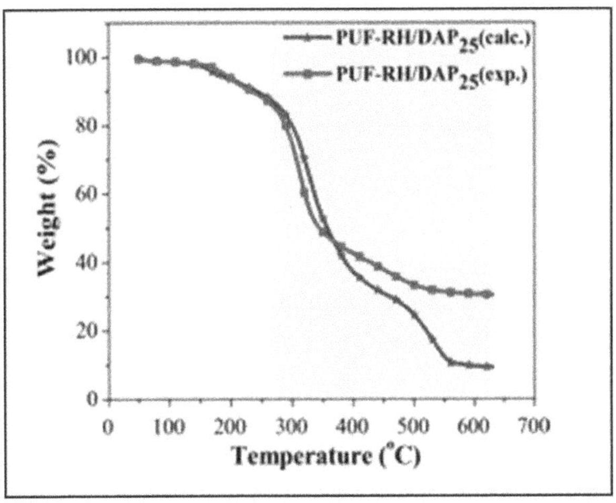

Figure 12. TG thermogram of PUFs. Reproduced with permission from reference (46). Copyright 2021 Springer.

Sykam et. al. studied the initial decomposition temperature of all samples, 25% and 50% weight loss temperatures, weight loss at 500 °C, and weight residue percentage at 700 °C. Negligible weight loss was observed at 50–100 °C. The decomposition temperature values of PU triazole foams (PUTFs) were found to be low compared with the decomposition temperature of PUF. Given that the phosphorus segment found in PUTFs can be easily decomposed, it is less thermally stable than pure PU. The findings show that PUTFs have a significant increase in coal residue compared with PUF, and they support the FR performances of PTF (35).

Wang et. al. observed that both the initial decomposition temperatures (i.e., $T_{5\%}$) and the maximum weight loss temperatures (i.e., T_{max}) of FR-RPUFs decreased compared with that of neat RPUFs. Because the urethane bonds formed between TDHTPP and MDI are less thermally stable than the urethane bonds between polyether polyol 4110 and MDI, the maximum weight loss of FR-RPUFs was reduced with respect to pure RPUFs. Also, the residual coal at 700 °C for RPUF-15 was 34.0% higher than that of pure RPUF (26.5%) (32).

Wang et al. has also observed that PDNP starts to thermally degrade at 179.3 °C and that the maximum decomposition temperature is 196.3 °C. The final amount of coal residue was 31.7 wt% at 600 °C. The formation of the high coal residue proved PNDP to be a good coal-forming agent (*47*).

Zhang et. al. noted that TG analysis and DTG graphs show that the degradation of RPUFs involves two stages. The first weight loss is likely due to evaporation of moisture absorbed by the foams and occurs at approximately 90 °C. Other weight loss that occurs between 200 and 350 °C is due to the degradation of the urethane bonds and the ether bonds of the polyol. Thanks to the addition of DOPO-BA, the maximum decomposition temperature increased from 294 °C to 312 °C for RPUF-FR5. The decomposition of urethane bonds and phosphorus groups create a phosphorus-rich layer between the PU matrix and the heat (*34*).

Evaluation of Combustion Behavior

P/N-synergistic FR improves combustion endurance. The reason is that the inclusion of the P/N FR increases the first thermal decomposition temperature of the PU and then the thermal pyrolysis of alcohols and the labile structures in the carbon layer emit a greater amount of nonflammable gases and create more carbon residues due to thermal oxidation (*11*, *47*). In this section, LOI, UL-94 vertical or horizontal burning tests, and cone calorimeter measurement results are discussed for evaluating the burning behavior of some synthesized P/N FRs.

LOI and Vertical and Horizontal Burning Tests

The burning behavior of FRs can be analyzed with the LOI test. FR-PU can reach a high LOI value and become fire resistant (*10*, *20*, *48*). P/N compounds begin to decompose at low temperatures and release phosphoric acid. A phosphorus-based FR changes the pyrolysis route of the substrate by reducing its degradation temperature, lowering the formation of gaseous combustors and facilitating the formation of coal. The high nitrogen content of P/N compounds increases the synergic effect of P/N (*20*). Depending on the subtilization of ammonia, more nonflammable gas is produced, thus allowing the gas phase that is formed to act as an FR and provide decomposition for the oxygen concentration surrounding the PU. The LOI value generally increases as the P/N content increases and may decrease when it exceeds a certain level. This may be because the initial degradation of PU is further increased with the increased FR content as the FR starts to degrade at low temperatures. In addition, some of the FR decomposes completely before the depolymerization of PU, and the FR takes no part in delaying the flame (*20*). Lin et al. obtained PUFs with 3, 5, 10, and 15% HAPO content and measured LOI values to analyze flame retardancy. It was determined that as the amount of HAPO included in PU increases, the LOI value improves and reaches 23.8 (*1*). Chen et al. measured LOI values of FPUF with different CMA contents (Table 4) and found that the LOI of FPUFs increases as the CMA content increases. When the CMA loading exceeded 10 parts per hundred of polyol by weight, foams with a self-extinguishing level were obtained (*10*).

Gu et al. tested LOI values in a study on DPWPU modified with ODDP. As the ODDP content increased up to 15% by weight, the LOI value of DPWPU increased up to 30.6%; however, when the ODDP content reached 20%, the LOI value decreased to 29.8% (*20*).

Ma et al. measured the LOI values of FPUF samples and found the LOI value for pure FPUF to be 17.5%. As the additive amount of FR increased, the LOI of the FPUF increased. D-D had the best flame retardancy and when 20 parts per hundred resin was added to FPUF, the LOI value reached 24.5% (*9*).

Table 4. The LOI values of FPUFs

Sample	*CMA content (php)*	*LOI(%)*
1	0	18.2
2	5	21.2
3	10	22.6
4	15	23.6
5	20	24.1
5	20	25.0
5	20	25.6

Source: Reproduced with permission from reference (*10*). Copyright 2012 American Chemical Society.

In another study, Li et al. measured the LOI values of FPUFs that included TDCP, MEL, or TDCP/MEL as FRs, and the LOI values were found to be 24, 23.5, and 25.5, respectively. The LOI of pure FPUF is 18, and it can be easily ignited under atmospheric conditions. Due to the P/N synergic effect, the highest LOI value was reached with the addition of TDCP/MEL (*11*).

Another indicator used to evaluate the combustion properties of materials is the UL-94 vertical burning test, which is a universal evaluation method (*1*, *2*, *9*). Evaluating the UL-94 test results of DPWPU modified with ODDP, Gu et al. observed that V-0 classification was achieved due to the rise of phosphorus content in polymers containing 10, 15, and 20% ODDP. This is evidence of a remarkable enhancement in the fire resistance of the DPWPU polymer by the presence of ODDP (*20*).

In a study conducted to examine the flame retardancy of PU/HAPO, Lin et al. tested the samples by UL-94 and determined that PUs with 10% and 15% HAPO had a V-0 rating. Furthermore, it was determined that, with its high cross-linking density, HAPO enhances the formation of viscous melt and thus provides non-drip properties to the polymer (*1*).

In a study about adding DOPO derivatives to FPUF, Ma et al. applied the vertical burning test to the samples. Digital photos are shown in Figure 13. Pure FPUF burned rapidly after ignition; the velocity of fire propagation was high, and melt drip was observed. After the inclusion of DOPO derivatives, the burning of FPUF was suppressed, the flame did not expand, and the drip was importantly decreased. As a result, the FR FPUFs were able to pass the vertical burning test (*9*).

Lin et al. photographed the coal residues after the UL-94 test for six RPUFs. The composites are shown in Figure 14 and the results are summarized in Table 5. Neat RPUF did not pass UL-94 V-0 because of its high flammability, and very little material remained after combustion. With the addition of EG and P/N compounds to RPUF, the UL-94 rating was improved, the self-extinguishment phenomenon was created, and robust coal morphologies were obtained. Thus, it is understood that the additives can act as barriers by forming compact coal layers that prevent heat and mass transfer. In addition, RPUF-2 had the longest t_1 time, which decreased with the addition of EG, confirming the synergic effect between EG and P/N compounds. It was also observed that RPUF-3 and RPUF-5 reached UL-94 V-0 classification. This situation demonstrates the outstanding synergic effect that exists between EG and MP compared with the effect between EG and MPi (*2*).

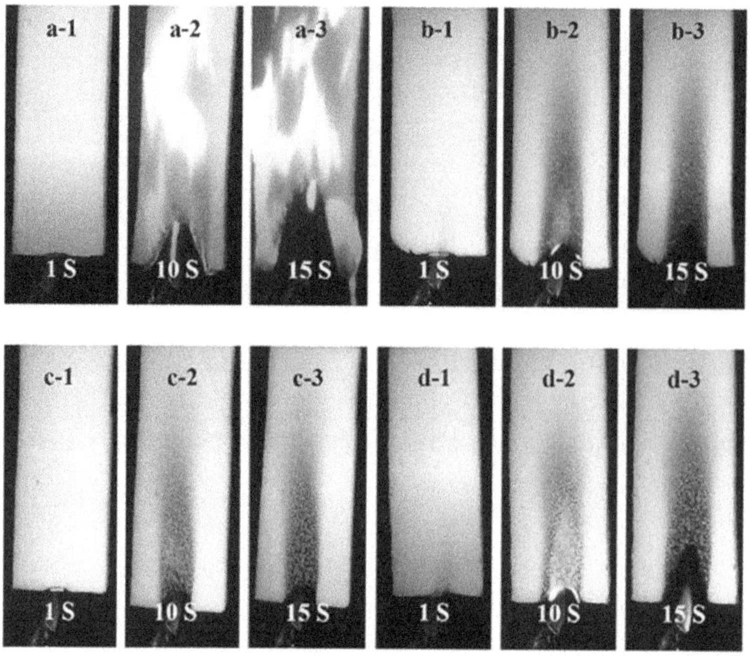

Figure 13. The photographs of (a) FPUF, (b) FPUF/D-M-20, (c) FPUF/D-D-20, and (d) FPUF/D-U-20 during the UL-94 test. Reproduced with permission from reference (9). Copyright 2020 Elsevier.

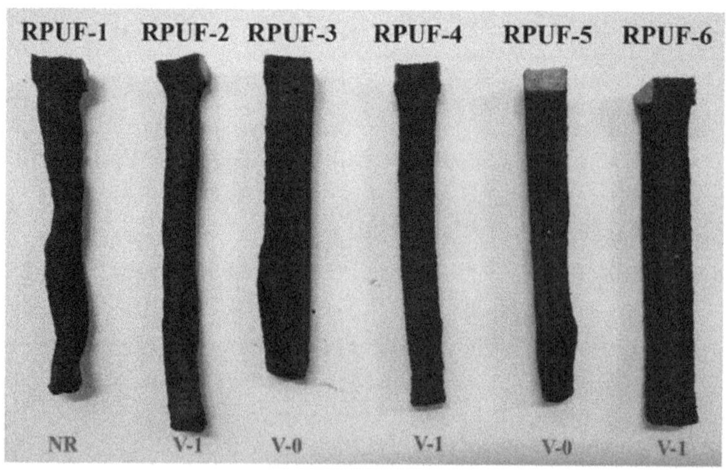

Figure 14. The photographs of coal residues of RPUF and its composites. Adapted with permission from reference (2). Copyright, some rights reserved; exclusive licensee [MDPI]. Distributed under a Creative Commons Attribution License 4.0 (CC BY) https://creativecommons.org/licenses/by/4.0.

In addition to the vertical flame test, the horizontal flame test has also been used in some studies to analyze the combustion behavior of materials. Li et. al. applied a horizontal flame test to FPUF samples containing FRs. The test results and residue morphologies after combustion are shown in Figure 15. All three types of FRs had reduced burning length and burning time (Figure 15a). The lowest values were achieved by adding TDCP and 5% MEL together. As can be seen in Figure 15(b), after the inclusion of the FR, the residual quantity and the amount of molten drop during combustion increased significantly, and the highest values were achieved in FPUF with N/P FR. A considerable amount of heat is exposed as the molten material drips, thus decreasing the temperature of the burning material. In addition, the melted drops cover the outside of the materials, obstructing the burning gas from getting into the materials and burning more (*11*).

Table 5. UL-94 data of RPUFs

Sample No.	$t_1/t_2\,(s)$	*UL-94*
RPUF-1	—	NR
RPUF-2	19.7/0.0	V-1
RPUF-3	6.9/0.0	V-0
RPUF-4	15.7/0.0	V-1
RPUF-5	2.4/0.0	V-0
RPUF-6	17.8/0.0	V-1

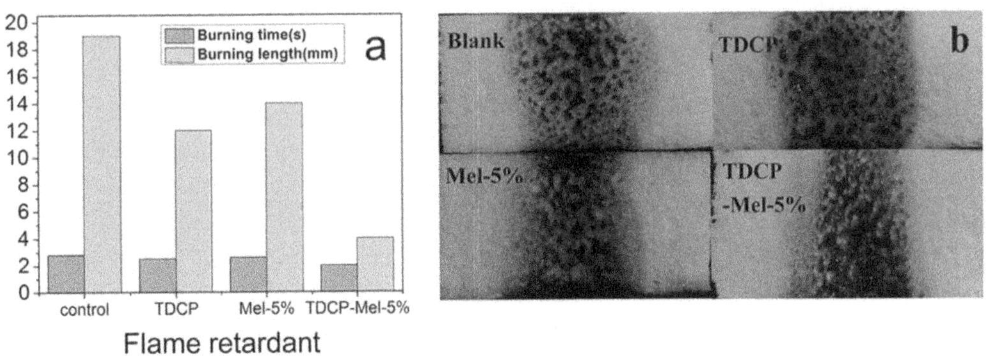

Figure 15. (a) Burning time and length and (b) residue morphology of FPUFs. Reproduced with permission from reference (11). Copyright 2019 Royal Society of Chemistry.

Yuan et al. observed that when DOPO-BA material is added to RPUF, the LOI value increases depending on the amount of material added. For example, by adding 10 and 20 php DOPO-BA to the formulations, the LOI increased from 20% to 24.8% and 28.1%, respectively, for FR-RPUFs (34).

In another study, Yuan et. al. found that the LOI of pure RPUF was 20.0% and that it increased to 27.0% when 15% EG by weight was included in the RPUF. When the EG mass value was kept constant and BHPP/MADP was added to RPUF in different proportions, the LOI for the FR-RPUF increased from 27% to 33.5%. From these results it can be concluded that there is an excellent synergistic effect between BHPP and MADP (33).

Sykam et al. found the measured LOI values to be 23% and 27% for PUTFs, corresponding to PUTF-50 and PUTF-100, respectively. This shows that the FR property of the material can be improved depending on the amount of PTFM added to the PUTFs (35).

Cone Calorimeter Test

The cone calorimeter test is a widely used method for evaluating the combustion behavior of materials (*1, 2, 9, 10*). The test environment is similar to the actual combustion atmosphere and the obtained test data can be used to form an estimate of the combustion behavior of materials in fire (*1*). As a result of this test, parameters such as time to ignition (TTI), peak heat release rate (pHRR), heat release rate (HRR), total heat release rate (THR), total smoke production (TSP), CO_2 production (CO_2P), CO production (COP), average effective heat of combustion (Av-EHC), average mass loss rate, and burn rate index (FIGRA) can be reached. Thus, the burning performance of materials can be analyzed quantitatively (*9, 10, 21, 23*). For example, HRR, which represents an important value in determining the combustion tendency, can be associated with flashover and fire spread in the case of a real fire. The HRR curve of pure RPUF exhibits a two-stage process. After heat is applied to the RPUF, it burns rapidly and the first maximum HRR peak occurs within 15 s. A decrease in HRR is observed after the formation of the protective char layer. Then, because of the strong heat source, cracks occur on the RPUF surface, leading to a second HRR peak (*21*).

Table 6 presents the cone calorimetry data for the study by Ma et al., where this test was performed. The TTI of neat FPUF was about 2 s, and the pHRR and THR values were 317.5 kW/m^2 and 24.2 MJ/m^2, respectively. When DOPO derivatives were added to the FPUF, TTI was increased to over 6 s, and the HRR and THR were considerably reduced. When the amount of DOPO derivatives included in the FPUF was increased, the HHR also decreased. The pHRR and THR values decreased by 55.7% and 52.9%, respectively, for FPUF/D-D-20. D-D appears to have the best FR property for the same loading. However, the DOPO derivatives caused an increase in density and TSP of FPUF because of the DOPO derivatives releasing products into the gas phase and improving the burning behavior. In relation to this, the Av-EHC also decreased. In addition, the decrease in the FIGRA reflected the maximum combustion effect of the material. As a result, the cone calorimetry results show that the DOPO derivatives can limit the combustion behavior of FPUF (*9*).

The results of the cone calorimeter test performed by Liu et al. for RPUF and its composites are shown in Figures 16 and 17, and the corresponding data are listed in Table 7. The TTI of RPUF decreased with the addition of P/N compounds and EG. This situation is associated with the dissociation process of the P–O–C bond in PU. When the HRR curves of RPUF and its composites are analyzed, it can be seen that the first peak that occurs is attributed to the presence of the shielding coal that prevents heat and mass transfer, and the second peak occurs when the shielding coal fails (Figure 16a,b). With neat RPUF, however, the first pHRR was observed later and the second pHRR was observed earlier. This difference is due to the FR additives in the pHRR curve and is related to the formation of the solid coal. The pHRR decreases with the addition of CEPPG, P/N compounds, and EG. The THR values showed the same trend as pHRR (*2*).

The SPR (smoke production rate) curves of RPUF and its composites are presented in Figure 17. Liu et al. demonstrated that RPUF composites have higher SPR values than neat RPUF because the P-including compounds thermally degrade into more smoke during combustion. The TSR values increased with the inclusion of CEPPG, MPi, and EG to RPUF, whereas they decreased with the addition of CEPPG, MP, and EG. The composite with the highest fire safety was RPUF-3, which had a TSR value of 32.7% lower than that of neat RPUF. This result shows the outstanding synergic impact of EG and MP regarding smoke emission. Consequently, the combustion behavior of RPUF can be decreased by adding the appropriate EG and P/N compounds (2).

Table 6. Cone calorimetry data of FPUFs

Sample	*TTI (s)*	*pHRR (kW/m²)*	*THR (MJ/m²)*	*TSP (m²)*	*Av-EHC (MJ/kg)*	*FIGRA*	*Char residue (%)*
Pure FPUF	2±1	318±17	24.2±2.1	1.58±0.10	24±3	3.20±0.13	2±1
FPUF/D-M-10	6±1	171±10	13.8±1.5	4.57±0.14	11±1	1.36±0.08	6±1
FPUF/D-M-20	7±1	166±8	12.5±0.8	5.30±0.25	12±2	1.32±0.05	9±1
FPUF/D-D-10	7±1	159±13	12.1±1.1	4.31±0.18	10±2	1.29±0.03	7±1
FPUF/D-D-20	7±1	141±9	11.4±0.6	4.91±0.13	9±1	1.14±0.06	10±1
FPUF/D-U-10	6±1	265±14	21.5±1.9	4.80±0.14	18±3	2.10±0.04	4±1
FPUF/D-U-20	6±1	232±11	18.4±1.4	5.46±0.31	16±2	1.84±0.11	7±1

Source: Reproduced with permission from reference (9). Copyright 2020 Elsevier.

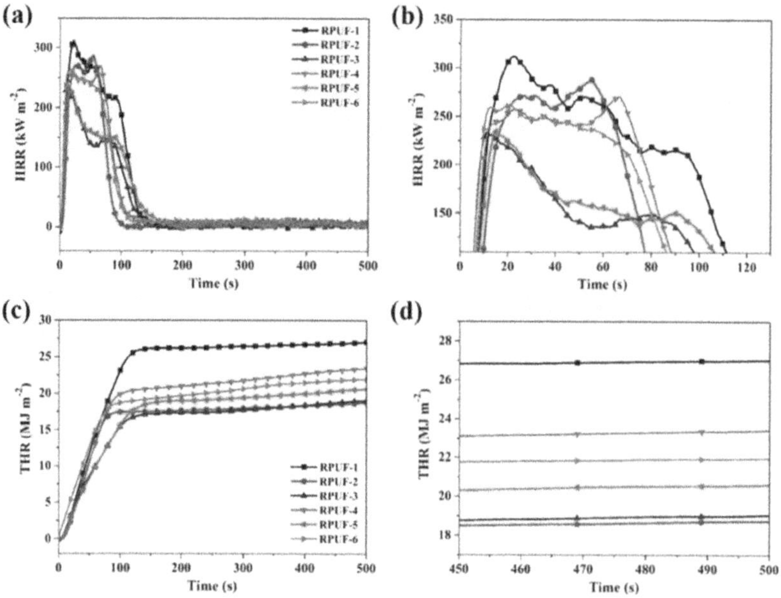

Figure 16. (a,b) HRR and (c,d) THR graphics of RPUF and its composites. Adapted with permission from reference (2). Copyright, some rights reserved; exclusive licensee [MDPI]. Distributed under a Creative Commons Attribution License 4.0 (CC BY) https://creativecommons.org/licenses/by/4.0.

Table 7. The data of cone calorimeter test for RPUF and its composites

Sample	*TTI (s)*	*pHRR (kW/m^2)*	*THR (MJ/m^2)*	*PSPR[a] (m^2/s)*	*TSR[b] ($m^2\ m^{-2}$)*
Error	±1	±21	±2	±0.02	±696
RPUF-1	5	312	27.0	0.24	3698
RPUF-2	6	287	18.8	0.42	3255
RPUF-3	4	233	19.0	0.38	2489
RPUF-4	4	270	23.6	0.31	3957
RPUF-5	3	235	20.6	0.35	3525
RPUF-6	3	262	27.0	0.34	4248

[a] PSPR: Peak of smoke production rate. [b] TSR: Total smoke release.

Source: Adapted with permission from reference (2). Copyright, some rights reserved; exclusive licensee [MDPI]. Distributed under a Creative Commons Attribution License 4.0 (CC BY) https://creativecommons.org/licenses/by/4.0.

Liu et al. evaluated cone calorimeter test results for HRR, THR, TSP, and smoke release rate values. PU-Neat, which had high values, as shown in the graphics, had both smoke and thermal hazards during burning. With the addition of HAPO, these values decreased significantly. According to the HRR curves, PU exhibited two peaks, 392 and 377 kW/m^2, whereas PU/HAPO/10% exhibited a single peak. Because of the FR, the formation of a compact coal layer on the surface of the composites and the dilution of the concentration of flammable gases prolonged the TTI of the PU. When the THR curves are examined, it can be seen that PU/HAPO/10% had a lower THR than did

PU-Neat. This decrease is related to the dilution of the oxygen concentration of ammonia and other nonflammable gases as a result of the decomposition of HAPO. With the addition of HAPO, the TSP value of PU-Neat also decreased by 38.8%, showing that effective smoke suppression is provided during burning hazards. The TTI value also increased owing to the addition of HAPO (*1*).

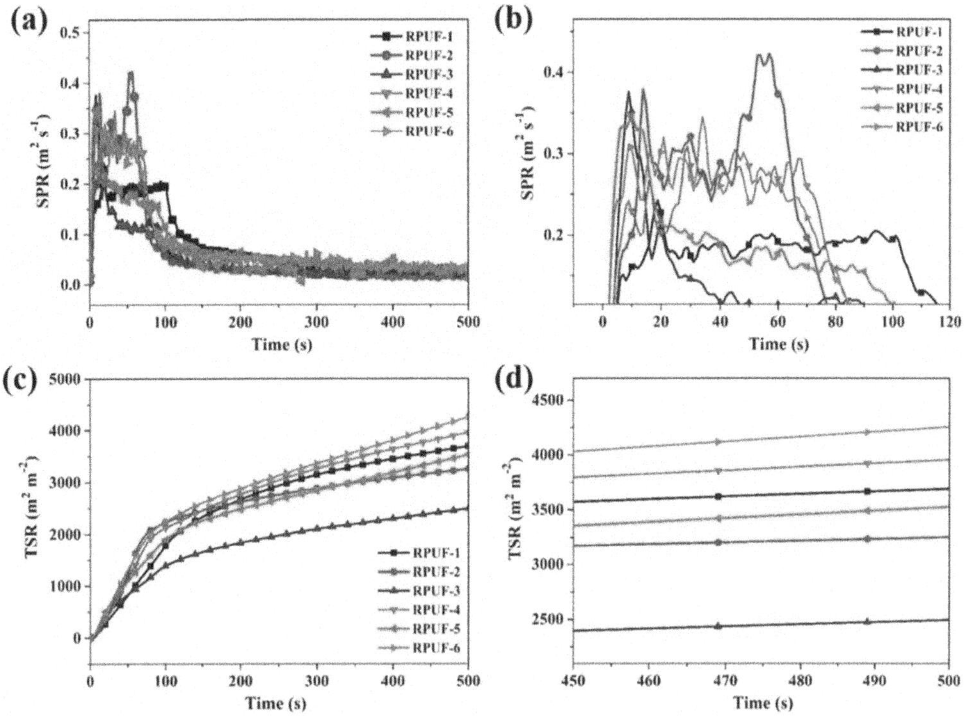

Figure 17. (a,b) SPR (smoke production rate) and (c,d) TSR graphics of RPUF and its composites. Adapted with permission from reference (2). Copyright, some rights reserved; exclusive licensee [MDPI]. Distributed under a Creative Commons Attribution License 4.0 (CC BY) https://creativecommons.org/licenses/by/4.0/

When A-EHC, A-MLR, and A-SEA values, which are among the parameters representing the smoke density, are examined, it can be seen that PU-Neat has higher values than PU/HAPO/10%. The decrease of the A-EHC for PU/HAPO/10% is due to the high amount of decomposed gas. For this composite, a decrease in the A-MLR value is also observed in connection with the formation of more coal. Reducing the combustion parameters to a great extent shows that PU's flame retardancy can be improved with P/N compounds (*1*).

The PUF composites were also tested by cone calorimeter and their flammability behaviors were evaluated by Ding et al. HRR and THR curves are presented in Figure 18. PHRR and THR values decreased with BHAPE and FR-PE addition. This reduction is possibly attributable to coal formation, which slows the heat and mass transfer between gas and condensed phases. The COP also decreased with the addition of BHAPE and FR-PE. The CO_2P peak was observed at the lowest FR-PE/BHAPE/PUF sample because FR-PE and BHAPE quicken COP and decrease CO_2P owing to the unfinished burning of PUF with FR (Figure 18). Because the P–O–C and P–C groups in FR-PE and BHAPE have lower thermal stability, FR-PE/BHAPE/PUF and 4110/BHAPE/PUF have lower TTIs compared with that of 4110/PUF. FRs also reduce Av-EHC and average mass loss rate values (*23*).

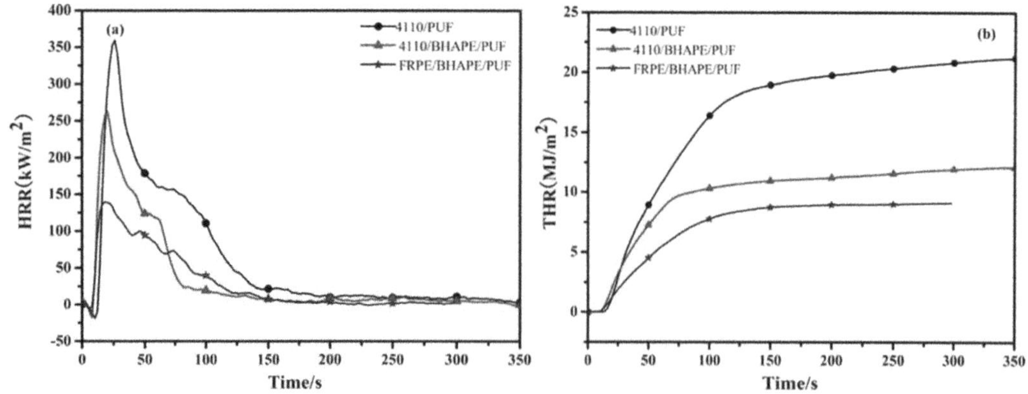

Figure 18. (a) HRR and (b) THR graphics of the PUF composites. Reproduced with permission from reference (23). Copyright 2017 Elsevier.

Chen et al. investigated the combustion behavior of CMA-filled FPUF and characterized the samples using cone calorimetry. The HRR curves of FPUF including 0, 10, and 20% CMA are given in Figure 19. The addition of CMA lowered the HRR of FPUF because noncombustible gases, such as ammonia decomposing from CMA, dilute the O_2 concentration surrounding the FPUF. The FPUFs that contained CMA, which has a higher HRR value for about 50 s, exhibited a low HRR value after 110 s that was due to coal formation (*10*).

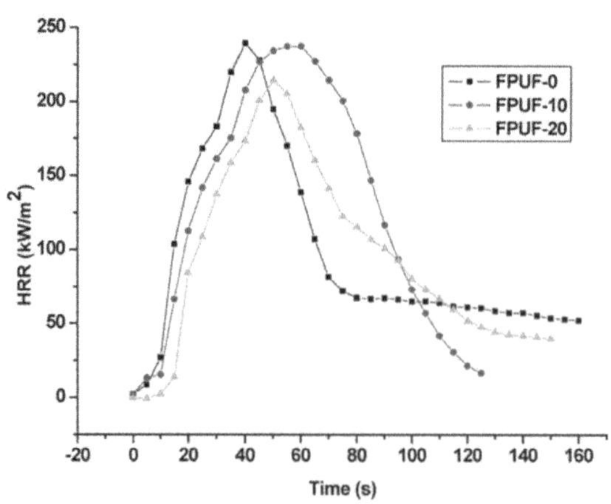

Figure 19. HRR curves of FPUFs. Reproduced with permission from reference (10). Copyright 2012 American Chemical Society.

Table 8 shows a rise in TTI and a decline in THR with increasing CMA loading. This rise is due to the P/N synergic effect of CMA that plays a part in both the gas and condensed phases (*10*).

Table 8. Cone calorimeter data of FPUFs

Sample	*CMA content (php)*	*pHRR (kW/m²)*	*THR (MJ/m²)*	*Char yield (%)*	*TTI (s)*
0	0	240	17	4.3	0
1	10	237	16	11.4	1
2	20	214	14	17.2	5

Source: Reproduced with permission from reference (*10*). Copyright 2012 American Chemical Society.

The cone calorimetry test was applied to samples in a study by Zhang et al. of the fire behavior of pure RPUF and FR-RPUF containing 20 php DOPO-BA. The time to fire materials increased from 2 s for pure RPUF to 4 s for FR-RPUF, indicating that the presence of FR delays the time to ignite the foam. The RPUF-FR20 extinction occurred within 296 s after ignition, whereas the pure RPUF flame took about 180 s to extinguish (*34*).

Yuan et al. investigated the synergistic effect of BHPP and MADP on the burning behavior of the RPUF matrix. When the data obtained from the cone calorimetry test is examined, it can be concluded that BHPP and MADP, which are reactive FRs, improve the flame retardancy of RPUF (33).

Coal Residues

P/N FRs accelerate the formation of metaphosphoric or poly(phosphoric acid) because of thermal decomposition. Thus, coal formation and dehydration of polyhydric alcohol is promoted. The poly(phosphoric acid) surrounding of the carbon residual layer inhibits the thermal oxidization of the coal layer. Therefore, a noncombustible gas is released, diluting the combustible gas concentration and delaying the burning propagation (*11*, *20*). This mechanism is presented in Figure 20 (*20*).

The FR mechanism can be interpreted most clearly by analyzing coal residues (2). The PUF coal can have a swollen and porous structure. With the addition of FR, the morphology of the coal can be changed and this can contribute to the burning behavior of the material. The morphology of DPWPU with and without FR is shown in Figure 21. Unlike the porous structure of DPWPU0 charcoal, DPWPU15 charcoal is smooth and dense. Because of this surface structure, it can serve as a barrier against air, heat, and pyrolysis products and reinforce flame retardancy (*20*).

Ding et al. investigated the structure of the FR-PUF coal by SEM and FTIR. According to the FTIR spectrum, FR-PE/BHAPE/PUF coal contains $-CH_3$, $-CH_2$ (2924 cm^{-1}) and aromatic groups (1580 cm^{-1}) and also P–O–C/ P–O–C groups (1077, 980 cm^{-1}). These results show that this coal primarily contains aromatic, phosphorus compounds and part of a methyl/methylene structure (23).

When SEM micrographs of external coal residues are examined, it can be observed that 4110/ PUF has a coal structure with fractures and great holes and does not have an efficient barrier layer. With the inclusion of FR-PE and BHAPE FRs, a more compact, stable, smoother coal structure is obtained. Thus, by significantly slowing the heat and mass transfer between the condensed and gas phases, the thermal stability and the combustion behavior of PUF can be improved (23).

The residual photographs of a study by Chen et al. examining the coal residues of FPUF that contain pure FPUF and CMA are presented in Figure 22. The intumescent char effectiveness improved with the increase of CMA loading. The enhanced flame retardancy of FPUF could be due to the coal's ability to serve as a barrier against air, heat, and pyrolysis products (*10*).

Lin et al. examined coal morphologies (Figure 23) and found that the coal of pure RPUF has many cracks and holes because the RPUF burns completely, releasing a large number of gaseous products. RPUF-2 had a smooth surface containing a few micro-sized particles but had a charcoal structure that presented very small cracks at some points. Owing to the swelling impact of EG and the catalytic carbonization of the P/N compounds, the inclusion of EG and P/N compounds results in coal residues with a more compact and continuous structure. RPUF-3 and RPUF-5 had smooth and continuous coal residuals because of the catalytic carbonization impact in the condensation phase, thus effectively preventing the production of volatile small molecules and heat transfer. In contrast, a large number of "wormlike" layers and fractures are shown in the coal residuals of RPUF-4 and RPUF-6. Therefore, RPUF-3 and RPUF-5 offered higher quality charcoal structure, indicating a more outstanding synergic effect between EG and MP than between EG and MPi (2).

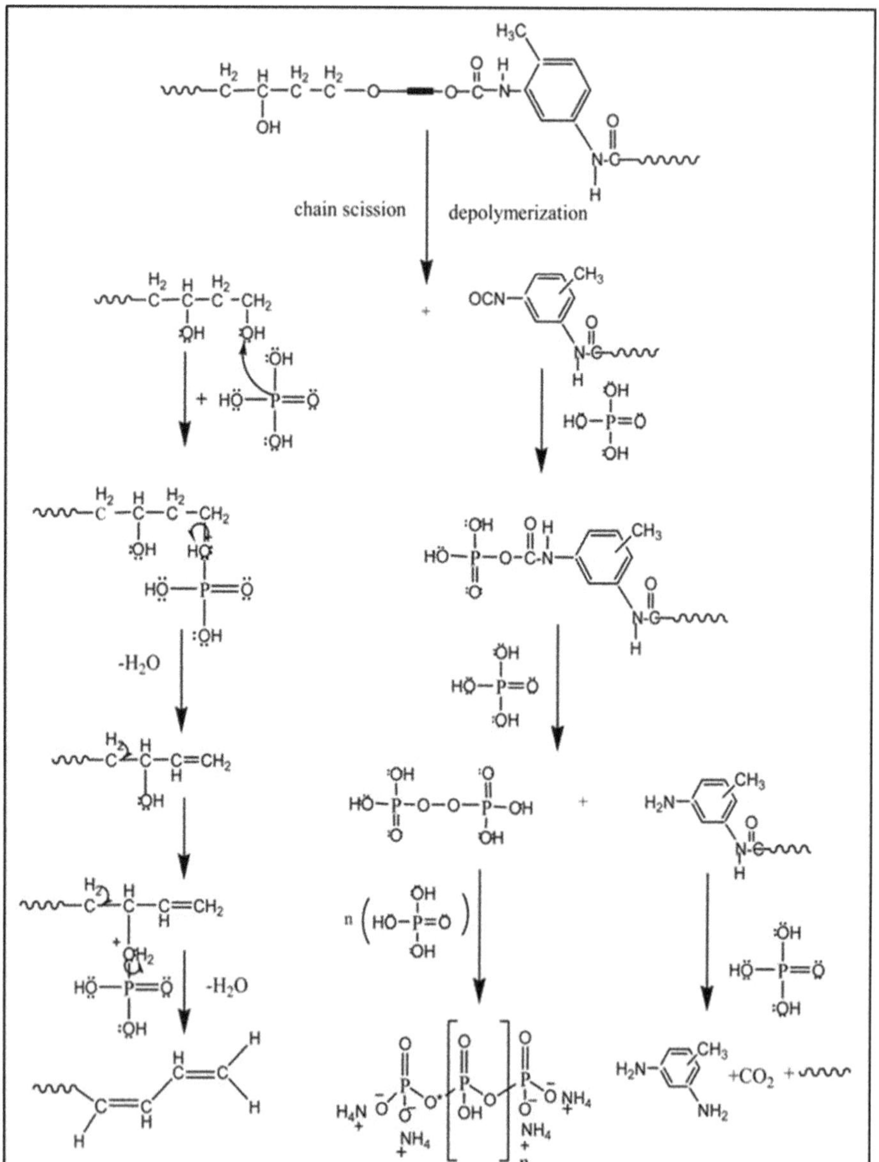

Figure 20. A mechanism of phosphoric acid dehydration and coal formation for PU. Reproduced with permission from reference (20). Copyright 2015 Wiley.

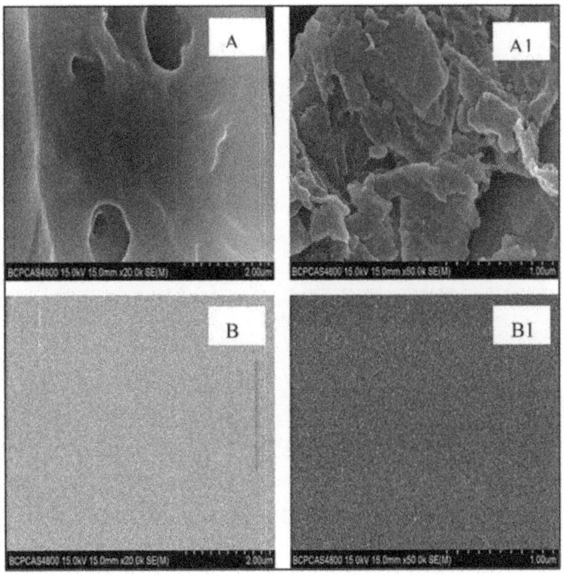

Figure 21. SEM micrographs of (A, A1) DPWPU0 and (B, B1) DPWPU15. Reproduced with permission from reference (20). Copyright 2015 Wiley.

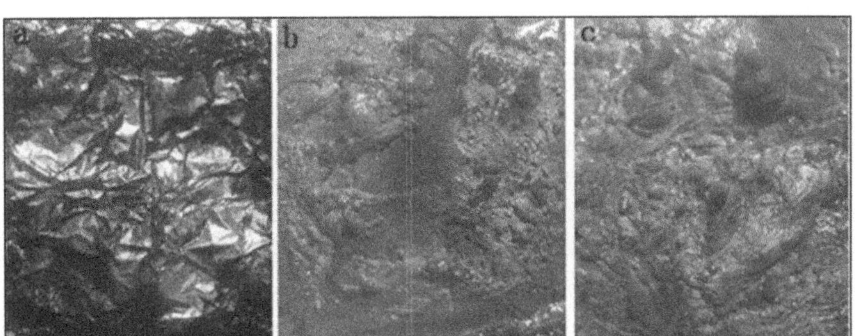

Figure 22. The charred residual photographs of (a) FPUF0 (b) FPUF10, and (c) FPUF20. Reproduced with permission from reference (10). Copyright 2012 American Chemical Society.

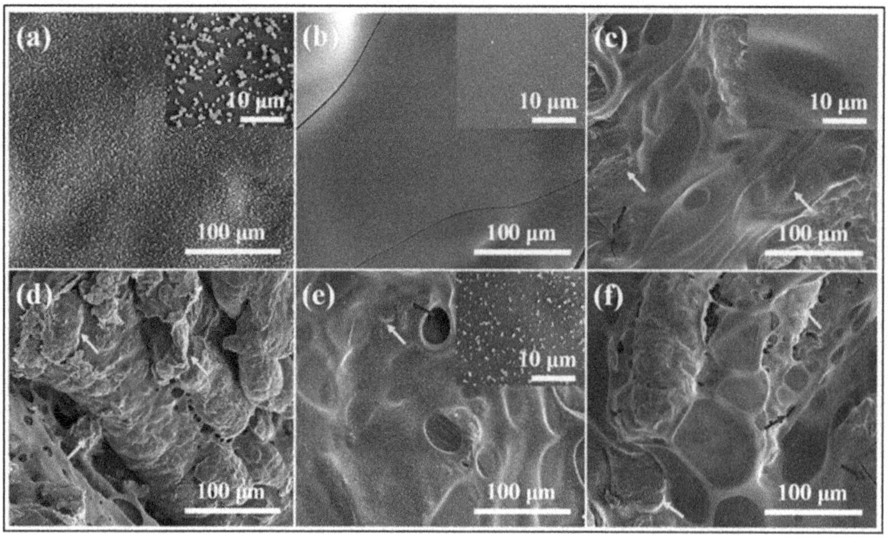

Figure 23. SEM micrographs of coal residuals of (a) RPUF-1, (b) RPUF-2, (c) RPUF-3, (d) RPUF-4, (e) RPUF-5, and (f) RPUF-6. Adapted with permission from reference (2). Copyright, some rights reserved; exclusive licensee [MDPI]. Distributed under a Creative Commons Attribution License 4.0 (CC BY) https://creativecommons.org/licenses/by/4.0.

Investigation of the FR Mechanism

The flammability of polymeric materials is associated with the thermal decomposition mechanism that occurs in the hyperthermal state. Hence, the degradation process of polymers plays an important part in understanding flame retardancy and the combustion process (2, 25). The FR should have a degradation temperature close to that of the FR-free polymer matrix (25).

Liu et. al. created a probable FR mechanism in accordance with the results of many tests (Figure 24). After igniting, RPUF thermally transforms into accendible gases, such as CO, HCN, and nitrogen oxide. P/N compounds, on the other hand, begin to degrade at low temperatures, with the formation of free radicals, such as NOx• and POx•, which can dilute the concentration of accendible gases. At the same time, POx• can delay heat dissipation and smoke generation by extinguishing flammable free radicals in the matrix. As the temperature increases, compact coal residuals are catalyzed. Meanwhile, the heat and oxygen transfer of the RPUF during the thermal decomposition process is isolated by the "wormlike" coal residues that are formed. These coal residuals serve as an effective barrier, further retarding the combustion of the RPUF (2).

Ding et. al. created a probable thermal decomposition process for FR-PE/BHAPE/PUF (Figure 25). In this process, thermal degradation mainly occurs in three stages. In the first step, P–O–C, P–N, and P–C bonds in FR-PUF are broken and free-radical chain reactions are inhibited by the formation of free-radical scavengers (PO•, P•, H•, and HO•). In the second step, aliphatic and aromatic compounds were formed with the increasing temperature and the phosphoric acid compounds form polyphosphoric acid, which accelerates coal formation as a catalyst. In the third stage, a constant coal layer that includes phosphorus species and polyaromatic structures is formed, and volatilized H_2O, CO, and CO_2 are revealed (23).

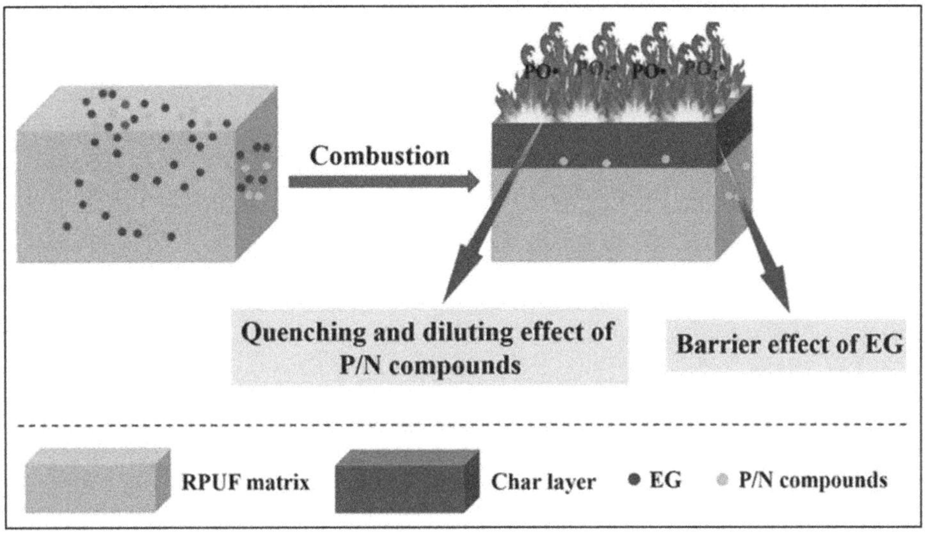

Figure 24. The FR mechanism of RPUF composites. Adapted with permission from reference (2).

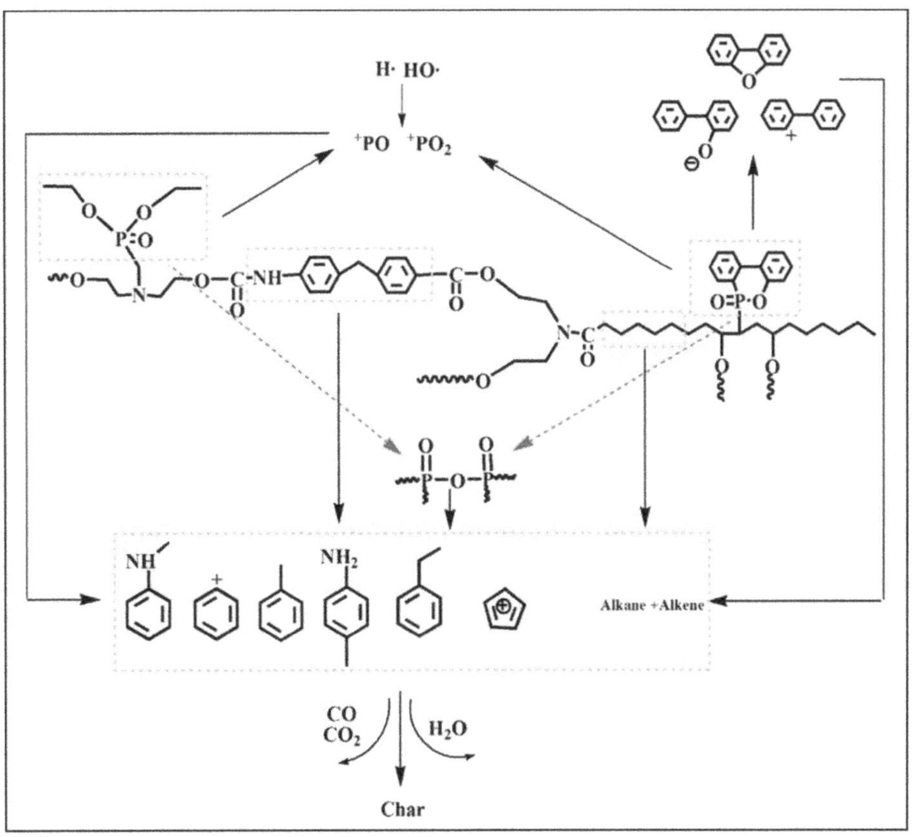

Figure 25. The thermal decomposition process of FR-PE/BHAPE/PUF. Reproduced with permission from reference (23). Copyright 2017 Elsevier.

Conclusion

FRs are extremely important for the PU industry because PU is one of the most flammable polymeric materials and releases a large amount of smoke and toxic gases when burned. The most used types of PU materials, especially in applications where there is a need for improvement of combustion properties, have been the foams. PUFs are easily ignited and have a high combustion velocity because of their air permeability, wide surface areas, and chemical structures. Additionally, the porous structure increases the surface area, thereby facilitating the heat exposure. Therefore, it is very important to develop the FR properties of PUFs. Halogen-based FRs are very effective and widely used for PUs, but they show some negative effects due to their chlorine content. Further, because of their toxic and acidic structure, they release toxic and dense smoke, making them harmful for human health and leading to environmental problems. Halogen-free FRs have received more attention. These FRs cause less harm to human health and the environment than other FRs. In particular, systems containing phosphorus and nitrogen are of great interest. Creating synergistic systems, such as P/N, is one of the most efficient methods to decrease the required FR content in materials. Synergic P/N FRs delay the burning process and help suppress smoke, thus improving the thermal stability of PUF improves. In addition, FRs with P/N content enhance the mechanical properties of PUFs, and these FRs form a higher quality coal residue on the surface that can protect the substrate from further thermal degradation.

References

1. Lin, Z.; Zhao, Q.; Fan, R.; Yuan, X.; Tian, F. Flame retardancy and thermal properties of rigid polyurethane foam conjugated with a phosphorus–nitrogen halogen-free intumescent flame retardant. *Journal of Fire Sciences* **2020**, 1–18, DOI: 10.1177/0734904119890685.
2. Liu, C.; Zhang, P.; Shi, Y.; Rao, X.; Cai, S.; Fu, L.; Feng, Y.; Wang, L.; Zheng, X.; Yang, W. Enhanced Fire Safety of Rigid Polyurethane Foam via Synergistic Effect of Phosphorus/Nitrogen Compounds and Expandable Graphite. *Molecules* **2020**, 25, 4741, DOI: 10.3390/molecules25204741.
3. Olcay, H.; Kocak, E. D.; Yildiz, Z. Sustainability in Polyurethane Synthesis and Bio-based Polyurethanes. In *Sustainable Textiles: Production, Processing, Manufacturing & Chemistry*; Muthu, S., Gardetti, M., Eds.; Sustainability in the Textile and Apparel Industries; Springer: Cham, Switzerland, 2020. https://doi.org/ 10.1007/978-3-030-38013-7_7
4. Price, D.; Horrocks, A. R. Polymer Degradation and the Matching of Flame Retardant Chemistry to Degradation. In *Fire Retardancy of Polymeric Materials*, 2nd ed.; Wilkie, C. A., Morgan, A. B., Eds.; CRC Press: Boca Raton, FL, 2010.
5. Eceiza, I.; Barrio, A.; Martín, L.; Veganzones, M. A.; Fernández-Berridi, M. J.; Irusta, L. Thermal and fire behavior of isophorone diisocyanate based polyurethane foams containing conventional flame retardants. *Journal of Applied Polymer Science* **2017**, *135*, 45944, DOI: 10.1002/app.45944.
6. Modesti, M.; Lorenzetti, A. Flame Retardant Design for Foam Materials. In *Fire Retardancy of Polymeric Materials*, 2nd ed.; Wilkie, C. A., Morgan, A. B., Eds.; CRC Press, Boca Raton, FL, 2010.
7. Singh, H.; Jain, A. K.; Sharma, T. P. Effect of phosphorus-nitrogen additives on fire retardancy of rigid polyurethane foams. *Journal of Applied Polymer Science* **2008**, *109*, 2718–2728. https://doi.org/10.1002/app.28324

8. Acuña, P.; Santiago-Calvo, M.; Villafañe, F.; Rodríguez-Perez, M. A.; Rosas, J.; Wang, D. Y. Impact of expandable graphite on flame retardancy and mechanical properties of rigid polyurethane foam. *Polymer Composites* **2019**, *40*, 1705–1715. https://doi.org/10.1002/pc.25127
9. Ma, S.; Xiao, Y.; Zhou, F.; Schartel, B.; Chan, Y. Y.; Korobeinichev, O. P.; Trubachev, S. A.; Hu, W.; Ma, C.; Hu, Y. Effects of novel phosphorus-nitrogen-containing DOPO derivative salts on mechanical properties, thermal stability and flame retardancy of flexible polyurethane foam. *Polymer Degradation and Stability* **2020**, *177*, 109160.
10. Chen, M.; Shao, Z.; Wang, X.; Chen, L.; Wang, Y. Halogen-Free Flame-Retardant Flexible Polyurethane Foam with a Novel Nitrogen−Phosphorus Flame Retardant. *Industrial & Engineering Chemistry Research* **2012**, *51*, 9769–9776. https://pubs.acs.org/doi/pdf/10.1021/ie301004d
11. Li, T.; Xing, M.; Wang, H.; Huang, S.; Fu, C.; Lou, C.; Lin, J. Nitrogen/phosphorus synergistic flame retardant-filled flexible polyurethane foams: microstructure, compressive stress, sound absorption, and combustion resistance. *RSC Advances* **2019**, *9*, 21192–21201.
12. Olcay, H.; Kocak, E. D. The mechanical, thermal and sound absorption properties of flexible polyurethane foam composites reinforced with artichoke stem waste fibers. *Journal of Industrial Textiles* **2020**. https://doi.org/10.1177/1528083720934193.
13. Olcay, H.; Kocak, E. D. Rice Plant Waste Reinforced Polyurethane Composites for Use as the Acoustic Absorption Material. *Applied Acoustics* **2021**, *173*. 10.1016/j.apacoust.2020.107733
14. Ugarte, L.; Saralegi, A.; Fernández, R.; Martín, L.; Corcuera, M. A.; Eceiza, A. Flexible polyurethane foams based on 100% renewably sourced polyols. *Industrial Crops and Products* **2014**, *62*, 545–551.
15. Joseph, P.; Ebdon, J. R. Phosphorus-Based Flame Retardants. In *Fire Retardancy of Polymeric Materials*, 2nd ed.; Wilkie, C. A., Morgan, A. B., Eds.; CRC Press: Boca Raton, FL, 2010.
16. Amado, J. C. Q. Thermal resistance properties of polyurethanes and its composites. In: Evingur, G. A.; Pekcan, O.; Achilias, D. S., Eds.; *Thermosoftening Plastics*; IntechOpen, 2019, DOI: 10.5772/intechopen.87039.
17. Kirpluks, M.; Cabulis, U.; Avots, A. Flammability of bio-based rigid polyurethane foam as sustainable thermal insulation material. In *Insulation Materials in Context of Sustainability*; Amjad, A., Asaad, A., Eds; IntechOpen, 2016; pp 87–111, DOI: 10.5772/62539.
18. Wilkie, C. A; Morgan, A. B. *Fire Retardancy of Polymeric Materials*, 2nd ed.; CRC Press: Boca Raton, FL, 2010.
19. Prager, F. H.; Rosteck, H. *Polyurethane and Fire: Fire Performance Testing Under Real Conditions*; Wiley-VCH: Weinheim, Germany, 2006.
20. Gu, L.; Ge, Z.; Huang, M.; Luo, Y. Halogen-Free Flame-Retardant Waterborne Polyurethane with a Novel Cyclic Structure of Phosphorus-Nitrogen Synergistic Flame Retardant. *Journal of Applied Polymer Science* **2015**, *132*, 3, DOI: 10.1002/app.41288.
21. Akdogan, E.; Erdem, M.; Ureyen, M. E.; Kaya, M. Synergistic effects of expandable graphite and ammonium pentaborate octahydrate on the flame-retardant, thermal insulation, and mechanical properties of rigid polyurethane foam. *Polymer Composites* **2020**, *41*, 1749–1762. https://doi.org/10.1002/pc.25494

22. Gaan, S.; Liang, S.; Mispreuve, H.; Perler, H.; Naescher, R.; Neisius, M. Flame retardant flexible polyurethane foams from novel DOPO-phosphonamidate additives. *Polymer Degradation and Stability,* **2015**, *113*, 180–188. https://doi.org/10.1016/j.polymdegradstab.2015.01.007.
23. Ding, H.; Huang, K.; Li, S.; Xu, L.; Xia, J.; Li, M. Synthesis of a novel phosphorus and nitrogen-containing bio-based polyol and its application in flame retardant polyurethane foam. *Journal of Analytical and Applied Pyrolysis* **2017**, *128*, 102–113. http://dx.doi.org/10.1016/j.jaap.2017.10.020
24. Wang, C.; Wu, Y.; Li, Y.; Shao, Q.; Yan, X.; Han, C.; Wang, Z.; Liu, Z.; Guo, Z. Flame-retardant rigid polyurethane foam with a phosphorus-nitrogen single intumescent flame retardant. *Polymers for Advanced Technologies* **2018**, *29*, 668–676. https://doi.org/10.1002/pat.4105
25. Akdogan, E.; Erdem, M.; Ureyen, M. E.; Kaya, M. Rigid polyurethane foams with halogen-free flame retardants: Thermal insulation, mechanical, and flame retardant properties. *Journal of Applied Polymer Science* **2020**, *137*, 47611, DOI: 10.1002/app.47611.
26. Bocchini, S.; Camino, G. Halogen-Containing Flame Retardants. In *Fire Retardancy of Polymeric Materials*, 2nd ed.; Wilkie, C. A., Morgan, A. B., Eds.; CRC Press: Boca Raton, FL, 2010.
27. Zhang, T.; Liu, W.; Wang, M.; Liu, P.; Pan, Y.; Liu, D. Synergistic effect of an aromatic boronic acid derivative and magnesium hydroxide on the flame retardancy of epoxy resin. *Polymer Degradation and Stability* **2016**, *130*, 257–263.
28. Levchik, S. V.; Weil, E. D. Thermal decomposition, combustion and fire-retardancy of polyurethanes—a review of the recent literature. *Polymer International* **2004**, *53*, 1585–1610. https://doi.org/10.1002/pi.1314
29. Zhang, W.; Li, X.; Yang, R. Study on the change of silicon and phosphorus content in the condensed phase during the combustion of epoxy resin with OPS/DOPO. *Polymer Degradation Stability* **2014**, *99*, 298–303.
30. Liao, F.; Zhou, L.; Ju, Y.; Yang, Y.; Wang, X. Synthesis of a novel phosphorus-nitrogen-silicon polymeric flame retardant and its application in poly(lactic acid). *Industrial Engineering Chemistry Research* **2014**, *53* (24), 10015–10023. https://doi.org/10.1021/ie5008745
31. Gaan, S.; Neisius, M.; Cuchere, O.; Liang, S.; Mispreuve, H. Flame retardant polyurethanes based on novel phosphonamidate additives. *Fire Safety Science* **2014**, *11*, 821–831.
32. Wang, S. X.; Zhao, H. B.; Rao, W. H.; Huang, S. C.; Wang, T.; Liao, W.; Wang, Y. Z. Inherently flame-retardant rigid polyurethane foams with excellent thermal insulation and mechanical properties. *Polymer* **2018**, *153*, 616–625.
33. Yuan, Y.; Yang, H.; Yu, B.; Shi, Y.; Wang, W.; Song, L.; Hu, Y.; Zhang, Y. Phosphorus and nitrogen-containing polyols: synergistic effect on the thermal property and flame retardancy of rigid polyurethane foam composites. *Industrial & Engineering Chemistry Research* **2016**, 55 (41), 10813–10822.
34. Zhang, M.; Luo, Z.; Zhang, J.; Chen, S.; Zhou, Y. Effects of a novel phosphorus–nitrogen flame retardant on rosin-based rigid polyurethane foams. *Polymer Degradation and Stability* **2015**, *120*, 427–434.
35. Sykam, K.; Meka, K. K. R.; Donempudi, S. Intumescent phosphorus and triazole-based flame-retardant polyurethane foams from castor oil. *ACS Omega* **2019**, *4* (1), 1086–1094.

36. Lu, S.; Feng, Y.; Zhang, P.; Hong, W.; Chen, Y.; Fan, H.; Yu, D.; Chen, X. Preparation of Flame-Retardant Polyurethane and Its Applications in the Leather Industry. *Polymers* **2021**, *13* (11), 1730.
37. Konig, A.; Fehrenbacher, U.; Hirth, T.; Kroke, E. Flexible polyurethane foam with the flame-retardant melamine. *Journal of Cellular Plastics* **2008**, *44* (6), 469–480.
38. Meng, X. Y.; Ye, L.; Zhang, X. G.; Tang, P. M.; Tang, J. H.; Ji, X.; Li, Z. M. Effects of expandable graphite and ammonium polyphosphate on the flame-retardant and mechanical properties of rigid polyurethane foams. *Journal of Applied Polymer Science* **2009**, *114* (2), 853–863.
39. Strąkowska, A.; Członka, S.; Konca, P.; Strzelec, K. New flame retardant systems based on expanded graphite for rigid polyurethane foams. *Applied Sciences* **2020**, *10* (17), 5817.
40. Wu, D. H.; Zhao, P. H.; Liu, Y. Q.; Liu, X. Y.; Wang, X. F. Halogen Free flame retardant rigid polyurethane foam with a novel phosphorus– nitrogen intumescent flame retardant. *Journal of Applied Polymer Science* **2014**, *131* (11)
41. Sałasińska, K.; Leszczyńska, M.; Celiński, M.; Kozikowski, P.; Kowiorski, K.; Lipińska, L. Burning Behaviour of Rigid Polyurethane Foams with Histidine and Modified Graphene Oxide. *Materials* **2021**, *14* (5), 1184.
42. Boult, M. A.; Gamadia, R. K.; Napier, D. H. Thermal degradation of polyurethane foams. In *I.Chem.E. Symposium Series*; No. 33, 1972.
43. Jiao, L.; Xiao, H.; Wang, Q.; Sun, J. Thermal degradation characteristics of rigid polyurethane foam and the volatile products analysis with TG-FTIR-MS. *Polymer Degradation and Stability* **2013**, *98* (12), 2687–2696.
44. Singh, H.; Jain, A. K. Ignition, combustion, toxicity, and fire retardancy of polyurethane foams: A comprehensive review. *Journal of Applied Polymer Science* **2009**, *111*, 1115–1143.
45. Liang, S.; Neisius, M.; Mispreuve, H.; Naescher, R.; Gaan, S. Flame retardancy and thermal decomposition of flexible polyurethane foams: structural influence of organophosphorus compounds. *Polymer Degradation and Stability* **2012**, *97* (11), 2428–2440.
46. Son, T.; Nguyen, H. T.; Phan, H. T.; Pham, C. T.; Nguyen, B. T.; Pham, L. H.; Do, T. V. V.; Vothi, H.; Kim, J.; Hoang, D. Q. Phosphorus/phosphorus-nitrogen flame retardants applied to polyurethane/rice husk eco-composites: thermal behavior, flame retardancy, and physico-mechanical properties. *Polymer Bulletin* **2021**, *78*, 2727–2743.
47. Wang, S. X.; Du, Z.; Cheng, X.; Liu, Y.; Wang, H. Synthesis of a phosphorus-and nitrogen-containing flame retardant and evaluation of its application in waterborne polyurethane. *Journal of Applied Polymer Science* **2018**, *135* (16), 46093.
48. Chattopadhyay, D. K.; Webster, D. C. Thermal stability and flame retardancy of polyurethanes. *Progress in Polymer Science* **2009**, *34* (10), 1068–1133.

Chapter 13

Future Aspects of Flame-Retardant Polyurethanes

Takashiro Akitsu*

Department of Chemistry, Faculty of Science, Tokyo University of Science, Tokyo 162-8601, Japan

***Email: akitsu2@rs.tus.ac.jp**

As we have studied flame retardants (FRs) (e.g., Br-containing metal complexes) for some polymers at the Center for Fire Science and Technology at Tokyo University of Science for many years, the materials chemistry of FRs has become an important issue from the viewpoint of the social safety of fire accidents. For example, in 2017, a train with a roof made of (or coated with) polyurethane materials was involved in a fire in downtown Tokyo. Despite the damage to the train, all of the passengers were able to evacuate safely, and no one was injured. In this chapter, the author briefly reviews the characteristics of polyurethane as a FR and combustible material, as well as previous studies on FRs for other conventional polymers. The author also discusses expected future aspects of fire considering conditions of low oxygen concentration and zero gravity in analogy with the present knowledge about polymers and combustion.

Introduction

Polyurethanes (PUs) are currently among the most common and prominent polymer materials in industrial products. However, fire involving PUs can also sometimes occur. For example, in 2017, a fire broke out in a three-story building in Yoyogi 5-chome, Shibuya-ku, Tokyo, Japan, and spread to the roof of the second car of the regular train (eight-car train) from Honatsugi to Shinjuku that had stopped between Sangubashi and Yoyogi-Hachiman on Odakyu Electric Railway's Odawara Line (Figure 1). According to the Tokyo Fire Department and Odakyu Electric Railway, about 300 passengers got off the railroad tracks and evacuated, and no one was injured during this fire. In fact, the roof of the train was made of (or coated with) PU materials. After this fire, a photograph of the damaged roof made of PU covered with char (carbon after burning) was presented on a news website. Odakyu Electric Railway reported that a train fire spread and a deformed burn mark remained on part of the roof. On the day after this fire accident, the reporter of a news website provided the following description of the train car involved in the fire (translated from the original Japanese by Google) (*1*):

The vehicle that left a burn mark is exactly the vehicle that put the flame on it. Almost half of the second car heading for Shinjuku, from the connection between cars 7 and 8 toward car 7, is burnt. The white roof of car 8 is black and scorched [on the side toward] car 7, and it undulates unnaturally in some places. The ditch where rainwater flows on both sides of the roof is also deformed, and fine charcoal is accumulated in it.

According to Odakyu Electric Railway, the stainless steel body does not burn, but the entire roof is coated with polyurethane resin to a thickness of 2 to 3 cm. It [polyurethane] is a roofing material [that is used] to improve insulation and waterproofness to avoid the effects of lightning strikes. However, a noncombustible material is mixed in this resin, and the flame retardant that meets the railway standards is maintained.

"Because it is flame-retardant, it [the fact that the roofing material is flame-retardant] does not mean that it will not burn at all, but the fire will go out after a certain amount of time" (driving vehicle section). It is presumed that the partially carbonized burn marks may be this roofing material.

Certainly, the inside of the train was as usual with flames on it, so I couldn't believe [the fire spread] slowly. It was when I entered car No. 7 with a burning roof that made me think that it was different from a normal car. When I opened the door of the connecting part, there was a chemical pungent odor of burning resin only in that vehicle.

However, looking at the air-conditioning outlet with the ceiling, windows, and panels removed, there is no dirt such as soot from the fire. Does [this] indicate that the inside of the car was isolated from the fire on the roof?

The on-site inspection by the fire department on [Sept 11] is over, and the company waits for the result. "The impact of fires along railway lines on trains is unprecedented. We will use this as a lesson to verify what we can do," said Odakyu's CSR [Corporate Social Responsibility] and Public Relations Department.

Figure 1. Normal roof of Odakyu Electric Railway's trains coated with PU. (Photograph courtesy of Takashiro Akitsu.)

The properties and characteristics of PU materials (hard and soft) are summarized here in terms of flame retardancy (2). For the production of PU foams (sometimes also called urethane foams), a polyisocyanate with isocyanate (NCO) groups and a polyol with hydroxyl (OH) groups are mixed together with a catalyst, foaming agent, and foam stabilizer, as well as other components, if needed, and foaming is carried out at the same time. The product is a uniform plastic foam. PU foams are roughly divided into soft (flexible) and hard (rigid) varieties. Soft foams are light, provide excellent cushioning and durability, and have a wide hardness range; they are mainly used in fields that require cushioning. On the other hand, hard foams have excellent heat-insulating properties and are mainly used as industrial materials that make the best use of these properties.

Preparation methods for hard PU foams result in the following features: It is essentially impossible for hard PU foams to be completely noncombustible. However, coating noncombustible materials on PU stops the radical reaction of oxygen with volatile compounds. To improve the thermal stability properties of hard PU foams, techniques such as (1) adding fire-resistant isocyanate-based polyimide foams, (2) forming inorganic polymer flame retardants (FRs) by heat, and (3) adding radical scavengers have been employed so far (3, 4).

Similarly to the popular poly(vinyl chloride) (PVC), PU acts as a FR itself. Widely used FRs for hard PU foams include (1) phosphates and halogens (to reduce flame propagation speed); (2) antimony trioxide, aluminum hydroxide, and hydrogen bromide (to decrease smoke); and (3) diisocyanate and polyol (included in the PU structure). It should be noted that features of thermosetting resins should also be considered. However, toxicity and durability as metal composites are also potentially important issues. In other words, emerging toxic gases (CO, CH_3Cl, HCN, NH_3, and NO_x) could be nonnegligible and cause serious issues in many cases (3, 5, 6).

On the other hand, commonly used methods for increasing the flame retardancy of soft PU foams (quick mold) are similar to those used for hard foams: (1) use of phosphate and halide compounds, N-containing polyols, and isocyanates; (2) direct inclusion of isocyanates, carbodiimides, and imides in the chemical structure; and (3) addition of fillers (e.g., $BaSO_4$ and $CaCO_3$). (4) Soaking soft PU foams in solutions of a FR may also be useful. However, decreasing the performance of the foams as functional materials through the addition of flame retardancy can occur by any of these methods, which may be a disadvantage of the selection of materials.

However, it should be noted that some compounds containing halogens, phosphorus, or nitrogen are considered dangerous materials according to Japanese fire laws (7). Such materials are specified in Article 2, Paragraph 7, Appendix 1, of the Fire Service Act and are classified into six categories. All of them have high risk of fire occurrence and spread, and the fire is difficult to extinguish in each case. They can be stored in the laboratory up to 0.2 times the sum of each designated quantity.

For phosphorus, in an air-collecting bottle filled with water, burn red phosphorus (class 2) on a burning spoon, cover it, dissolve the white smoke in water, and confirm the red discoloration of methyl orange. A class 2 material (flammable solid) is a solid that is easily ignited by a fire or that is easily ignited at a relatively low temperature (less than 40 °C); it easily catches fire and is difficult to extinguish quickly:

$$4P + 5O_2 \rightarrow P_4O_{10} \quad (1)$$

$$P_4O_{10} + 6H_2O \rightarrow 4H_3PO_4 \quad (2)$$

Regarding nitrogen, class 5 materials (self-reactive substances) are solids or liquids that generate a large amount of heat at a relatively low temperature due to hydrolysis or a similar reaction or that react explosively. Because combustibles and oxygen supplies coexist, class 5 materials are self-combustible and cannot be extinguished through suffocation, even if the air supply is shut off. Generally, the fire must be extinguished with a water or foam fire extinguisher. A typical class 5 compound is nitrobenzene, obtained from benzene and nitric acid (class 6). Nitrobenzene is then used with tin and concentrated hydrochloric acid to form aniline and aniline hydrochloric acid. These reactions are often mentioned in chemistry textbooks.

Diversity makes it difficult to judge what is dangerous. In a lecture on environmental energy chemistry, the author introduced the Fischer–Tropsch method for producing artificial petroleum from coal. According to this process, carbon monoxide obtained from coal and hydrogen are reacted at high temperature and high pressure in the presence of a thorium(IV) oxide catalyst to produce hydrocarbons (artificial petroleum) and hydrogen. The author was surprised to see all of the dangerous substances involved. That is, carbon monoxide easily binds to the heme iron of hemoglobin and causes carbon monoxide poisoning, high-pressure hydrogen gas easily reacts explosively with oxygen, and thorium is a radioactive element. However, the author noticed that, among these substances, only liquid hydrocarbons (class 4) are actually classified as dangerous goods (including only solids and liquids) under the Fire Service Act in Japan. It appears that determining a safe substance is just a matter of viewpoint (regulating law) and quantity (representing a storable standard amount according to danger, designated quantity, etc.) (*7*).

Tests and requirements for FRs have been established as follows: (1) burn-through and thin-material test methods (ASTM; results of different tests sometimes conflict), (2) lower emissions of volatile compounds upon the heating of polymers, and (3) char production before decomposition into lower compounds. In this context, (4) low heat of combustion and high heat of decomposition, (5) coating with air shut -off (suffocation) and FR materials, and (6) radical scavengersare also important (*8*).

In terms of fire extinguishing, however, if these elements are contained in PU materials, there is no use for ideal discussions or comventional fire-exinguishing methods. For example, principle 5 of shutting off air can be supported as follows: The three requirements for combustion are (1) combustibles, (2) an oxygen supply, and (3) an ignition source (heat source). Of these, the suffocation fire-extinguishing method works by removing oxygen.

Moreover, principle 4 regarding a high heat of decomposition may be important for materials containing both FR and dangerous elements. Allotropes of red phosphorus (class 2) and yellow phosphorus (class 3) are known. (Sulfur, which has orthorhombic, monoclinic, and rubber-like allotropes, is also class 2.) When yellow phosphorus is heated at about 250 °C for several hours in a nitrogen atmosphere, reddish-brown red phosphorus is obtained. The color of red phosphorus is due to the inclusion of impurities. Red phosphorus is applied to the sides of the match boxes and ignites when rubbed. (The enthalpy produced by burning phosphorus pentoxide is $\Delta H = -17.6$ kJ/mol.) Water injection and fire extinguishing are possible. On the other hand, white phosphorus is obtained by igniting a mixture of silica sand and coke with a phosphorus mineral containing calcium phosphate as the main component. White (yellow) phosphorus has a tetrahedral structure with a side of 225 pm. Because it spontaneously ignites in air, white phosphorus is stored in water to break contact with the air. Because it has a low melting point (44 °C) and is fluid, its fires can be extinguished with water and sand (*7*).

Aside from elements acting as FRs, preferable chemical structures may be proposed by including the following factors: (1) polyethers (hydrogen bonds) for volatile compounds; (2) three-dimensional bridging networks; (3) ring structures for high flash points; (4) char production by $(NH_3)_3PO_4$; (5) C=O and C–O bonds for low heats of combustion; (6) C=O, C=C, O–H, and

P compounds for high heats of decomposition; (7) surface coatings by P, Cl, and Br compounds; and (8) Br and Cl compounds for trapping radicals. In addition to halogen compounds, non-halogen compounds with the same capabilities are being developed to avoid toxicity issues, and such materials have been expanded for graphite-based materials as well (*9*, *10*).

Thus, known principles of FRs for PU can be summarized as follows: For halogen-containing phosphoric acid ester, phosphorus decomposes to produce phosphoric acid and polyphosphoric acid in the solid phase, and the film blocks oxygen. This has the effect of stopping combustion through the carbonization effect of the dehydration of phosphorus. For halogens alone, in the gas phase, halogens trap radicals generated by the decomposition of the resin. This releases an inert gas such as hydrochloric acid to dilute the oxygen concentration and stop the fuel.

Chemical Structure-Based FRs for (Other) Polymers

The combustion process generally proceeds in the steps of (1) heating, (2) decomposition, (3) ignition, (4) combustion, and (5) flame propagation. However, in the development of FRs that has been carried out to date, the target is mainly combustion (step 4). Research has been conducted on methods for removing the three essential conditions of combustion substances, energy, and oxidants. The mechanism works by decomposing and burning an additive added to the polymer near its decomposition temperature, depressing the combustion rate of the polymer, gaining a time lag before flame propagation (step 5), and reducing the degree of fire. This works to keep flame propagation to a minimum.

However, the significance of the initial steps leading up to combustion, namely, heating, decomposition, and ignition, is also noted. They lead to the irreversible deterioration of the material (softening, decomposition, vaporization, etc.) (reduction), and there is no precedent for compounds that have the function of blocking or alleviating such consequences. In addition, FR technology that reduces or does not use halogens is required because of the impact of halogen-containing materials on the human body and the environment, but the additives currently being developed have low decomposition temperatures and their target polymers are limited. There are various problems such as a large amount of additives and a decrease in polymer recyclability, and no practical alternatives have been identified.

On the other hand, the technology conventionally used for the heat stabilization of polymers was developed for the purpose of capturing active decomposition products that are desorbed in advance by heat, inactivating metals, and other causes. The emphasis is on stopping the promotion of the decomposition reaction. For example, a combination of a zinc salt and a barium salt of higher fatty acids is used as a heat stabilizer that captures and detoxifies the hydrogen chloride desorbed by the heat of vinyl chloride. However, the actions of these heat stabilizers are all irreversible reactions, and chemical changes in the materials and the accumulated effects of long-term irreversible reactions are unavoidable, even if they do not lead to visible physical changes or decomposition. Further, additives such as FRs and heat stabilizers often exhibit antagonistic effects when used in combination, and there is a need for a single highly functional and safe additive.

The thermophysical properties of PU and other polymers are in the category of polymer chemistry. Indeed, in principle, the flashover of PVC wall panels can be analyzed by detailed measurements of test polymers. Furthermore, the emission of gases (e.g., CO) after the burning of polymers can be a serious problem in fires (Figure 2). However, because research into a more general FR and fire science requires an interdisciplinary approach, joint research and human resource education face unique difficulties. The use of partnerships may be one breakthrough for this issue. Toward this end, we have investigated chemical structure–based FRs of other polymers (not only

PU) having amide or halogen groups (Figure 3). For example, this section describes a report of international joint work on testing FRs for (other) polymers (*11–13*).

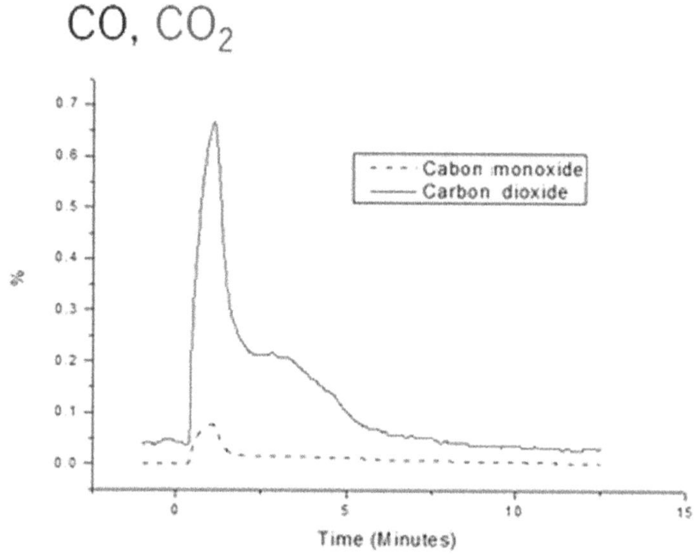

Figure 2. Concentrations of carbon monoxide (dashed line) and carbon dioxide (solid line) versus time in air during the burning of certain PVC-based polymers.

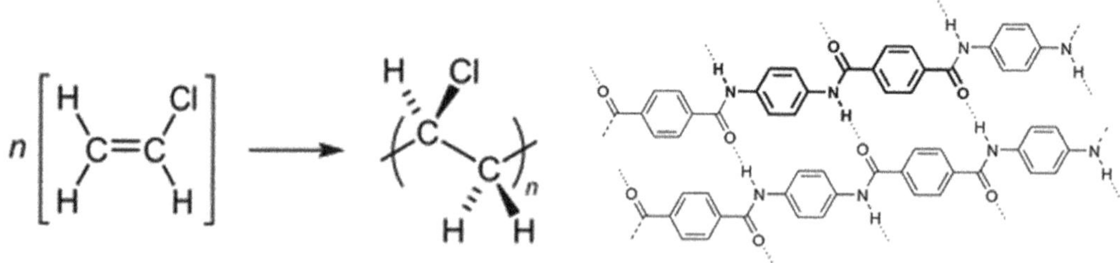

Figure 3. Chemical structure–based FR polymers: (left) PVC and (right) a typical polyamide (Kevlar).

In research on PVC, Professor R. K. Soni of the Department of Chemistry, Chaudhary Charan Singh University, Meerut, Uttar Pradesh, India, who visited Japan under a joint research agreement with the Center for Fire Science and Technology and performed cone calorimeter measurements there. Cooperation with polymer chemists (Professor Soni and his colleagues) contributed greatly to our work. Around the time of his first visit to Japan, a partner school agreement was signed between Charan Singh University and the Graduate School of Science at Tokyo University of Science (TUS). Professor Soni developed a technology for synthesizing amide compounds using poly(ethylene terephthalate) (PET), which is the raw material for PET bottles (similar to Figure 3, right). If it were a PVC film, it is expected that the FR would at least be thermally stable, although it would depend on the country's standards, according to calculations based on thermal analysis. A PVC film containing a bromine-substituted metal complex described later was also measured as a sample for comparison. Therefore, the researchers have been accumulating data by repeating thermal analysis and cone calorimeter measurements after visiting Japan several times.

A summary of thermophysical experiments for polyamide-based polymers as FRs is as follows: (1) Synthesis of *N,N'*-substituted terephthalamides or PVC-based polymers was carried out by the depolymerization of PET waste (ammonolysis or aminolysis) or a conventional film formation method. (2) Simultaneous thermogravimetric analysis (TGA)/differential thermal analysis (DTA) thermograms were recorded on a PerkinElmer Diamond TG/DTA system at 20.00 °C/min under a nitrogen atmosphere from 40 to 700 °C . (3) In detail, for the experiments, plasticizer was mixed at 60–70 °C (heated to 110 °C). The formulation comprised PVC resin, 100 parts per hundred resin (phr); dioctyl phthalate, 50 phr; $CaCO_3$, 15 phr; stearic acid, 1 phr; and terephthalamide, 10 phr. Two cycles of roll milling were applied (144 °C for the first, 146 °C for the second), followed by compression molding at 142–145 °C (holding for 1.5 min, cooling for 1 min). (4) Cone calorimeter data and TGA were used to investigate the thermal behavior of terephthalamide-stabilized PVC sheets under an irradiation heat flux of 50 kW/m^2 (some examples are listed in Table 1). A PVC sample was subjected to single-step thermal degradation with a peak maximum at 285.56 °C. The sample exhibited a high value for flashover propensity, which indicates a high level of risk; however, the total heat released suggests an intermediate level of risk. The PVC sheet was well stabilized thermally, and hence, efforts should be made to overcome the problem of heat release to make PVC sheets safe for construction applications (Table 1).

Table 1. Summary of Heat Measurements and Analyses for Some FR PVC Samples

Sample	*Parameter*	*Value (MJ/m^2)*
1	Total heat released	44.01
2	Average heat generation rate in 60 s	273.23
3	Average heat generation rate in 60 s	169.42
4	Average heat generation rate in 60 s	130.00
5	Peak heat generation rate	366.03

The following description is mainly based on the report of international joint work by Professor Soni.

Outline of Academic Activities during Professor Soni's Stay at TUS

During the period of the visit (15 days, from November 15 to November 30, 2018), different academic activities were discussed, and appropriate actions were taken. The different samples brought by Professor Soni were handed over, and the testing procedure was discussed on November 15, 2018. We performed some experiments on cone calorimetry (Figure 4) at the Noda campus of TUS and received the data for analysis. The carbon samples produced during the experiments were preserved by Professor Pal of Chaudhary Charan Singh University for further studies on solid-state materials. We also analyzed some samples for TGA and differential scanning calorimetry (Figure 5) at the central research facilities of TUS, Kagurazaka campus, Tokyo, Japan. Three samples were also tested by proton nuclear magnetic resonance spectroscopy.

In addition, we submitted two research proposals to Professor Akitsu for future mutual research collaboration. As scheduled, Professor Soni delivered a lecture on the topic of UV curing kinetics investigations of photocurable acrylic formulations on Nov 28, 2018, at the Kagurazaka campus, Tokyo, Japan.

Figure 4. Equipment for cone calorimetry (top left). Sample of PVC sheet, (top right) setting to a container, (bottom left) before heating, and (bottom right) after burning. (Photographs courtesy of Takashiro Akitsu.)

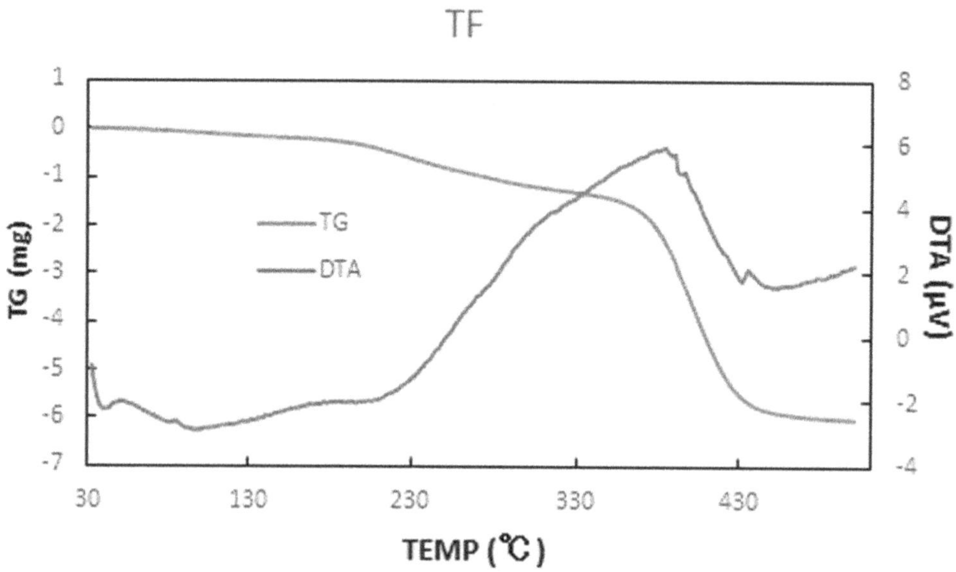

Figure 5. Typical example of TG/DTA data for the thermal behavior of a certain amide polymer to record temperature during burning.

Future Plans for Academic Collaboration with the TUS Host Researcher

For the future, Professor Soni submitted two research proposals to Professor Akitsu. The two institutions also plan to support student exchanges, write a joint book, and organize a joint international conference.

Metal Complexes as FRs for (Other) Polymers

In this section, metal complexes as FRs for polymers are discussed. Initially, the direct reason for developing complex chemistry in fire-related research is that the author remembered the phenomenon of structural change (structural phase transition) due to temperature changes in trans-type Schiff base metal complexes, which has been known for a long time. This is a phenomenon in which a complex with a planar structure at room temperature undergoes an endothermic structural phase transition when heated and changes to a tetrahedral structure (*14*). Therefore, we performed the TGA/DTA of the metal complex alone. However, because it was in an organic–inorganic composite material, that is, a poly(methyl methacrylate) (PMMA) film, the temperature rise of PMMA was slowed by the endothermic heat accompanying the deformation of the complex.

Before submitting a patent application, we investigated previous studies using metal complexes as polymer FRs and found that there were no examples of trans-type Schiff base metal complexes, but there were examples of cis-type (so-called salen-type) Schiff base metal complexes. To show inventiveness and superiority, it was necessary to add additional functions to the salen-type metal complex.

Taking advantage of the photochemical properties of organic–inorganic composite materials including salen-type Schiff base complexes, we have been conducting research to apply such composites to new types of solar cell dyes (close to dye sensitization and organic thin films). With the spread of renewable energy, solar cells are being installed not only in utility-scale solar power plants (Figure 6) but also in individual buildings (Figure 7). However, once a fire breaks out, there are new issues in terms of safety measures such as electric shock during fire-extinguishing activities. Module materials that make up new types of solar cells that are expected to be put into practical use in the future include organic dyes (metal complexes), polymers, electrolytes (iodine) in the dye-sensitized type and organic compounds in the organic thin-film type. Semiconductors (metal complexes, fullerenes) and conductive polymers are among the materials that can potentially burn. Therefore, it is desired to synthesize a metal complex that performs both as a dye that absorbs light of as long a wavelength as possible to effectively use sunlight and as a FR that makes flammable materials contained in components, especially polymers, hard to burn.

For example, safety measures during firefighting activities (problems of water discharge and electric shock) have been identified. Solar cells of several types and modules may encounter significant issues because of certain polymer materials. Moreover, because modules combine several materials, the resulting problems may be complicated.

A related Br-substituted metal complex was further synthesized, PVC was selected as the polymer, and combustion products were analyzed by TGA/DTA/mass spectrometry (MS) (*15*). The results showed that chlorine detection was characteristic. We tried to simulate the results by dynamics calculations but were unsuccessful. If a supercomputer equipped with computational chemistry software suitable for this purpose could be used, this is an approach that we would like to pursue again in the future.

There are still problems to be solved, and theoretical research toward this end is also underway. These efforts include identifying FRs for flammable module components (e.g., polymer FRs for film and lens components), investigating the combustion mechanism (e.g., molecular dynamics calculations of plastics combustion, Fourier transform infrared spectroscopy and electronic state calculations of bond cleavage in dye complexes), and analyzing combustion products (e.g., Fourier transform infrared spectroscopy, MS).

Figure 6. The world's first airport fully powered by solar energy in India. (Photograph courtesy of Takashiro Akitsu.)

Figure 7. Solar cell panels on the roof of a building in France. (Photograph courtesy of Takashiro Akitsu.)

As a solution, a metal complex that serves as a FR for plastics is desired. The current situation can be summarized as follows:

(1) The endothermic deformation of the structure of a Br complex has a FR effect on PMMA.
(2) There are patents for salen-type complexes.
(3) A metal complex that doubles as a dye for dye-sensitized solar cells and a FR is desired.

Consequently, we have investigated dual-purpose (dye for solar cells and FR for polymers in solar cells) metal complexes (Figures 8 and 9). Increasing the number of Br atoms introduced into a molecule results in both a shift to longer wavelengths of the light absorbed and an increase in the decomposition temperature for such metal complexes.

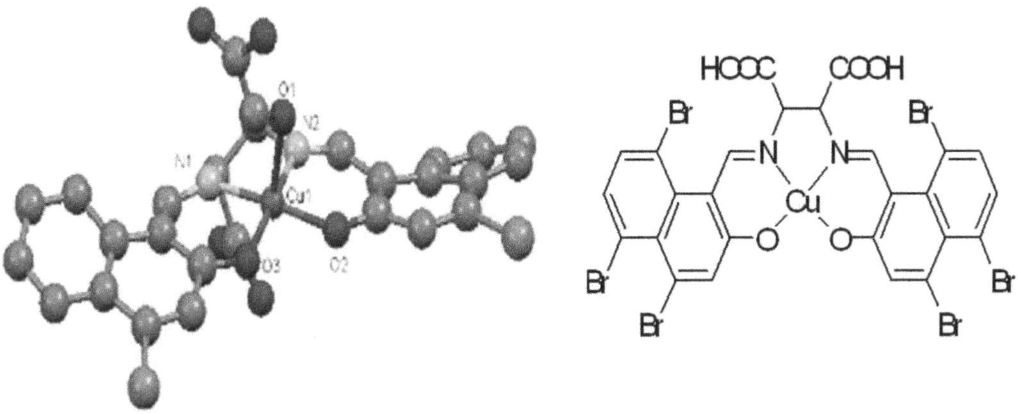

Figure 8. (Left) steric structure of a Br-containing metal complex, (right) chemical structure of a Br-containing metal complex.

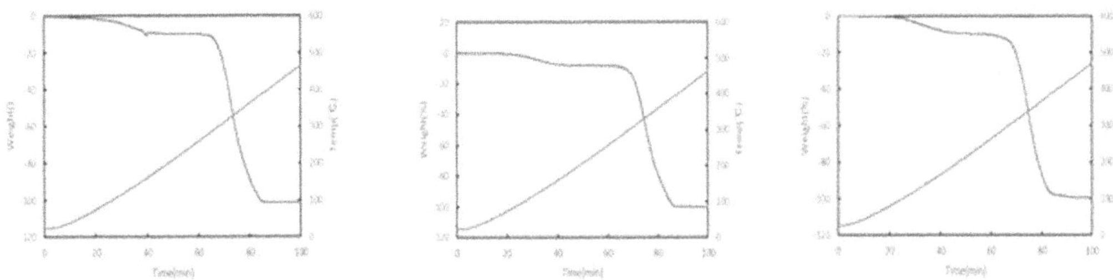

Figure 9. TGA/DTA scans of PMMA containing metal complexes having (from left to right) two, four, and six Br atoms. A gradual increase of the decomposition temperature can be observed upon the addition of multiple Br atoms to a molecule.

The following sentences are a part of a research proposal for dual-purpose (dye for solar cells and FR for polymers in solar cells) metal complexes:

Continuing from last year, in this year, we conducted thermal analysis and combustion product identification (TGA/DTA/MS) of PVC films containing PET recycled heat stabilizers and PVC films containing a brominated copper complex FR. This brominated copper complex FR is a complex that can be expected to have an applied function as a dye for a dye-sensitized solar cell. It has been pointed out that, in dye-sensitized solar cells and lithium batteries, not only the plastic of the module but also the electrolytes and organic solvents that are components of the device material can ignite or burn, making FRs important. From the viewpoint of the spread of renewable energy materials and safety measures in the future, plastic materials that have both a dye function and a FR (heat stabilizer) function are required. In anticipation of such research developments next year, we plan to introduce a manual printing machine for manufacturing solar cell materials and consumables for dye evaluation.

It was shown that, as the number of Br substituents with strong electron-attracting properties increased in the complex molecule, the peak near 280 nm in the UV–vis spectrum shifted to slightly longer wavelengths and the absorption intensity increased. It was confirmed by TGA/DTA that the decomposition temperature of the PMMA film containing the complex increased as the number of bromine substituents increased. Therefore, it was clarified that the effect of the Br-containing complex as a FR was enhanced.

Because the electrochemical properties of all of the complexes satisfied the conditions for use in dye-sensitized solar cells, we created cells in which titanium oxide and glass substrates were layered and measured the photoelectric conversion efficiency with a solar simulator. However, unfortunately, only very low values were measured. Improving the performance of these solar cells could entail improving the cell itself (improving the method of applying the dye to the substrate, using another glass material, etc.) or employing a combination with other dyes (using aggregates, photoelectron transfer reactions). We are still continuing research into other physicochemical approaches, such as incorporating a site that can be used for glass, and utilizing nonlinear optical effects.

Other Strategies for FRs

Because the introduction of elements into the chemical structure is effective for PU, the same strategy is also used for fire materials other than FRs. Keep in mind, however, that halogens (bromine, chlorine, and fluorine) are both good FRs and environmentally harmful elements. Therefore, the following research was conducted: A fluorine-based surfactant was used as a component of an alcohol-resistant foam fire-extinguishing agent to extinguish a water-soluble liquid typified by alcohol. Because fluorine-based surfactants have a low surface tension peculiar to fluorine compounds, foam fire-extinguishing agents containing them quickly flow and accumulate on the flammable liquid surface when extinguishing a fire. Through this action, a foam fire-extinguishing agent that is not able to extinguish fires on a water-soluble liquid can improve the liquid resistance against a water-soluble liquid by containing a fluorine-based surfactant as a component. Based on this principle, we obtained a foam fire-extinguishing surfactant from a vendor and conducted research to prevent foam rupture by adding metal ions that are difficult to oxidize (*16*). We want to apply this strategy to PU as well.

In the near future, there will be more opportunities to use energy at high density, and it is predicted that the frequency of fires will increase. Specifically, dealing with new types of dangers such as runaway lithium-ion batteries and fires at utility-scale solar power plants is an urgent issue. To solve this problem, from the viewpoint that combustion is an interface phenomenon, the goal is "to process and design the interface of materials well and to develop next-generation materials with excellent heat resistance and fire resistance without impairing characteristics and performance." It is also necessary to establish a research base for the development of FR materials and fire-extinguishing agents or to incorporate or cooperate with a current research base that has such a goal.

If innovation is promoted in collaboration with companies that contribute to society through these research results, the following scenarios can be foreseen: We are developing new inorganic FRs that are completely different from conventional organic FRs, such as halogen-based and phosphorus-based FRs and inorganic FRs such as metal oxides and antimony. In particular, these materials are intended for engineering plastics such as materials used to coat electric wires (PVC, polyethylene, PMMA, polystyrene, etc.); transportation equipment such as aircraft and automobiles; electrical, electronic, optical, and medical equipment; bulletproof glass materials (polycarbonate); and even electronics. The purpose is to develop new forms of FRs such as equipment bases (plastic bases for central processing units, solar panels, etc.). The FRs to be developed are metal complexes called organic metals, and to further enhance their FR properties, these complexes should contain hydrocarbons with halogen elements in their side chains.

Among fire-extinguishing agents, alkali metals (lithium, sodium, potassium, etc.) used in batteries, for example, are characterized as class 3 dangerous substances, react with oxygen in water and air, spontaneously ignite, and sometimes even explode. The containers of these alkali metal batteries are covered with a plastic-based thin-film coating material, and a FR is added so they do not burn with heat. Finally, we will develop a self-extinguishing fire-extinguishing agent and a chemical that prevents the combustion of coating materials such as those used on fuel cells. On the other hand, it would be a great success if could be developed a powder-based fire-extinguishing agent for small-scale alkali metal fires, as well as a mobile or packaged powder fire-extinguishing device.

Future Aspects of Gravity

The frontier of fire science will extend into extreme environments, and methods for such situations should be established in advance. For example, space and marine plans and the corresponding technologies are being developed by the government of Japan (*17*, *18*). Regarding the future plans of the Japanese government in fire science, Professor Y. Matsubara (Center for Fire Science and Technology, TUS) had multiple exchanges with the author. In this exchange of opinions, two important aspects of fire environments that will be important in the near future (gravity and oxygen concentration) were discussed. This discussion was based on materials for the Council for Science, Technology and Innovation and the "Follow-up on the Growth Strategy" announced in July 2020.

This issue is not unrelated to fires involving PU and macromolecules (Figure 10). In general, the model of polymer combustion can be summarized as follows: In the gas phase, combustion occurs in the presence of oxygen and involves the radiation of heat. In the solid phase, the process also involves the continuous conduction of heat, the thermal decomposition of the polymer, and the diffusion of flammable gas. When the gas is given heat and oxygen, combustion continues (*19*). Are the inner structures of flames and the behavior of the oxygen supply similar under low-gravity conditions to those observed under normal conditions? Normally, the concentration of oxygen in the air is approximately 20%, even during the burning polymers (Figure 11), and flames are formed in the direction opposite to the force of gravity (i.e., above the combustion points). Whether current fire science and technology and FR mechanisms are also valid under such environments is an important question from this viewpoint. Herein, we state the influences of gravity and oxygen concentration in analogy to the cases for PU.

Figure 10. The burning of a PVC-based polymer sample and its flame during a cone calorimetry test. (Photograph courtesy of Takashiro Akitsu.)

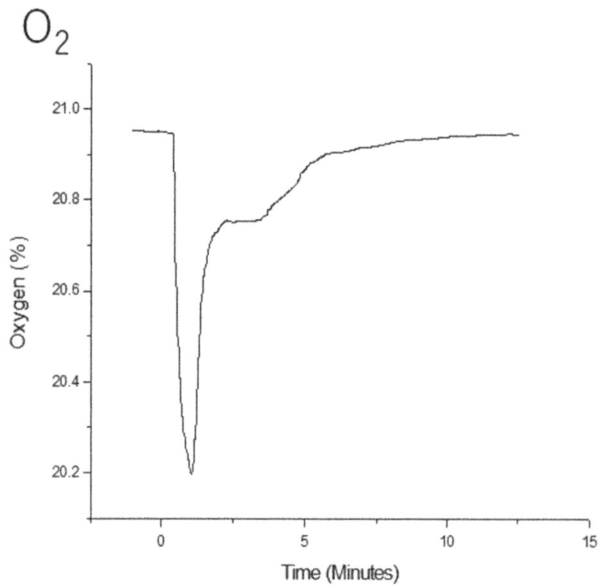

Figure 11. Concentration of oxygen versus time in air during burning a certain PVC-based polymer.

It is well known that the flames of candles on Earth (Figure 12) are different from those in space. Without gravity, flames become round-shaped (no convection of flammable gases) and blue in color (insufficient oxygen supply). The familiar candle flame shape under normal conditions has a thin core and a blue part on the outside. The upper part of the flame is bright orange. The shape of the flame is the shape of a vertically long teardrop with a thin top.

Figure 12. Candles on a birthday cake, showing the typical flame shape under normal conditions. (Photograph courtesy of Takashiro Akitsu.)

Dr. Akira Yoshino (awarded the Nobel Prize in Chemistry in 2019) summarized the important points of the science of candles from Faraday's famous lecture (*20*). Candles continue to burn because they are supplied with oxygen. However, there is no doubt that there is no air convection, so oxygen is brought about by a phenomenon other than convection, namely, diffusion. Diffusion for a candle flame should be considered as follows: "When the concentration of a substance varies from place to place, the concentration of the substance becomes uniform over time".

In other words, the burning of a candle consumes the air and oxygen around it, reducing the air and oxygen concentrations in that region. The temperature also decreases. Because the concentrations of air and oxygen now vary from place to place, diffusion tries to make the concentrations uniform, with new air entering the region around the candle and the oxygen concentration increasing. As this process repeats, oxygen is continuously supplied to the candle, and it continues to burn. However, there are conditions for this diffusion to occur, and it does not occur if the space where the candle is burning is narrow or there are obstacles that interfere with the diffusion, and it also depends on the thickness and shape of the candle. It is said that it does not always keep burning.

Next, he explained the color of flames. The candle flame in a weightless environment is dim blue. Why does it exhibit such a color that is completely different from its color on Earth? (This may be straightforward question for many people.) Supplying oxygen by diffusion is much slower than supplying it by convection. The amount of oxygen that can be used is so small that there is only barely enough oxygen to burn. There is only enough oxygen to disappear quickly if the conditions are bad.

Based on the behavior on Earth, the temperature of the flame in space is predicted as follows: In space, the temperature of the flame is lower than that on Earth. The color of the candle flame on Earth is orange because the wax is heated to a high temperature and undergoes thermal decomposition, and the carbon atoms are heated. However, in zero gravity, the temperature of the flame is low, so the candle does not become hot enough to be pyrolyzed. That is why the flame is a dim blue, as seen around the flame core on Earth.

Previously, combustion under microgravity was also investigated by other researchers. Under Earth's gravity, high-temperature combustion gas rises by convection, and oxygen is supplied to the flame. Under microgravity, there is no convection, and the transport of combustibles is controlled by diffusion. Then, the flammable material burns incompletely with a round flame. Under microgravity, the slow, nonflaming oxidation of the porous material PU was observed (*21*).

In addition to the combustion reaction, if it is related to gravity, which is difficult and almost unknown, a chemical experiment assuming the universe can be a "model" experiment at present. For example, in consideration of the effect of rotation, we once proposed a project to search for the origin of biochemistry and identifying functional materials using the angular momentum of cosmic light as follows:

By monitoring space radiation, it is possible to measure lunar radiation and electromagnetic waves. There is a strong theory that light with the angular momentum of the universe explains the origin of chirality in materials such as amino acids. For comparison with the conditions of conventional experimental facilities, we will extend our experiments to the conditions of the lunar environment and further proceed with the goal of elucidating the origin of biochemistry and searching for functional materials.

The contributions of such work could lead to the establishment of asymmetric synthesis and reaction methods using circularly polarized light and their application to chiral functional materials. This could, in turn, increase the supply of biologically related chiral substances (pharmaceutical products) and enable the selective reaction (treatment) of biological substances. This basic research could also lead to the creation of new organic–inorganic composite functional materials that take into consideration the conditions of the lunar environment (strong UV light, anoxia, low gravity), abundant resources, and elemental composition.

Future Aspects of Oxygen Concentration

High concentrations of oxygen were investigated to establish standards for hyperbaric oxygen therapy rooms. At that time, safety for hyperbaric oxygen therapy rooms was studied in the field of neither buildings nor health. Ignition temperatures may be lower than under normal conditions, and ignition energy phenomena under high pressure (due to oxygen as well as helium or nitrogen) may be different than under normal condition. Hazards arise when the partial pressure of flammable oxygen is high. Combustion occurs more intensely than at atmospheric pressure and spreads quickly.

Extinguishing fires at high pressure and high oxygen concentration is difficult but has been partly established as follows (22): (1) Hot gas, electrical wires, and electric sparks may be ignition sources. (2) Under microgravity, the ignition point is low, and flame propagation is generally slow. (3) Extinguishing fires by suffocation is difficult, and cooling must be effective. (4) Water, foam, and evaporation liquid fire extinguishers are all suitable. (5) Rapid decompression leads to decompression sickness.

A simple comparison of known methods and extreme conditions can be summarized as follows. (Note that the comments in parentheses represent important questions that should be considered and overcome later.)

(1) There is less emission of volatile compounds upon the heating of polymers. (What is the effect of the pressure of the atmosphere?)
(2) Char is produced before decomposition into lower compounds. (Is combustion incomplete?)
(3) The heat of combustion is low, and the heat of decomposition is high. (Is combustion easy and heat absorption difficult?)
(4) Suffocation and the use of FR materials are effective in extinguishing fires. (Does the effectiveness of fire-extinguishing approaches depend on oxygen pressure?)
(5) Radical scavengers can also be used to establish fires. (Are the spatial features of reactants suitable? Would they be suitable in the gas phase?)

In the case of fire accidents in space, we have proposed a study to clarify the aspects of current research in the field of fire safety that can be utilized and those that need to be further changed to the environment of space. In fact, most of the accidents in space development occur during launch and, therefore, in Earth's atmosphere. A total of three fires broke out on the Russian space station, and a total of six fires occurred on the American space shuttle. In the 1997 Russian accident, a fire was caused by an oxygen tank. Although the number is smaller than for general fires, some research reports on countermeasures are available. Unique conditions are required in terms of both gravity and oxygen concentration. Fire extinguishers at the International Space Station in Japan use carbon dioxide as a fire extinguishing material. It also uses a respiratory system. On the other hand, Russian space stations use foam fire-extinguishing agents containing distilled water and nitrogen. Fire experiments that occur in a microgravity environment are also being conducted on the International Space Station. The main purpose of these experiments is to investigate fire protection standards in space. Although the oxygen concentration and material fire safety standards are observed, the story changes if there is no oxidant supply to the combustion reaction because there is no natural convection under microgravity (the effects of gravity and oxygen are related each other). Many of the causes of fires in space have been electrical. In addition, research on solid fuels is required.

In this context, unexpected results under normal reaction conditions can occur even in a fire accident in a chemical experiment that is usually seen, which was investigated in our previous studies (*23, 24*). Is it possible to extract potentially dangerous substances and experimental operations in terms of fire from the text "General Chemistry Student Experiment" text used at our university and predict the fire risk mechanically from the dangerous substance data? We attempted to clarify what kind of danger would arise if it could not be done, through a comparison with accident cases. From the results, it was found that the types of fires caused by human carelessness (explosion during experimental operation; electrostatic charge; conditions under which a machine was used; proximity to fire; abnormal reaction or explosion due to impurities in the reaction raw materials; effects of scale-up, particle size, or surface conditions of metal reagents) could occur even if the substances handled in the chemical experiment and the experimental operation were familiar. If chemical experiments are conducted without following appropriate rules, not only fires but also accidents and environmental pollution can result. At the same time, if appropriate measures are taken once the properties of chemical substances are understood, it should be possible to effectively avoid the danger of fires.

Moreover, PU is known to be a porous material, but research on hard porous materials and oxygen is also active in other fields. Many researchers have hybridized metal complexes with other functional materials. Organic–inorganic composite materials for this purpose are further adding functions and improving performance. For example, various functions such as gas occlusion in nanopores and control of associated catalytic reactions and physical properties control are expected, as well as metal–organic frameworks and coordination polymers. We have been studying the control of changes in the crystal structure of metal oxides due to surface adsorption modification of metal complexes by compounding metal oxides that are widely used as battery materials and catalysts. Both metal–organic frameworks and metal oxides (or simple metals and nanoparticles) use elemental materials that are expected to have oxygen storage capacity. Further function addition is required for an oxygen-concentrating material (e.g., gas selectivity). Adsorption of oxygen onto one part of a material, such as the stepwise oxygen adsorption of hemoglobin, causes structural changes in other parts of the material.

Conclusion

This chapter has described examples of research into the FR properties of polymer materials and prospects for their use in future special environments. For fire safety in future material systems, it is not sufficient to rely on expectations based on past data (*25–28*). Rather, some type of model experiments may be a useful tool if difficult situations are included (*29*). Alternatively, the game-changing idea of investigating systems, rather than individual components, will become increasingly important for the future (*30*).

Temporarily, such an approach may be useful for carrying out such cross-disciplinary research (in the future) within a certain university, aiming at efficiency and risk avoidance, using "big data" and simulations in addition to model experiments, with only experimental time and costs. However, the author would prefer to implement a research system that reduces the danger and environmental burdens. As a concrete project, the author will call for project participation from related fields under the theme of model experiments and try to reduce or predict actual experiments using artificial intelligence (AI) and big data simulations.

According to the *2019 White Paper on Fire Service* by Japan's Fire and Disaster Management Agency (*31*), enhancing the firefighting and disaster prevention system using AI and robots is a priority research topic. Toward this end, the Cabinet Office of Japan decided on an integrated innovation strategy for science and technology based on Japan's *Science and Technology Basic Plan*. To realize the goals of Japan's Society 5.0 initiative, which is proposed in Japan's Science and Technology Basic Plan, the government will promote research and development that contributes to innovation in AI and robots and promote social implementation of research results. First, fire science requires cooperation across various specialized fields. For example, as a fundamental technology for research and development of a firefighting robot system utilizing AI technology, this effort will require:

(1) technology that utilizes AI to carry out effective firefighting activities against large-scale fires by sharing information, distributing functions to multiple robots, and ensuring that the robots cooperate with each other;
(2) technology that utilizes AI to achieve image recognition and space recognition even under harsh conditions so that the robots can operate semiautonomously in areas far from firefighters; and
(3) heat-resistant technology that conducts firefighting activities in high-temperature areas close to large-scale fires.

Other technologies will also be needed. To develop equipment and materials while comprehending risks with a small number of personnel, cooperation at least among fields of science and engineering such as construction, machinery, mathematical information, physics, medical engineering, health, and materials is required. For efficient cross-disciplinary integration, big data and simulations should be leveraged to provide ways to reduce experimental time, cost, risk, and environmental impact.

References

1. Nakajima, M. Odakyu Line reveals vehicle due to fire spread on train. *Automotive Media Response*, Sept 11, 2017. https://response.jp/article/2017/09/11/299630.html (accessed March 9, 2021) [in Japanese].
2. *Polyurethane Foam*. Japan Urethane Industry Association website. http://www.urethane-jp.org/english/ (accessed March 9, 2021).
3. Kurita, S.; Tagaya, M. Hard Polyurethane Foams. *Nippon Gomu Kyokaishi (J. Soc. Rubber Ind., Jpn.)* **1982**, 55, 159–167[in Japanese].
4. Kurachi, Y. The Concept of Nanotechnology in Flame-Retardant Polymer. *Nippon Gomu Kyokaishi (J. Soc. Rubber Ind., Jpn.)* **2002**, 75, 344–347[in Japanese].
5. Itoge, O.; Irie, Y.; Suzuki, H.; Koura, K.; Ando, T.; Tsutsumi, T.; Minami, S. Fundamental Study on the Combustibility of Cellular Plastic Insulation. *J. Environ. Eng.* **2010**, 75, 1–8.
6. Miyahara, Y.; Watanabe, Y.; Ueno, T.; Mizuno, K.; Ishikawa, T.; Takeda, K. Study on the Combustion Product Gases of Polyurethane and Other Polymers and Their Safety. *Zairyo (J. Soc. Mater. Sci., Jpn.)* **2009**, 58, 747–752[in Japanese].
7. Akitsu, T. Chemical Experiment/Fire Service Act Dangerous Goods and Science. *Chem Educ.* **2018**, *66*, 564–567[in Japanese].

8. Sato, S.; Sakai, M. Flame retardant of rigid urethane foam. *Kogyo Kagaku Zasshi [J. Chem. Ind. (Jpn.)]* **1967**, *70*, 1592–1598[in Japanese].
9. Kakinuma, Y. (2005) Recent Trends of Polyurethane Elastomers. *Nippon Gomu Kyokaishi (J. Soc. Rubber Ind., Jpn.)* **2005**, *78*, 85–87[in Japanese].
10. Nakagawa, Y. Recent Development of Flame Retardant Polymeric Materials Containing Expandable Graphite. *Nippon Kasai Gakkai Ronbunshu (Bull. Jpn. Assoc. Fire Sci. Eng.)* **2006**, *56*, 37–43[in Japanese].
11. Teotia, M.; Verma, A.; Akitsu, T.; Tanaka, S.; Takahashi, K.; Soni, R. K. TGA Decomposition and Flame Profile Measurement of Terephthalamide Stabilized PVC by Cone Calorimeter. *J. Sci. Ind. Res.* **2017**, *76*, 438–441.
12. Teotia, M.; Verma, A.; Akitsu, T.; Tanaka, S.; Haraguchi, T.; Soni, R. K. Salen Type Copper (II) Complexes as Flame Retardants for PVC Sheets. *J. Sci. Ind. Res.* **2019**, *78*, 46–49.
13. Soni, R. K.; Sharma, A.; Akitsu, T.; Teotia, M. Flame Profile Measurement of Cu(II) based Salen Complex Filled Thermally Stabilized PVC Sheets by Cone Calorimeter. *J. Sci. Ind. Res.* **2020**, *79*, 582–585.
14. Aota, Y.; Akitsu, T. Conformational calculations of trans-Schiff base metal complexes for endothermal organic/inorganic fillers in flame retardant polymers. *Asian Chem. Lett.* **2012**, *16*, 19–32.
15. Takahashi, K.; Tanaka, S.; Yamaguchi, M.; Tsunoda, Y.; Akitsu, T.; Sugiyama, M.; Soni, R. K.; Moon, D. Dual purpose Br-containing Schiff base Cu(II) complexes for DSSC dyes and polymer flame retardants. *J. Korean Chem. Soc.* **2017**, *61*, 129–131.
16. Wakisaka, T.; Morita, M.; Akitsu, T. Metal containing surfactants of fire extinguishing compositions for bioethanol blended gasoline. In *Advances in Petroleum Engineering-II: Petrochemical*; Pant, K. K., Sinha, S. Bajpai, S., Govil, J. N., Eds.; Chemical Technology Series; Studium Press LLC: Houston, TX, 2015; Chapter 16, pp 321–336.
17. *Space Policy*. Cabinet Office of Japan, Tokyo, Japan. https://www8.cao.go.jp/space/index.html (accessed March 9, 2021) [in Japanese].
18. *Revised "Marine Energy and Mineral Resources Development Plan" with the aim of utilizing marine resources*. Agency for Natural Resources and Energy of Japan, Tokyo, Japan, Feb 21, 2019. https://www.enecho.meti.go.jp/about/special/johoteikyo/kaiyokaihatukeikaku.html (accessed March 9, 2021) [in Japanese].
19. Okoshi, M. Flame Retardants of Thermosetting Resin. *Nettowaku Porima Ronbunshu (J. Network Polym., Jpn.)* **2019**, *40*, 200–205[in Japanese].
20. Yoshino, A. *Special lecture "Science of Candle"*; NHK Publishing: Shibuya, Tokyo, Japan, 2020.
21. Ito, K. Combustion in Microgravity Environment. *Materia Japan* **1994**, 33, 1039–1043[in Japanese].
22. Nakakuki, A. High-pressure, high-concentration laboratory fire and safety measures. *Saf. Eng.* **1972**, *11*, 98–105.
23. Mizuno, K.; Ueki, K.; Akitsu, T. (2018) Case studies on predictability in university chemistry experiment accidents. *Int. J. Fire Sci. Technol.* **2018**, 37, 1–16.
24. Akitsu, T. Laboratory relay. *Nippon Kasai Gakkai Ronbunshu (Bull. Jpn. Assoc. Fire Sci. Eng.)* **2019**, *69*, 50–54[in Japanese].

25. Akitsu, T.; Onami, Y.; Katsuumi, N. Do thermally mild chemical reactions (for avoiding fire accidents) give rise to unexpected products? *Edelweiss Chem. Sci. J.* **2020**, 3, 15–16.
26. Kida, J.; Akitsu, T. Changes and developments in hydrogen energy as seen in TV programs. *Int. J. Fire Sci. Technol.* **2019**, *38*, 21–25.
27. Kida, J.; Akitsu, T. Fire Hazard Assessment of New Automotive Battery Materials Using SDS Information. *Int. J. Fire Sci. Technol.* **2019**, *38*, 1–19.
28. Date, N.; Sato, A.; Takeuchi, K.; Mori, T.; Yokosuka, K.; Ito, Y.; Shima, K.; Suzuki, K.; Taguchi, H.; Chiyozaki, T.; Yanagawa, K.; Giles, T. K.; Akitsu, T. Fire at Notre Dame Cathedral and Lead Materials in the Environment. *Int. J. Fire Sci. Technol.* **2020**, *39*, 17–37.
29. Akitsu, T.; Mitani, Y. Potential Fire Hazard and Photochemical Tests of Azobenzene of Several Lightning Sources. *Edelweiss Chem. Sci. J.* **2020**, 3, 6–7.
30. Mizuno, K.; Akitsu, T. Environmental Chemistry Lectures Based on System Thinking for University Students. *Edelweiss Chem. Sci. J.* **2019**, 2, 14–16.
31. *Overview of the 2019 White Paper on Fire Service*; Fire and Disaster Management Agency (FDMA): Tokyo, Japan, 2019; translated by International Fire Service Information Center, Tokyo, Japan. https://www.kaigai-shobo.jp/files/fireserviceinjapan/20210301_White_paper02_eng.pdf (accessed June 25, 2021).

Editor's Biography

Ram K. Gupta

Dr. Ram K. Gupta is an associate professor at Pittsburg State University. Dr. Gupta's research focuses on biobased polymers; flame-retardant polyurethanes; conducting polymers and composites; and green energy production and storage using two-dimensional materials, optoelectronics and photovoltaics devices, organic–inorganic heterojunctions for sensors, biocompatible nanofibers for tissue regeneration, scaffold and antibacterial applications, and biodegradable metallic implants. Dr. Gupta has published more than 230 peer-reviewed articles; has made more than 280 national, international, and regional presentations; has chaired many sessions at national and international meetings; has edited many books; and has written several book chapters. He has received more than $2.5 million for research and educational activities from many funding agencies. He is serving as editor-in-chief, associate editor, and editorial board member of numerous journals.

Indexes

Author Index

Subject Index

P

Printed in the USA/Agawam, MA
November 7, 2022

800877.003